# Nanofibers - Production, Properties and Functional Applications

# Nanofibers - Production, Properties and Functional Applications

*Editor*

**Hua Fen Han**

scitus academics

**Nanofibers - Production, Properties and Functional Applications**

Edited by **Hua Fen Han**

ISBN: 978-1-68117-528-7
Library of Congress Control Number: 2016942928


www.scitusacademics.com
Box No. 4766, 616 Corporate Way,
Suite 2, Valley Cottage,
NY 10989



**Notice**



Printed in United States of America on Acid Free Paper

# Preface

In recent years, with the rapid development of nanotechnology and nanoscience, great progress has been made not only in the preparation and characterization of nanomaterials, but also in their functional applications. Nanofibers have an extremely high specific surface area due to their small diameters as crucial one-dimensional nanomaterials and nanofibers membranes are particularly porous with excellent pore interconnectivity. These special features and functionalities of the materials themselves impart nanofibers with a selection of novel properties for applications in areas as diverse as biomedical engineering and storage, energy collection and optics, sensors, composite reinforcement, protection of the environment, filtration, enzyme carriers, catalyst and release control, drug delivery and wound healing. Exclusive nanofibers are ideal for a wide range of applications, from medical to consumer goods and manufacturing to high-tech applications for aerospace, condensers, transistors, drug delivery systems, fuel cells, energy storage, battery separators and information technology. Nanofibers have significant applications in the field of filtration because their surface area is significantly better and have smaller micropores than webs of melt blown (MB). High porous structure with a high surface area makes it ideal for many applications for filtration. Nanofibers are ideal for filtering water or air submicron particles.

This compilation provides up-to-date coverage of preparation, properties and functional applications for nanofibers. The information in this book will be of immense use to anyone associated with the nanofibers discipline.

# Table of Contents


Chapter 1 Preparation of Functionalized Nanofibers and their Applications 1

Chapter 2 Hierarchically Structured Electrospun Fibers 25

Chapter 3 Preparation of Cellulose-based Nanofibers Using Electrospinning 63

Chapter 4 Physical Properties of Polypeptide Electrospun Nanofiber Cell Culture Scaffolds on a Wettable Substrate 77

Chapter 5 Keratin-based Nanofibres 101

Chapter 6 Fabrication of Pd Doped $WO_3$ Nanofiber as Hydrogen Sensor 129

Chapter 7 Nanofiber Reinforced Composite Polymer Electrolyte Membranes 145

Chapter 8 Preparation of Ag/HBP/PAN Nanofiber Web and Its Antimicrobial and Filtration Property 181

Chapter 9 A Review of Electrospun Conductive Polyaniline Based Nanofiber Composites and Blends: Processing Features, Applications, and Future Directions 201

Chapter 10 Tailoring Crystallinity of Electrospun Plla Fibres by Control of Electrospinning Parameters 249

**Chapter 11** **Synthesis of Alumina Nanofibers and Composites** **271**

**Index** **289**

# Chapter 1

# Preparation of Functionalized Nanofibers and their Applications

**Young-Seak Lee[1] and Ji Sun Im[1]**

[1] Chungnam National University, Republic of Korea

## 1. INTRODUCTION

Recently, nano-scaled materials have been investigated with amazing increased interest due to their many advantages, such as large surface area and many active surface sites. Among different nano-scaled materials, nanofibers have been widely applied in industry due to the ease in production processes compared to other nano-materials. Nanofibers can be identified as fibers having diameters between tens and hundreds of nanometers. This nano-scaled diameter of fibers can give an enormous surface area per unit mass.

Nanofibers are usually fabricated by the electrospinning method as a non-woven mat. Non-woven fabrics are textile materials consisting of randomly oriented fibers connected together by physical entanglements or bonds between individual fibers, without any knitting or stitching. These non-woven nanofibers are the primary alternative for traditional textiles as filtration media, energy storage media, hygienic and health/personal care textiles, thermal and sound insulating materials, ecological materials, building materials, geo-textiles and automotive textiles.

# 2. FORMATION OF NANOFIBERS BY ELECTROSPINNING

## 2.1. Theory of Electrospinning

New applications require fibers with increasingly smaller diameters. Since the surface area is proportional to the fiber diameter and the volume is proportional to the square of the diameter, the specific surface area is inversely proportional to the fiber diameter, leading to high specific surface areas for small fibers. In addition, pore size depends on the fiber diameter; therefore, small fibers produce non-wovens with a small pore size. There are several methods for producing small diameter fibers using high-volume production methods, such as fibrillation, island-in-sea, and the novel melt-blowing system, in addition there are highly accurate methods such as nano-lithography and self-assembly. However, the usefulness of above methods is restricted by combinations of narrow material ranges, high costs and low production rates. In comparison, electrospinning is a simple and low cost process and has an intermediate production rate. Electrospinning is a process for sub-micron scale polymer-based filament production (usually called nanofibers) by means of an electrostatic field. Due to these forces, the meniscus of a liquid flowing out of a capillary nozzle elongates, forming a fine jet, that is later atomized into fine droplets. The droplets obtained by this method are electrically charged. A basic electrospinning setup consists of three elements: an electrical generator (high voltage supply), a capillary (jet source) and a metal collector (target). Depending on the flow rate and potential of the capillary, the droplets can be of submicron size, with a narrow size distribution. The solution is usually electrically charged by the generator, and the collector is grounded, but it is also possible to invert the process by electrically charging the collector and grounding the solution. When the electrostatic forces exceed the surface tension force, the pendent droplet at the capillary tip is stretched into a cone (called Taylor cone), and a solution stream is ejected (Dzenis, Y.A., 2004).

Whether the jet will form a continuous fiber or disperse into droplets depends on polymer molecular weight, polymer chain entanglement, and the solvent applied to the process (specifically, its evaporation rate). It is known from the literature that smooth fibers are produced when the product of intrinsic viscosity (η) and polymer concentration (c), known as the Berry's number, Be=ηc, is greater than a certain critical value Becr, which is characteristic of the polymer. Specific viscosity of a polymer solution is determined as the ratio:

$$\eta_{sp}=(\eta_0-\eta_s)/\eta_s \tag{1}$$

where $\eta_0$ is the zero shear rate viscosity of the polymer solution at concentration (c), and $\eta_s$ is the solvent viscosity. From this equation, the intrinsic viscosity (η) of the polymer is determined as a linear extrapolation of specific viscosity $\eta_{sp}$ measured for various concentrations to the concentration at c=0:

The intrinsic viscosity (η) is also related to the molecular weight ($M_W$) of a linear polymer by the Mark–Houwink equation:

$$\eta = KM_W{}^a \qquad (2)$$

where the constants (*K*) and (*a*) depend on the polymer, solvent and temperature.

## 2.2. Electrospinning Setups

A basic electrospinning setup consists of three elements: an electrical generator (high voltage supply), a capillary (jet source) and a metal collector (target) (Fig. 1).

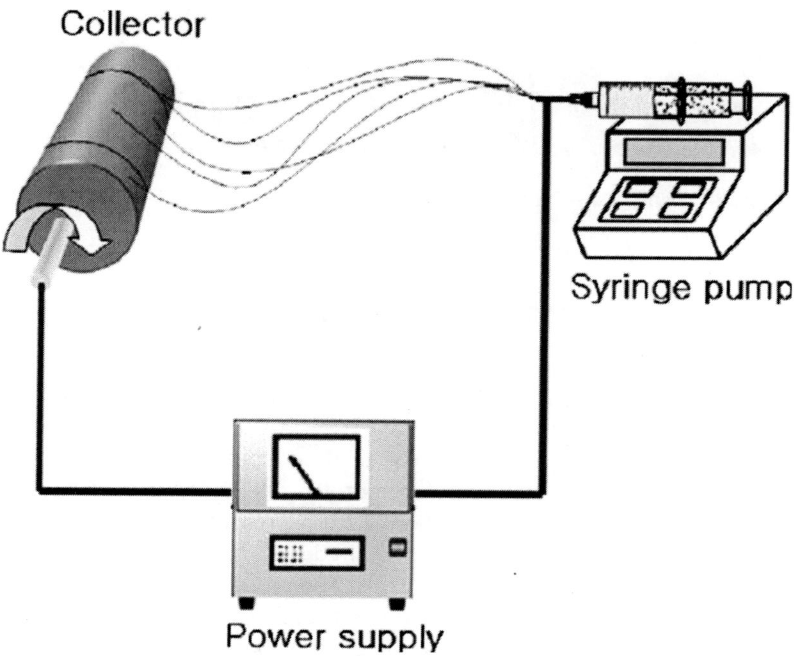


**Figure 1.** Diagram of the electrospinning apparatus (J.S. Im et al., 2008).

In our group, polyacrylonitrle (PAN)-based nanofibers were obtained by the electrospinning method. The electrospinning was carried out under the following conditions: [10 wt% of PAN/DMF polymer solution, 15 kV] (J.S. Im et al., 2008).

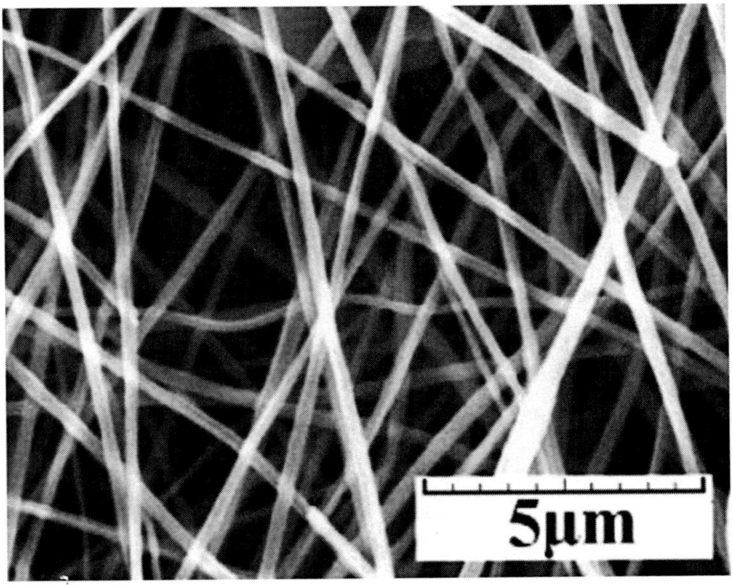


**Figure 2.** PAN-based electrospun carbon fibers.

# 3. FUNCTIONALIZATION OF NANOFIBERS

## 3.1. Activation Process For Porous Nanofibers

Nanofibers have many benefits because of their large surface area for active reaction sites. Activation processes for improving these active sites have been applied by chemical/physical activation. As a representative case of physical activation, the activation agent was embedded into the fibers and then removed by physically removing the agents. The process for PAN-based porous nanofibers physically activated by silica is presented in Fig. 3 (J.S. Im et al., 2009). The silica-activated carbon nanofibers are shown in Fig. 4 (J.S. Im et al., 2009). The pores generated by physical activation are clearly observed.

## 3.2. Heat Treatment Effect For Physical Properties

Heat treatment has been carried out widely because it can change physical properties. The electrical enhancement of PAN-based carbon fibers was presented by (J.H. Huang et al., 2006) in Fig. 5. A stabilization treatment was carried out at 280 C for a duration of 4 h, and samples with the same stabilization process parameters were used as the starting material of the carbonization treatment. Cyclic voltammetry (CV) measurements were carried out in a potential window of 0 to 1 V with different scan rates ranging from 100 to 1000 mV/s. It is noted that the curve for electrospun PAN-based carbon fibers shows a zigzag-like behavior, which indicates the hydrogen ions in the electrolytes are not easily adsorbed on the surface pores of PAN-CFs. In contrast, after thermal treatments (especially after the carbonization process), the capacitive performance is still stable even at the scan rate of 1000 mV/s. This result reveals that the capacitive performance of PAN-based carbon fibers is improved by the carbonization process.

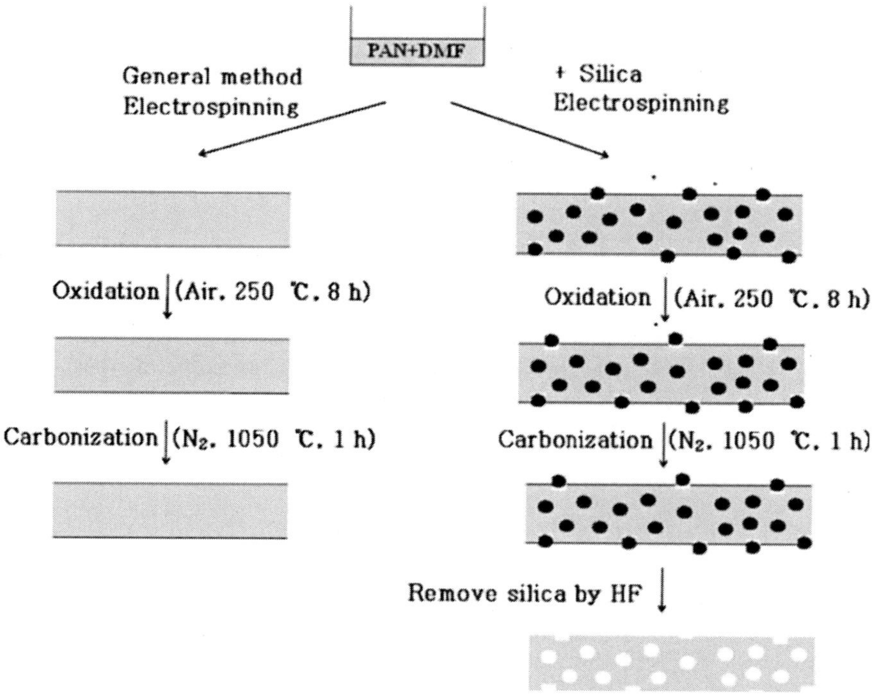


**Figure 3.** The procedure of manufacturing carbon fibers and silica-activated carbon fibers.

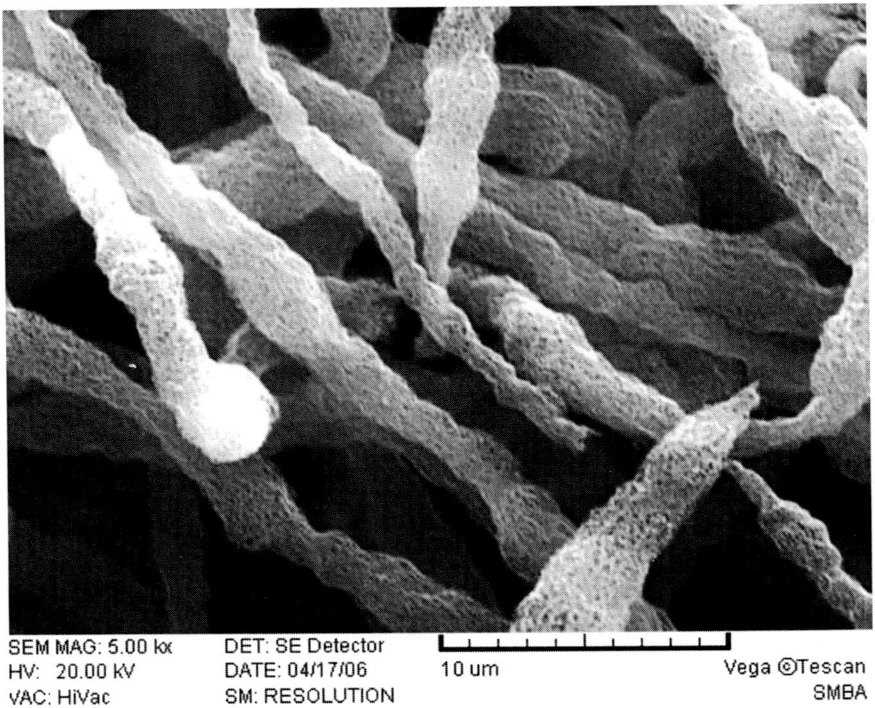


**Figure 4.** Silica-activated carbon nanofibers.

This enhancement of electrical properties has been attributed to the improved electrical conductivity caused by the orientation of carbon structures. It was presented that the electrical conductivity can be improved significantly by well-oriented carbon structures through heat treatment (J.S. Im et al., In Press). The test was carried out on a variety of heat treatment temperatures and amounts of carbon nanotube additives, as shown in Fig. 6. The electrical conductivity was improved over three fold by the effects of heat treatment.

## 3.3. Modification Of Nanofibers

The modification of nanofibers has been applied widely to give them improved properties. The result, which shows that the removal of multi-metal ions was improved by surface modification using acid treatment, is shown in Fig. 7 (S.J. Park et al., 2004). They explained that the surface functional groups containing oxygen content by acid treatment played an important role for the removal of metal ions.

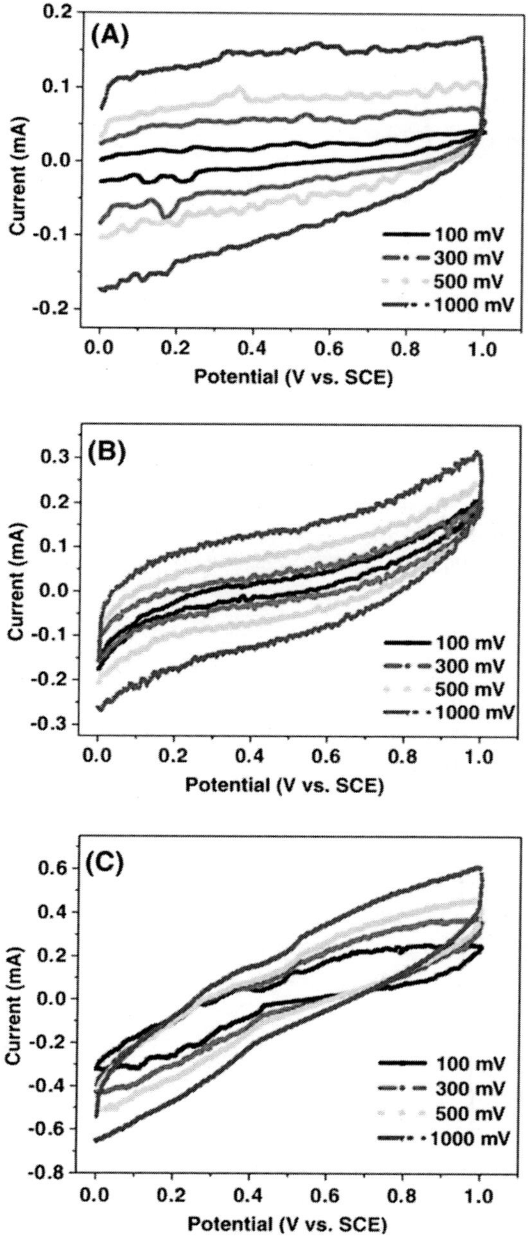


**Figure 5.** Capacitive behavior of (A) electrospun, (B) stabilized, (C) carbonized PAN-based carbon fibers.

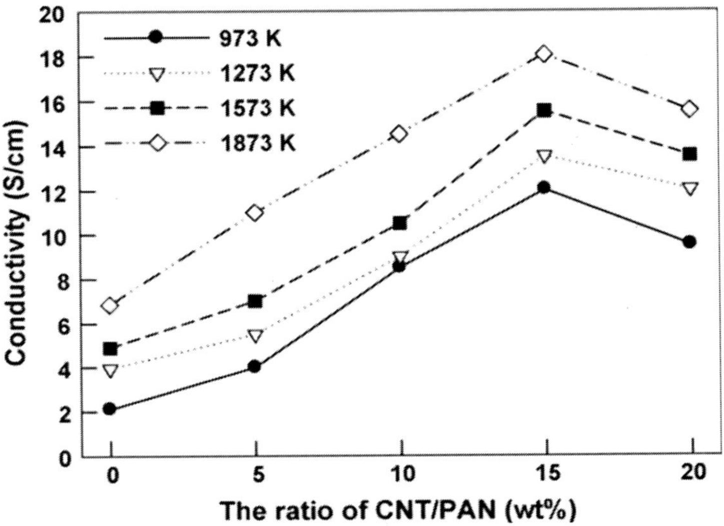


**Figure 6.** Electrical conductivity of carbon fiber/carbon nanotubes composite.

A plasma treatment was carried out to enhance the surface energy of carbon nanofibers for good bonding to the matrix (V. Bruser et al., 2003). The gas-state plasma treatment was carried out by using the fluidized bed plasma reactor depicted in Fig. 8. The surface energy increased over two fold after 5 min of plasma treatment (V. Bruser et al., 2003).

The effects of functionalized surface groups on nanofibers were investigated based on improved gas adsorption by our group (J.S. Im et al., 2009). The fluorination treatment of porous nanofibers was carried out in the gas state for 5 min. The capacity of methane storage was increased by attraction effects of fluorine on porous nanofibers.

## 3.4. Nanofibers Complex

The use of nanofibers as the reinforcing filler and conducting additive in polymers to improve their mechanical and electrical properties is generally encountered in polymer technology. Nanofibers, such as carbon and glass fibers, are routinely used in composites of a range of different polymers. Improvement in modulus and strength, achieved by using nanofibers in a composite, has been presented by many researchers. The improved storage modulus of epoxy resin by addition of carbon nanofibers is presented in Fig. 11 (Derrick R.

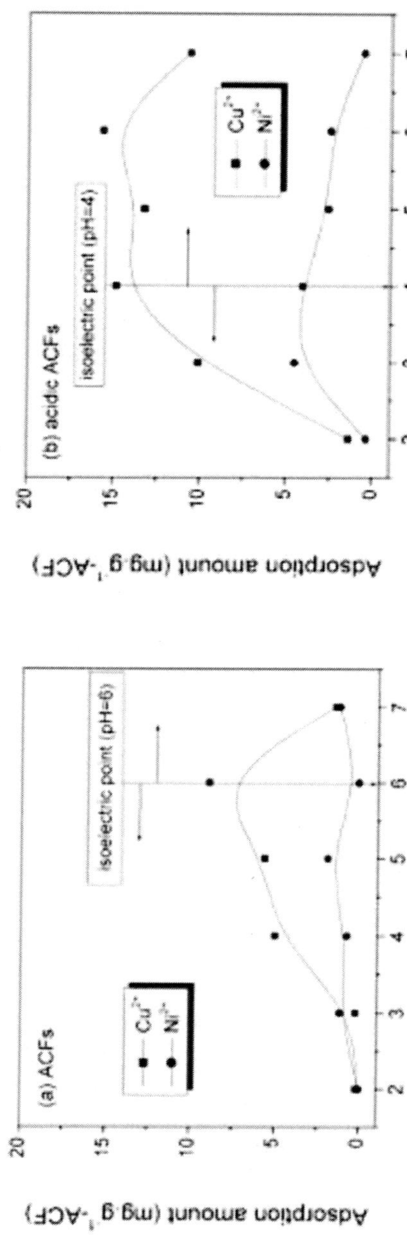


**Figure 7.** The removal of multi-metal ions on (a) activated carbon fibers and (b) acid-treated activated carbon fibers as a function of pH at 25 C.

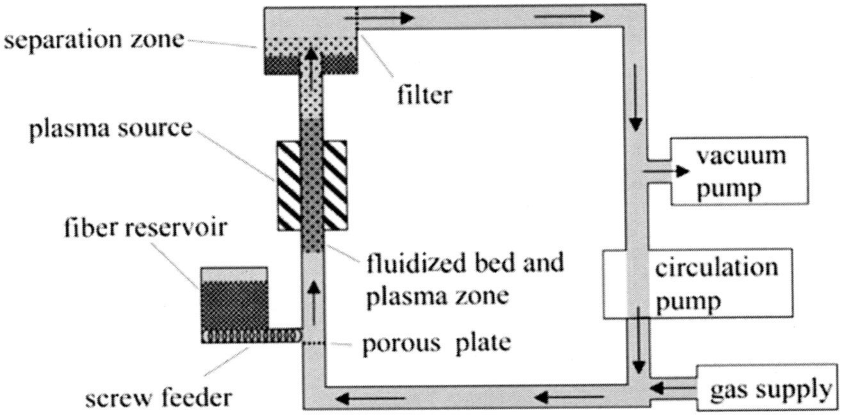


**Figure 8.** Fluidized bed plasma reactor (V. Bruser et al., 2003).

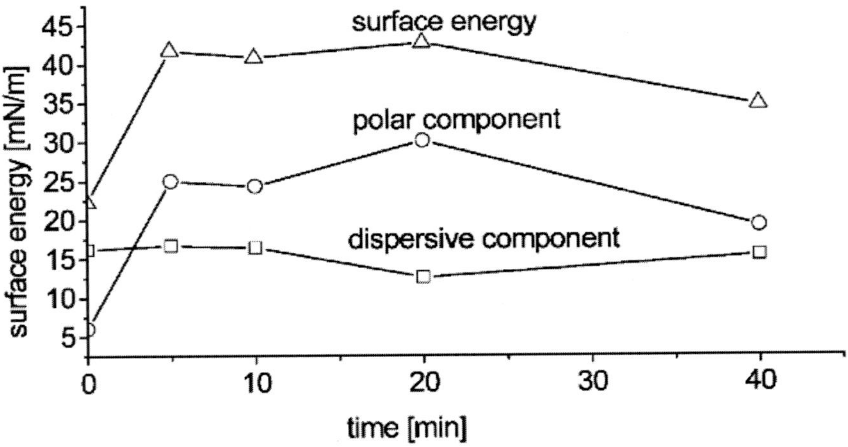


**Figure 9.** Surface energy at 1.2 mbar and 80 W; gas composition: Ar:$O_2$=1:1 (V. Bruser et al., 2003).

Dean et al., 2009). The transition temperature shifted significantly higher, indicating that the epoxy/carbon nanofiber composite can be stained up to higher temperatures.

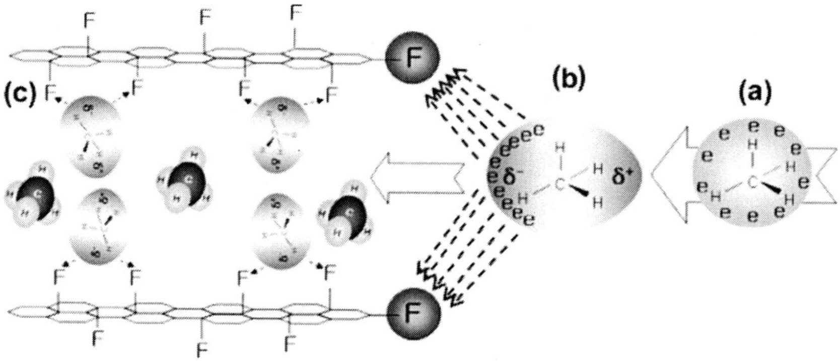


**Figure 10.** Mechanism of methane storage; (a) methane molecule, (b) electron-attracted molecule and (c) adsorbed methane molecule in the carbon silt pore (J.S. Im et al., 2009).

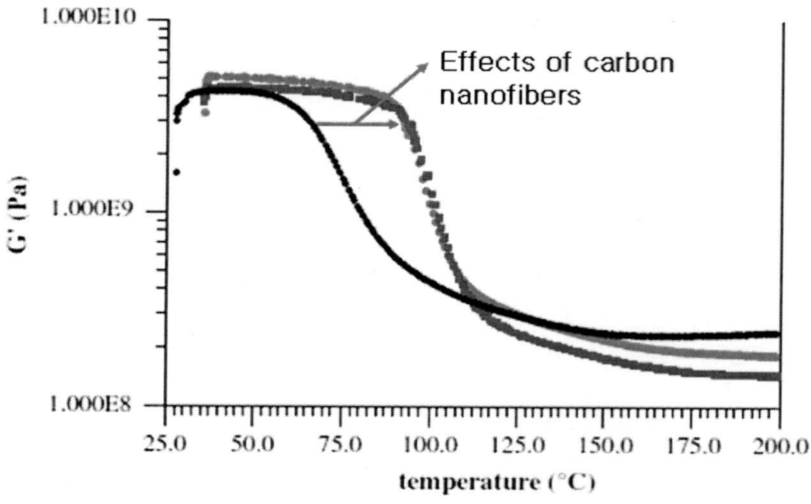


**Figure 11.** Storage modulus of epoxy resin based on the effects of carbon nanofibers (Derrick R. Dean et al., 2009).

This mechanical improvement is attributed to the properties at the fiber/matrix interface and are therefore dependent on the surface area of the interface. S.G. Prolongo's group synthesized Epoxy/carbon nanofiber composites by suing functionalized nanofibers with amino groups (S.G. Prolongo et al., 2009). They found that the dispersion of carbon nanofibers is improved by the functionalization process up to a nanofiber content of 1 wt%. They also found that improved dispersion of carbon nanofibers markedly affects the physical and thermo-dynamical mechanical properties of the epoxy nanocomposites. The addition of functionalized carbon nanofibers causes an important increase in the coefficient of thermal expansion and glassy storage modulus of nanocomposites.

The use of carbon nanofibers for enhanced electrical properties has also been reported widely. The group of Torsten presented results showing that a reduced specific resistance of epoxy resin was obtained by the addition of carbon nanofibers (Torsten Prasse et al., 2003). They also found that the orientation of carbon nanofibers were important for the electrical properties of the epoxy complex. When the carbon nanofibers were oriented along the direction of the applied electric field, the measured electrical properties were improved more efficiently.

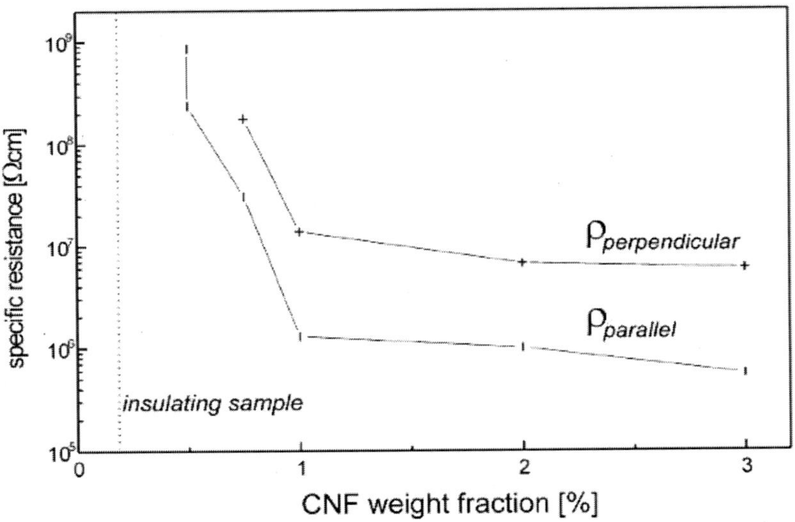


**Figure 12.** Specific resistance of epoxy resin as a function of carbon nanofiber weight fraction with different orientations of carbon nanofibers (Torsten Prasse et al., 2003).

Metal/nanofiber complexes were also investigated to maximize the advantages of both metals and nanofibers. The self-assembly method of metal-complexed bolaamphiphiles was introduced by Masaki Kogiso's group (Masaki Kogiso et al., 2004). The illustration is shown in Fig. 13.

For example, a $TiO_2$ photocatalyst was embedded in electrospun fibers to optimize the photocatalytic ability shown in Fig. 14 (J.S. Im et al., 2008). Due to the low density of electrospun nanofibers, it was possible to make the photocatalysts floating on water for enhanced absorbance of UV light.

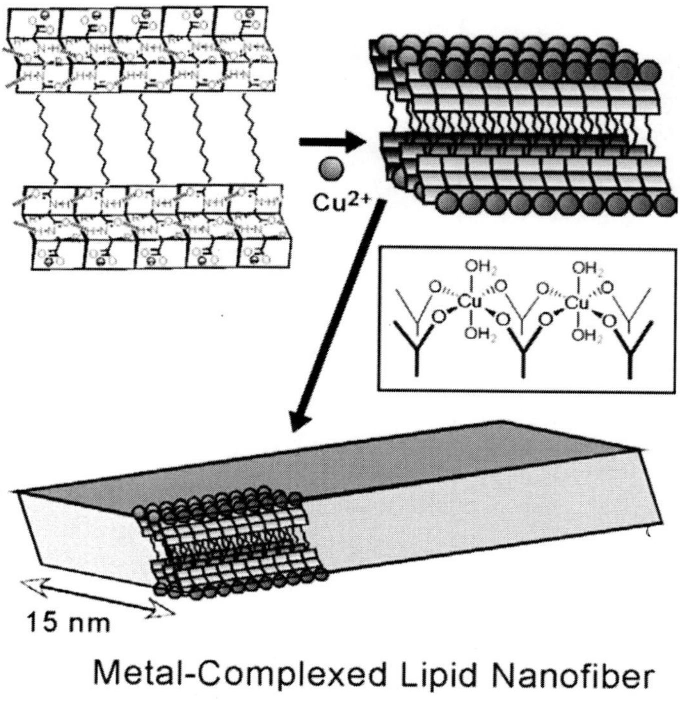


**Figure 13.** A proposed structure of the copper(II) acetates complex and a schematic illustration of the molecular packing within the nanofibers (Masaki Kogiso et al., 2004).

# 4. APPLICATION OF NANOFIBERS

## 4.1. Energy Storage Materials

Nanofibers have been applied as a storage media for alternative energy sources such as hydrogen and natural gases. Porous carbon nanofibers have especially been investigated widely due to their large specific surface area and high pore volume. Hydrogen and natural gases can be stored by physical adsorption, indicating that the use of these gases is easy. The superior storage capacity of porous carbon nanofibers was presented through comparison with other porous carbon materials such as graphite, carbon nanotubes, and activated carbon, as shown in Fig. 15 (J.S. Im et al., 2009).

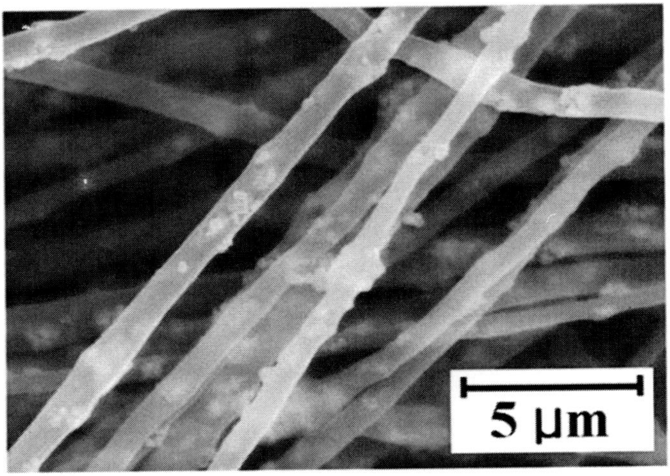


**Figure 14.** SEM images of electrospun PAN-based nanofibers containing $TiO_2$ (J.S. Im et al., 2008).

The main reason suggested for improved hydrogen adsorption was that electrospun activated carbon nanofibers might be expected to have an optimized pore structure with controlled pore size. This result may come from the fact that the diameters of electrospun fibers can be controlled easily, and optimized pore sizes can be obtained with a highly developed pore structure.

To find the optimized activation conditions, carbon nanofibers were activated based on varying the chemical activation agents, reaction time, reaction temperature, and the rate of inert gas flow. The role of the type of chemical activation and inert gas glow rate was explained by recent paper (A. Linares-Solano et al., 2009). When comparing potassium hydroxide and sodium hydroxide, a higher developed pore structure was observed with potassium hydroxide. A quicker flow rate of inert gas was also beneficial for an optimized pore structure for hydrogen adsorption of carbon nanofibers.

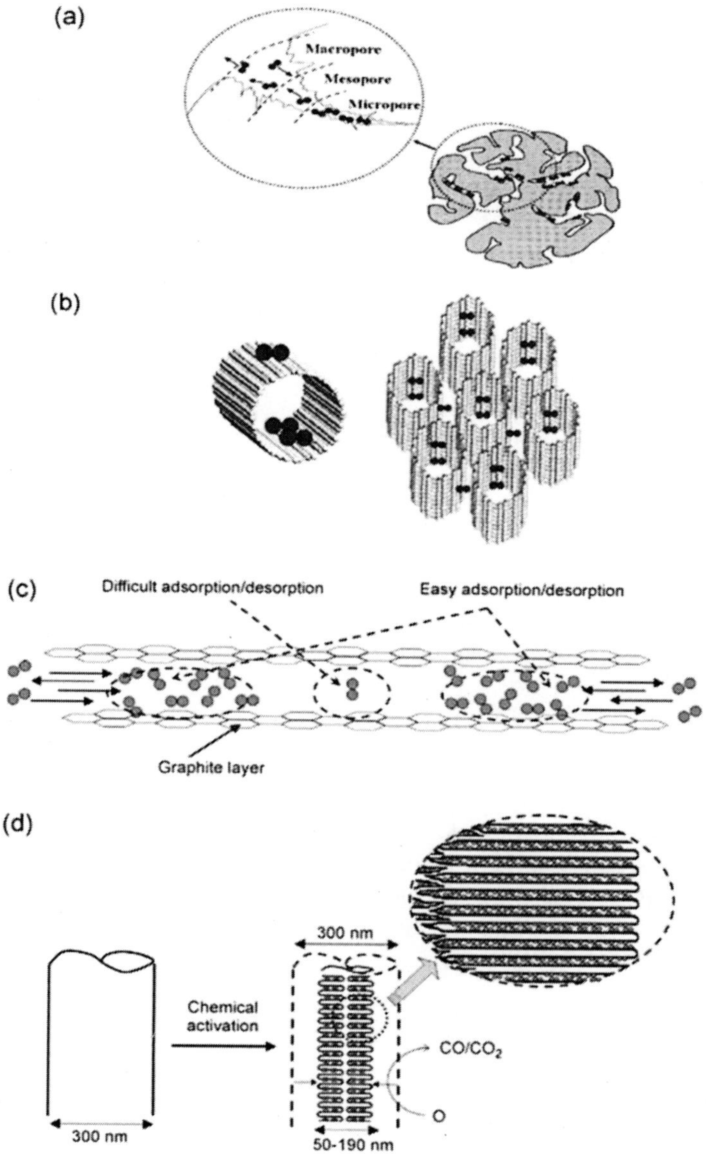


**Figure 15.** The mechanism of hydrogen adsorption using various carbon materials; (a): activated carbon, (b): single walled carbon nanotube, (c): graphite, (d): electrospun activated carbon nanofibers (J.S. Im et al., 2009).

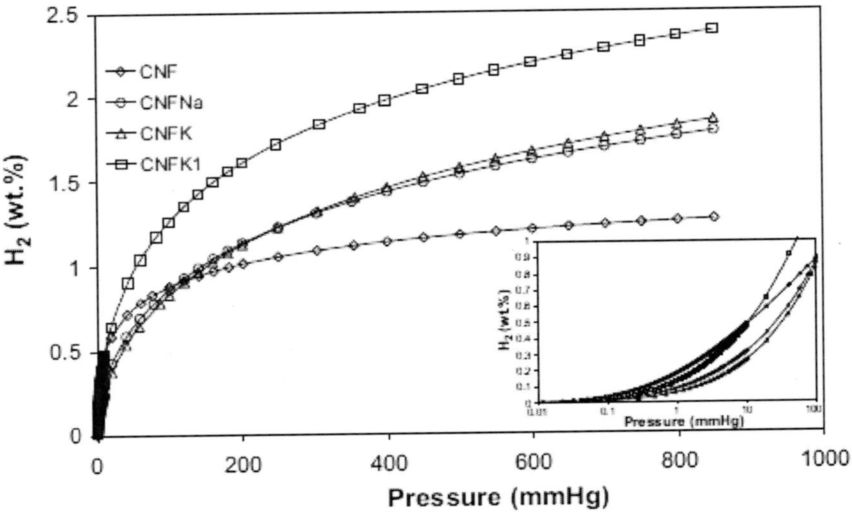


**Figure 16.** Hydrogen adsorption isotherms at 77 K; CNF: carbon nanofiber, CNF Na: NaOH activated carbon nanofiber ($N_2$ flow of 500 ml/min), CNF K: KOH activated carbon nanofiber ($N_2$ flow of 500 ml/min), CNF K1: KOH activated carbon nanofiber ($N_2$ flow of 800 ml/min) (A. Linares-Solano et al., 2009).

The MCF (metal-carbon-fluorine) system was introduced by using electrospun carbon nanofibers (J.S. Im et al., 2009). The transition metal and fluorine were introduced on the activated carbon nanofibers to use the gap of electronegativity between the metal and fluorine. As a result of this gap, the hydrogen molecules were attracted into the carbon pores effectively.

## 4.2. Ecological Materials

Photocatalysts (usually $TiO_2$) are used widely because of their high activity, chemical stability, robustness against photocorrosion, low toxicity and low cost. Nano-sized photocatalysts have been synthesized by many methods, such as the sol-gel process and thermal treatments, to enhance the photocatalytic ability with larger active sites. However, the recovery of nano-sized photocatalysts is not easy, making reuse difficult and secondary contamination likely. Therefore the immobilization method has been studied widely. Nanofibers have been investigated as one of the solutions using two methods: first, photocatalysts were fabricated as a non-woven mat by the electrospinning method; second, photocatalysts were embedded into a matrix of electrospun nanofibers. These materials can render large active sites for the photodegradation of pollutants. The TiO2 nanofibers were fabricated by C. Tekmen and A.K. Alves's groups (C. Tekmen et al., 2008, A.K. Alves et al., 2009). S.G. Lee's group synthesized $SiO_2/TiO_2$ nanofibers by electrospinning and presented their synergy effects (S.W. Lee et al., 2007). The $TiO_2$ embedded PAN-based carbon nanofibers were prepared and investigated based on the floating effect with a low density of nano-web (J.S. Im et al., 2008). The photocatalytic efficiency was maximized because photocatalysts can be activated by UV-light without any interruption such as UV absorption of dust water.

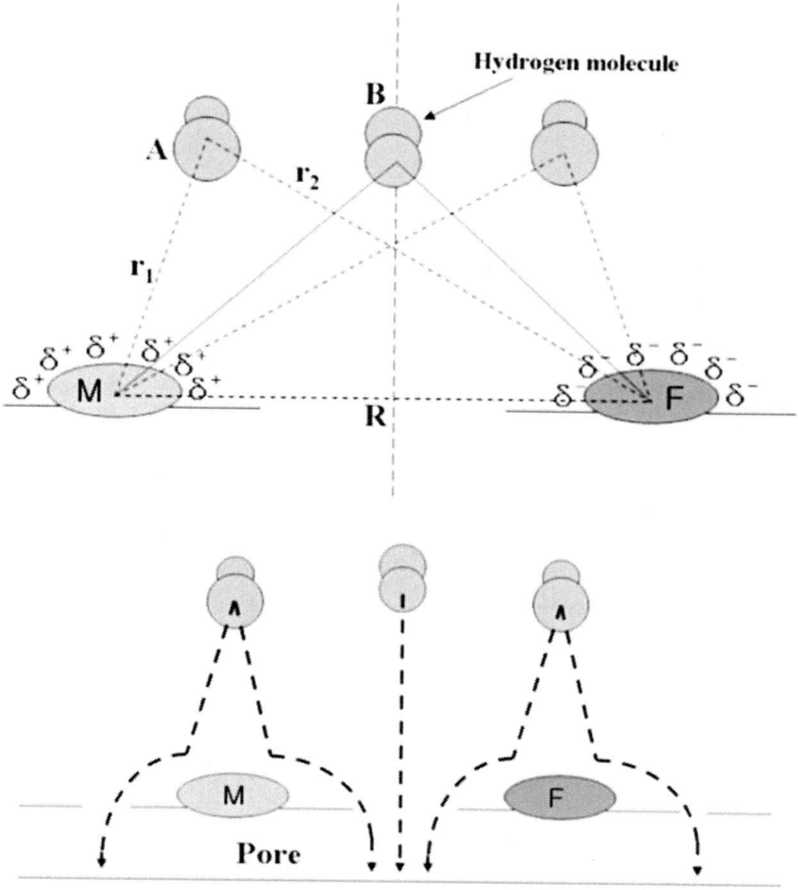


**Figure 17.** Mechanism of hydrogen adsorption enhancement by the MCF system.

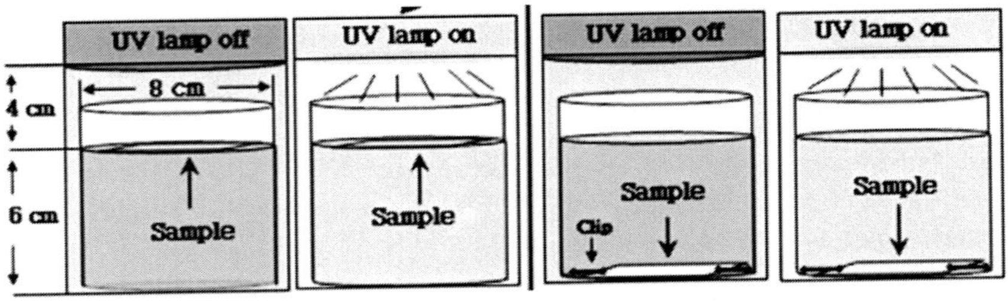


**Figure 18.** Photocatalytic effects of nanofibers involving $TiO_2$ based on the floating property.

Nanofibers have also been studied as a membrane for the filtration of heavy metals in groundwater (Y. Sang et al., 2008). The experimental setup was shown in Fig. 19. From static adsorption experiments, an efficient removal of $Cu^{2+}$, $Cd^{2+}$, and $Pb^{2+}$ was observed.

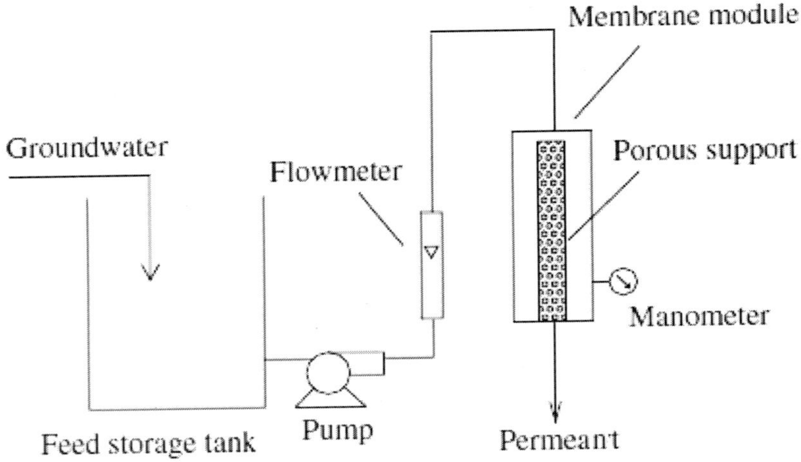


**Figure 19.** Nanofiber membrane test as a groundwater treatment (Y. Sang et al., 2008).

## 4.3. Biomaterials

The biomedical applications of electrospun fibers remain the most interesting research area. Nanofibers can contribute in diverse emerging medical areas such as drug delivery, organogenesis, genomic medicine, rapid bedside clinical tests, and smart wound dressings. Among these applications, nanofiber-based three-dimensional scaffolds for tissue engineering and the design of nanofiber devices for sustained delivery systems have been attracted by the many advantages of electrospun fibers. The majority of nanofiber anti- neoplastic agent delivery systems have been envisioned for the treatment of malignant gliomas (a type of brain tumor). The current drug delivery system of choice is post tumor-resection implantation of a drug-eluting wafer. Thus, numerous studies have tried to elucidate the benefits of implanting a nanofiber delivery system over a wafer-based system.

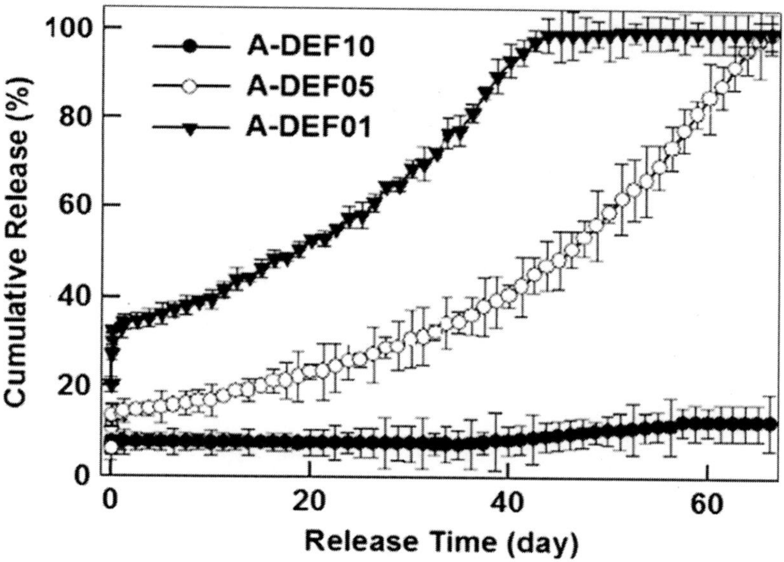


**Figure 20.** Controlled drug release of electrospun hydrogel nanofibers by fluorination treatment.

Recent results show that the released drug of electrospun nanofibers can be controlled by surface treatment (J.S. Im et al., 2009). Sustained drug release was observed by controlling the swelling rate of hydrogel nanofibers using a fluorination treatment.

The scaffolding application of electrospun nanofiber mats has already been applied in the industrial field because their size range approximates the structural features present in tissues surrounding animals' bodies. Based on the effects of various layered nanofiber matrices presented in Fig. 21, the ability of the scaffold was tested (S.H. Park et al., 2008).

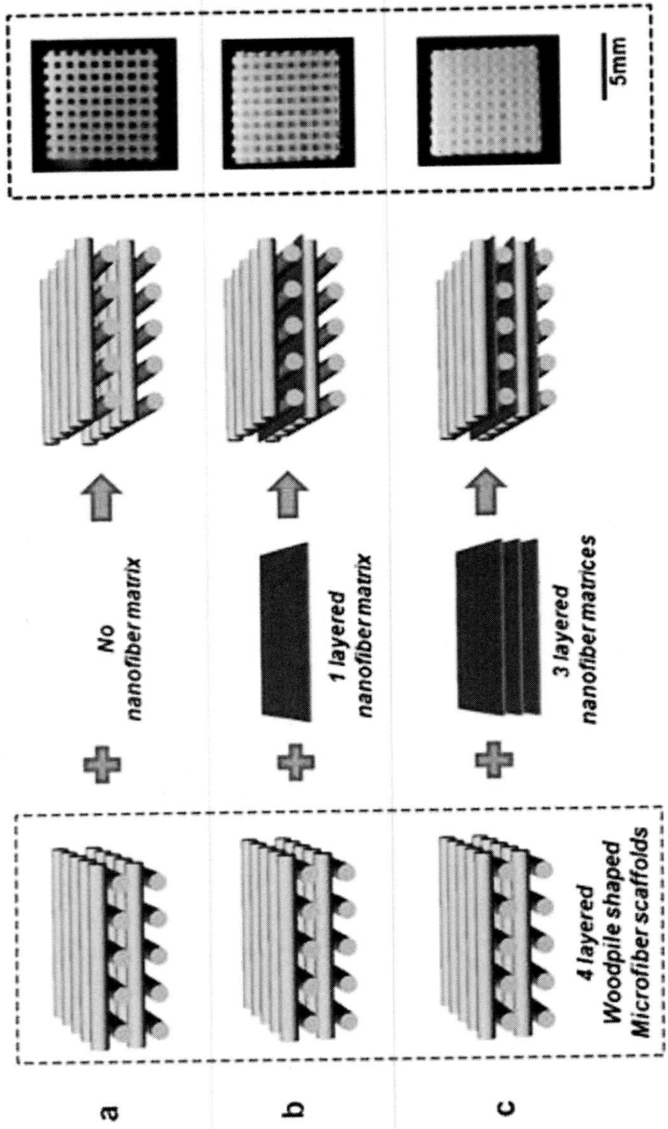


**Figure 21.** Schematic diagrams and photographs of the three types of hybrid scaffolds used in the cell culture experiments. (a) Type I has no nanofiber matrix and only a PCL scaffold as a control specimen. PCL scaffolds combined with (b) a one-layer PCL/collagen nanofiber matrix (type II) and (c) a three-layer PCL/collagen nanofiber matrix (type III) (S.H. Park et al., 2008).

The M.P. Lutolf group presented the concept of the stem cell niche and the role of the extracellular matrix in regulating stem cell survival and signaling. They used electrospun nanofibers as a basement membrane between two phases.

By controlling the ratio of the polymer sources, the scaffold can easily have different properties. InFig. 23, the lacunae in A1, B1 and C1 were enlarged and compared ((Hap/PLGA (%w/w); A1 (0/100), B1 (5/95) and C1 (10/90)) (H. Nie et al., 2009). Numerous osteoclast-like cells (identified by black circles) were observed to reabsorb the trabecular bone throughout the defects at this time-point. More isolated lacunae were observed in A1 and B1 than C1.

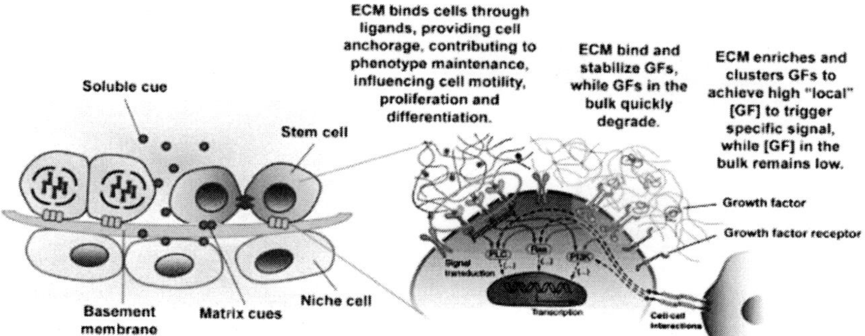


**Figure 22.** The concept of the stem cell niche and role of the extracellular matrix in regulating stem cell survival and signaling. ECM: extracellular matrix; GF: growth factor; [GF]: growth factor concentration. Matrix (M.P. Lutolf and J.A. Hubbell, 2005).

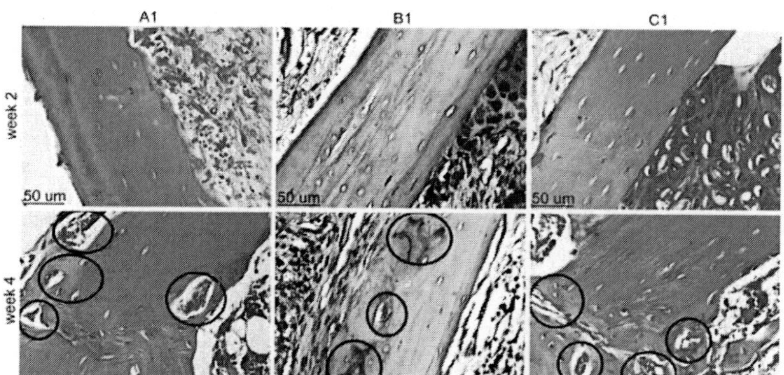


**Figure 23.** Histological specimens from nude mice tibias after 2 and 4 weeks of implantation of A1, B1 and C1. Original magnification is 400 × for all (H. Nie et al., 2009).

The wound healing ability of nanofibers is shown in Fig. 24 (J.S. Choi et al., 2008). The wound of mouse was effectively healed by electrospun nanofibers within 14 days.

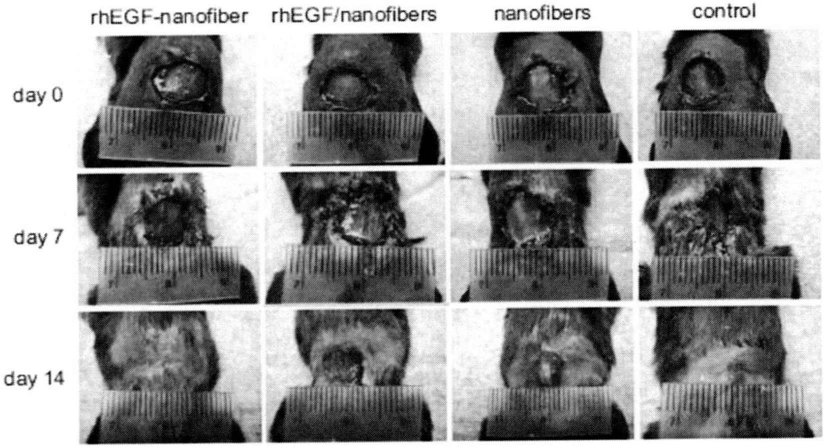


**Figure 24.** Extent of wound healing in diabetic C57BL/6 mice treated with various formulations (J.S. Choi et al., 2008).

# REFERENCES

1. K. Alves, F. A. Berutti, F. J. Clemens, T. Graule, C. P. Bergmann, 2009 Materials Research Bulletin, 44., 312-317, 0025-5408

2. Y. A. Dzenis, 2004 Spinning continuous fibers for nanotechnology. Science, 304(5679), 1917-1919, 0036-8075

3. S. F. Garcia, E. V. Ortego, M. Kunowsky, M. Kimura, A. Oya, A. L. Solano, 2009 International Journal of Hydrogen Energy, In Press, 0360 3199

4. K. J. Green, D. R. Dean, U. K. Vaidya, E. Nyairo, 2009 Composites: Part A, 40., 1470-1475, 1359-835X

5. M. Heintze, V. Bru¨ser, W. Brandl, G. Marginean, H. Bubert, S. Haiber, 2003 Surface and Coatings Technology, 174 175 ., 831-834, 0257-8972

6. J. L. Hu, J. H. Huang, Y. K. Chih, C. C. Chuang, J. P. Chen, S. H. Cheng, J. L. Horng, 2009 Diamond & Related Materials, 18., 511-515, 0925-9635

7. J. S. Im, J. S. Jang, Y. S. Lee, 2009 Synthesis and characterization of mesoporous electrospun carbon fibers derived from silica template. Journal of Industrial and Engineering Chemistry, In Press, 1226 086 X

8. J. S. Im, M. I. Kim, Y. S. Lee, 2008 Preparation of PAN-based electrospun nanofiber webs containing TiO2 for photocatalyst degradation. Materials Letters, 62(21-22), 3652-3655, 0167-577X

9. J. S. Im, S. J. Kim, P. H. Kang, Y. S. Lee, 2009 The improved electrical conductivity of carbon nanofibers by fluorinated MWCNTs. Journal of Industrial and Engineering Chemistry, In Press, 1226 086 X

10. J. S. Im, S. J. Park, Y. S. Lee, 2009 The metal-carbon-fluorine system for improving hydrogen storage by using metal and fluorine with different levels of electronegativity. International Journal of Hydrogen Energy, 34(3), 1423-1428, 0360-3199

11. J. S. Im, S. J. Park, Y. S. Lee, 2009 Superior prospect of chemically activated electrospun carbon fibers for hydrogen storage. Materials Research Bulletin, 44(9), 1871-1878, 0025-5408

12. J. S. Im, J. Yun, Y. M. Lim, H. I. Kim, Y. S. Lee, 2009 Fluorination of electrospun hydrogel fibers for a controlled release drug delivery system. Acta Biomaterialia, In Press, 1742 7061

13. J. S. Kam, K. W. Leong, H. S. Yoo, 2008 Biomaterials, 29., 587-596, 0966-0844

14. S. W. Lee, Y. U. Kim, S. S. Choi, T. Y. Park, Y. L. Joo, S. G. Lee, 2007 Materials Letters, 61., 889-893, 0167-577X

15. M. P. Lutolf, J. A. Hubbell, 2005 Nature biotechnology, 23., 47-55, 1087-0156

16. H. Nie, M. L. Ho, C. K. Wang, C. H. Wang, Y. C. Fu, 2009 Biomaterials, 30., 892-901, 0966-0844

17. S. H. Park, T. G. Kim, H. C. Kim, D. Y. Yang, T. G. Park, 2008 Acta Biomaterialia, 4., 1198-1207, 1742-7061

18. S. J. Park, J. S. Shin, J. W. Shin, S. K. Ryu, 2004 Journal of Colloid and Interface Science, 275., 342-344, 0021-9797

19. T. Prasse, J. Y. Cavaillé, W. Bauhofer, 2003 Composites Science and Technology, 63., 1835-1841, 0266-3538

20. S. G. Prolongo, M. Campo, M. R. Gude, R. Chaos-Moran, A. Urena, 2009 Composites Science and Technology, 69., 349-357, 0266-3538

21. Y. Sang, F. Li, Q. Gu, C. Liang, J. Chena, 2008 Desalination, 223., 349-360, 0011-9164

22. Tekmen, A. Suslu, U. Cocen, 2008 Materials Letters, 62., 4470-4472, 0167-577X

# Chapter 2

# Hierarchically Structured Electrospun Fibers

**Nicole E. Zander**

US Army Research Laboratory, Weapons and Materials Research Directorate, Aberdeen Proving Ground, MD 21005, USA

## ABSTRACT

Traditional electrospun nanofibers have a myriad of applications ranging from scaffolds for tissue engineering to components of biosensors and energy harvesting devices. The generally smooth one-dimensional structure of the fibers has stood as a limitation to several interesting novel applications. Control of fiber diameter, porosity and collector geometry will be briefly discussed, as will more traditional methods for controlling fiber morphology and fiber mat architecture. The remainder of the review will focus on new techniques to prepare hierarchically structured fibers. Fibers with hierarchical primary structures—including helical, buckled, and beads-on-a-string fibers, as well as fibers with secondary structures, such as nanopores, nanopillars, nanorods, and internally structured fibers and their applications—will be discussed. These new materials with helical/buckled morphology are expected to possess unique optical and mechanical properties with possible applications for negative refractive index materials, highly stretchable/high-tensile-strength materials, and components in microelectromechanical devices. Core-shell type fibers enable a much wider variety of materials to be electrospun and are expected to be widely applied in the sensing, drug delivery/controlled release fields, and in the encapsulation of live cells for biological applications. Materials with a hierarchical secondary structure are expected to provide new superhydrophobic and self-cleaning materials.

## KEYWORDS

electrospinning; hierarchical; core-shell; helical; nanoporous; patterned; nanofiber

## 1. INTRODUCTION

Self-assembled nanostructures are gaining attention in the fields of tissue engineering, drug delivery, sensing, catalysis and energy harvesting/storage [1,2,3,4,5,6,7,8]. In particular, nanofiber materials are of great interest because of their long length scales, resulting in very high aspect ratios, and the high degree of fiber orientation possible. Nanofibers are generally prepared using drawing with a micropipette, template synthesis, phase separation, chemical vapor deposition, melt blowing, and electrospinning [9,10,11]. Of the aforementioned techniques, electrospinning is the most versatile and simplest method of preparing such nanofibers. It is also potentially scalable and requires little equipment. Electrospinning is an electrostatically-driven process that generates fibers of nanometer to micrometer diameters. Specifically, when a high voltage is applied to a liquid droplet, electrostatic repulsion overcomes surface tension to stretch the droplet. If molecular cohesion or chain entanglement in the droplet is sufficiently high, the droplet does not break-up (electrospray), but instead continues to stretch to form fine fibers on a grounded collection target. During the electrospinning process, the jet undergoes a whipping instability, which is the primary mechanism for the formation of sub-micron fibers [12,13].

Collector geometry can influence mat morphology, such as fiber alignment and pattern. Collectors can be stationary, such as metal plates or parallel electrodes, or rotating-type disks or mandrels [14]. In general, a randomly oriented web of fibers is collected on static targets, whereas aligned fibers are collected on spinning substrates. Some collectors possess unique geometries to affect particular fiber orientation or patterns. Theron *et al.* used a knife-edge collector to generate highly aligned fibers, whereas others have used collectors with grids or charged needles to create patterned nanofibrous mats [15,16,17,18,19]. Several researchers have used a series of parallel electrodes to generate aligned fibers rather than rotating collectors [20,21,22,23]. Li *et al.* used a pattern of four electrodes to generate a cross-bar array of aligned nanofibers [24].

Polymers in the solution and melt form, as well as non-polymeric materials such as ceramics and metals, can be fabricated into nanofibers via electrospinning. An assortment of natural polymers, such as the proteins collagen, elastin and silk fibroin, has been electrospun [25,26,27,28,29,30,31]. In addition, a wide variety of synthetic polymers such as polystyrene (PS), polycaprolactone (PCL), poly(methyl methacrylate) (PMMA) and poly(vinyl alcohol) (PVA) have been electrospun from solution. Melt electrospinning requires that a heating system surround the reservoir and, despite its advantages of improved cost effectiveness and environmental safety due to the absence of solvents in the process, is generally more challenging. Polyethylene and polypropylene are two polymers that have been successfully fabricated from the melt [32,33]. Ceramic fibers such as $TiO_2$ have been fabricated from a sol of ceramic precursor and a host polymer such as polyvinylpyrrolidone (PVP) [34]. Metallic fibers such as copper fibers have also been generated by electrospinning a metal acetate or nitrate solution with a host polymer, followed by degradation of the host polymer and reduction [35].

The electrospinning field is extremely broad, and consequently there have been many useful reviews discussing various aspects from detailed fiber formation mechanisms to the formation of nanocomposite fibers to discussions on the wide range of applications [24,36,37,38,39]. On the other hand, the focus of this review is quite narrow, highlighting unique fiber morphologies. The review will briefly cover the factors impacting the morphological characteristics of nanofibers such as fiber diameter and porosity, patterning, and then focus on primary and secondary hierarchical structures of electrospun nanofibers.

# 2. GROSS MORPHOLOGY

## 2.1. Fiber Diameter

One of the most highly studied and fundamental areas in electrospinning is control over fiber diameter. Many factors that affect the fiber diameter, including solution concentration, viscosity, molecular weight, polymer feed rate, capillary size, solution conductivity, working distance and applied voltage [11,40]. Shin *et al.* found that the resulting fiber diameter was highly dependent on the stretching and acceleration of the fiber jet prior to solidification [41]. Fridrikh *et al.* developed a model to predict fiber diameter based on surface tension and electrostatic charge repulsion. In this model, solution conductivity and viscosity are also taken into account [42]. Of the many parameters discussed, concentration/solution viscosity is one of the most important factors affecting fiber diameter. Demir *et al.*found that fiber diameter of polyurethane fibers was proportional to the cube of the polymer concentration, following a power law relationship [43]. Figure 1 displays the effect of polymer concentration on the fiber diameter of PCL fibers prepared from dimethylformamide (DMF) and methylene chloride. As can be clearly seen from the images and chart, the fiber diameter increases significantly from 10–20 wt % PCL to 40 wt % PCL.

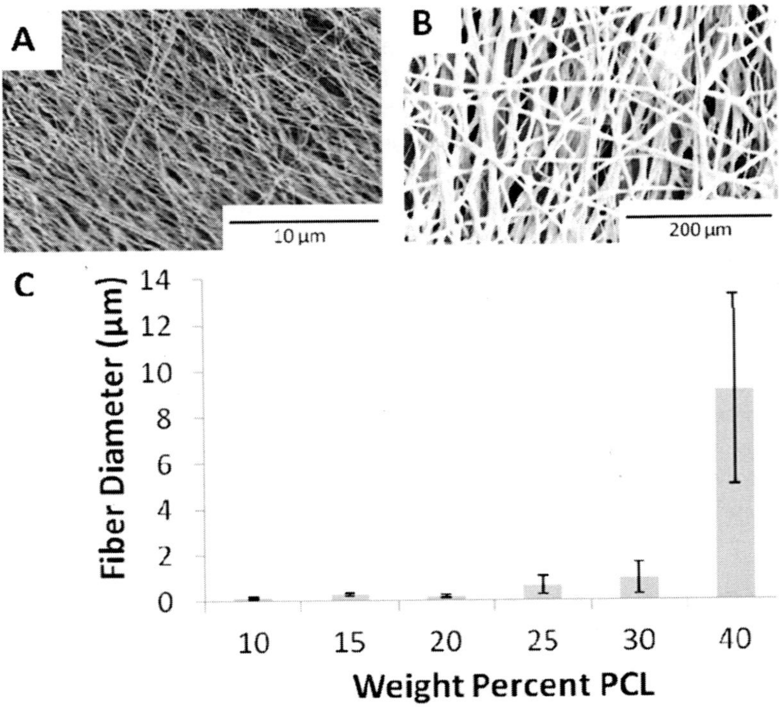


**Figure 1.** Scanning electron micrographs of electrospun polycaprolactone (PCL) fibers formed from solutions of varying PCL concentrations in methylene chloride and dimethylformamide (DMF) (50/50 w/w) solvent. (**A**) 20 wt % PCL; (**B**) 40 wt % PCL; (**C**) chart of PCL concentration (wt %) and fiber diameter. Data are expressed as mean ± standard deviation.

The extent of polymer chain entanglements, expressed as solution viscosity, plays a critical role in controlling not only the fiber diameter, but also the fiber quality or uniformity. If the density of chain entanglements is less than some critical value (as defined by each polymeric system), the electrospinning jet breaks up, and beads or electrosprayed droplets rather than fibers are formed. If the density of entanglements is sufficient but the extent of chain overlap is less than the critical value, beaded fibers are formed [44,45,46]. Interestingly, the morphology of the beads on the fibers changes shape as the solution viscosity changes from a rounded droplet-like shape with low viscosity solutions to stretched droplet or ellipse to smooth fibers when sufficient viscosity is achieved, as shown inFigure 2 [47,48,49]. Zong *et al.* found that not only does the shape of the beads change with viscosity, but the spacing between the beads increases. The researchers postulated that the drying of the fibers had a large effect on resulting fiber morphology. Wet fibers, formed from lower concentration solutions, undergo a solidification process governed by surface tension and a relaxation process controlled by the viscoelastic property. In contrast, fibers formed from higher concentrations are generally dried by the time they are collected [50].

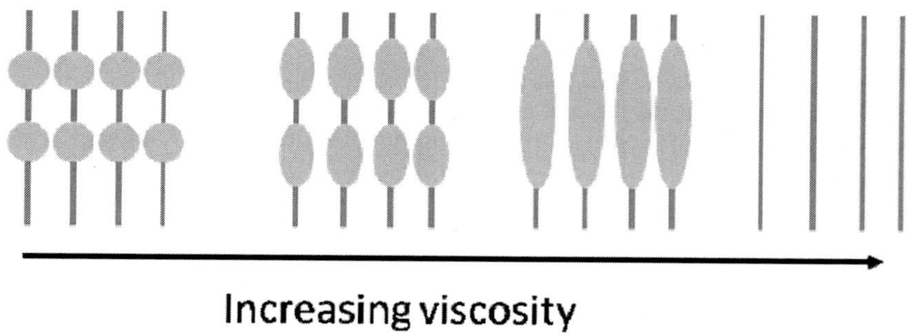


**Figure 2.** Schematic showing the effect of increasing the polymer solution viscosity on the morphology of the resulting electrospun fibers.

## 2.2. Porosity

Fiber mat porosity is another controllable parameter that has been extensively studied, in particular for tissue engineering (cellular infiltration) and filtration applications. There are many approaches to controlling the pore size in nanofiber mats including the addition of salt particles or ice crystals to the electrospinning solution, the formation of composites made of electrospun fibers and electrosprayed polymer beads that can be later removed, or the generation of micron-sized fiber diameter scaffolds [51,52,53,54,55,56]. The latter approach generates micron-sized inter-fiber spaces because pore size, in general, scales with fiber diameter. As a possible compromise to maintain the nanometer-diameter fibers and high surface area and micrometer-diameter fibers for sufficient pore-size for air/fluid diffusion, some researchers have explored forming composites of nanometer and micrometer fibers [57,58], electrospinning nanometer fibers onto micrometer fibers [59], and electrospinning nanometer fibers of two polymers, and then later removing the fibers from one of the polymers to increase the porosity [60].

## 2.3. Patterns

Patterning is another way to control the morphology/architecture of the electrospun fiber mat. The effect of collector geometry, particularly for the alignment of fibers, has been extensively studied [15,24,61,62,63]. Rotating collectors such as mandrels or disks, as well as electrodes separated by a gap, are facile ways to generate oriented fibers [64]. Unaligned or randomly oriented fibers are traditionally collected on a static substrate, such as a grounded metal plate. To generate micro- or macro-scale patterns of nanofibers, researchers have explored the use of conductive grids or a series of charged needles to guide the collection of fibers along the charged regions [17,18,19,65]. Li *et al.* developed a variety of aligned cross-bar patterns using a layer-by-layer approach with series of conductive grids with void gaps [66]. Zussman *et al.* attached an aluminum table to a rotating disk collector to generate cross-bar patterns in a one-step process [67]. Some researchers have developed collectors to generate patterns of fibers that mimic the natural fibrous patterns seen, for example, on bamboo leaves or goose feathers [68]. Figure 3 shows a variety of collectors and their effect on the nanofiber mat pattern. Ding *et al.* recently demonstrated the selective deposition of nanofibers on an elastomeric substrate containing gold-coated pyramidal spikes [69]. Star and radial fiber patterns have also been prepared using enamel patterned steel-sheet collectors [70]. Ye *et al.* developed a method to create a bird's nest pattern of fibers using an unpatterned plate collector. To do this, they doped chlorinated polystyrene (PS) and polypropylene with an ionic liquid that enhanced the charge accumulation of the collecting fibers. Thus, the fibers were able to form only thin layers of fibers aligned on top of one another before depositing in another location on the collector, which eventually created a nest-like pattern. The use of an insulating collector enhanced the pattern formation by increasing the charge accumulation at the collector [71].

Another method used to generate patterned fiber mats is the controlled deposition of fiber using direct-write methods, sometimes referred to as Near Field Electrospinning (NFES) or Scanned Tip Electrospinning Deposition (STED) [72,73,74,75,76]. This process generally uses conductive tips rather than needles with extremely small tip-sample separations [8,77]. At such small working distances, the fibers do not undergo the bending instability and therefore are collected in a more stable region of the jet, allowing for highly controlled fiber collection, generally on a moving substrate. The benefit of this technique is that micro-meter-sized patterns can be generated within the resolution limits of the linear motion stage utilized for collection. Kameoka *et al.* was able to use the STED method to deposit aligned nanoparticles in a polymeric fiber [76].

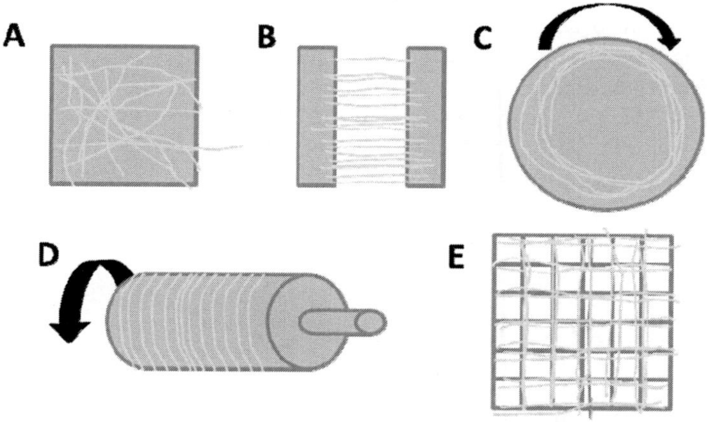


**Figure 3.** Different collector geometries for electrospinning. (**A**) static plate; (**B**) parallel electrodes; (**C**) rotating disk; (**D**) rotating mandrel; (**E**) grid.

Other methods to generate micro-scale patterns include magnetic electrospinning with magnetic nanoparticles doped into the fibers and subsequently aligned in a magnetic field, micro-contact transfer of fiber patterns using a patterned elastomeric stamp, and photolithography [78,79,80,81]. Carlberg *et al.* incorporated the photoinitiator benzoin methyl ether into a polyurethane solution, formed fibers and then exposed the fiber mat to ultraviolet light using a mask to create patterns [80]. Sharma *et al.* used a similar procedure with electrospun fibers composed of a photoresist to generate micro-sized patterns of carbon nanofibers after post-processing steps [81].

# 3. HIERARCHICAL PRIMARY STRUCTURES

## 3.1. Introduction

In addition to the well-studied gross morphology of electrospun fibers, such as fiber diameter, porosity, and the newer area of micro- to macro-scale patterns, electrospun materials exhibit various other fascinating morphologies, such as helical, ribbon-like, beads-on-a-string, multi-channel tubular, porous honeycomb, nanoporous, nanopillared, core-shell and hollow structures, many of which are also observed in nature [82,83,84,85,86]. Much of the focus in generating nanofibers with hierarchical structures is to impart new properties to the materials, such as superhydrophobicity, self-cleaning, or unique optical properties. Figure 4 displays various natural hierarchical structures side-by-side with bio-mimetic versions created from electrospun fibers. These particular bio-mimetic fibers possess the unique property of superhydrophobicity, as observed in the natural lotus and silver ragwort leaves.

## 3.2. Internally Structured Fibers

### 3.2.1. Core Shell Fibers

Core-shell fibers typically are bi-component fibers consisting of a core of one material type and an encapsulating shell of another material. These types of fibers can be very useful in the drug-delivery or tissue engineering fields by allowing the controlled release of a drug or growth factors [91,92,93,94]. In addition, a structurally supportive but non-biocompatible core can be encapsulated with a biocompatible sheath [95,96]. Non-electrospinnable materials, such as low molecular weight or conjugated polymers with limited solubility, can be fabricated into fibers via formation as the fiber core and subsequent removal of the shell [24,97,98,99,100]. Core-shell fibers are traditionally made by electrospinning of the core material, which serves as a template for the deposition of the shell layer via chemical vapor, plasma or solution deposition, or through use of a co-annular needle [101,102,103].

Core-shell fibers, particularly with the core removed to make hollow tubes, also have many applications in the sensing field. Choi *et al.* generated hollow ZnO fiber sensors by coating a sacrificial electrospun fiber template with ZnO followed by calcination. The hollow ZnO fiber membrane showed enhanced sensitivity for $NO_2$ compared to a ZnO thin film [104]. Gao *et al.* prepared polyaniline tubes using electrospun PVA as a template. The nanotubes showed a higher selectivity and quicker response to trimethylamine gas [105]. Park *et al.* made $TiO_2$-ZnO core-shell fibers using a $TiO_2$ template and atomic layer deposition. The core-shell fibers showed good oxygen sensing capability [106]. Zhan *et al.* generated hollow $TiO_2$ fibers using a coaxial spinneret and machine oil as the core fluid. By adjusting the sol and oil pressures, tunable wall thicknesses and inner diameters were achieved [107].

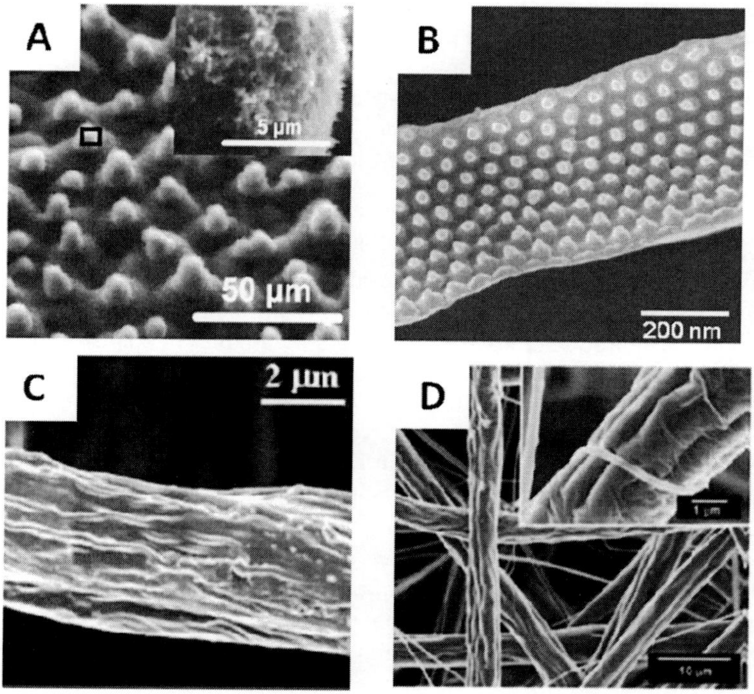


**Figure 4.** Examples of electrospun bio-mimics of natural hierarchical structures. (**A**) Lotus leaf; (**B**) pillared poly(methyl methacrylate) (PMMA) electrospun fiber mimic; (**C**) silver ragwort leaf; (**D**) electrospun fiber mimic made from nylon 6 and polystyrene (PS). (Figure (**A**) reprinted with permission from Wang *et al.* [87]. Copyright 2009 American Chemical Society; Figure (**B**) reprinted with permission from Chen *et al.* [88]. Copyright 2012 American Chemical Society; Figure (**C**) reprinted with permission from Miyauchi *et al.* [89].Copyright 2006 IOP Science; Figure (**D**) reprinted with permission from Li *et al.* [90]. Copyright 2009 American Chemical Society).

As mentioned above, core-shell fibers have many applications in the biological fields. Zhang *et al.* prepared core-shell PCL-collagen fibers using a co-annular needle. The growth of human dermal fibroblasts on core-shell fibers was compared to PCL fibers soaked in collagen. Cellular proliferation was improved up to 200% on the core-shell fibers compared to the surface treated fibers [96]. Zhang *et al.* encapsulated the model protein fluorescein isothiocyanate-conjugated bovine serum albumin (FITC-BSA) within a polyethylene glycol core and PCL shell. Variation of the inner and outer flow rates controlled the loading of FITC-BSA and fiber diameter, respectively [108]. Townsend-Nicholson *et al.* used a coaxial spinneret to encapsulate living mammalian cells in polydimethylsiloxane fibers. The post-cultured cells were viable and showed no evidence of evidence cellular damage from the electrospinning process [109]. Klein *et al.* encapsulated three different strains of bacteria into polyethylene oxide (PEO)/PCL core-shell fibers for bioremediation applications. The encapsulated cells maintained some of their phosphatase, β-galactosidase and denitrification activity [110].

Recently, superhydrophobic core-shell fibers have been fabricated. Han *et al.* used coaxial electrospinning to generate superhydrophobic and oleophobic fibers for the first time. The electrospinning of normally non-electrospinnable Teflon was facilitated with a PCL core using a co-annular needle set-up. Water contact angles of 158° and dodecane contact angles of 130° were reported [111].

There are several drawbacks using the main two methods, template deposition and a co-annular needle, to form core-shell fibers. The former requires two or more steps and may result in a non-uniform shell coating. The latter method has parameters that must be carefully considered, including solution viscosity, miscibility, and interfacial tension requirements between the two solutions to obtain continuous entrainment of the core material within the shell [95]. Recently, core-shell fibers have also been fabricated by electrospinning certain emulsions, a process that avoids many of the drawbacks associated with the co-annular approach [112,113,114].Figure 5 displays a schematic of a co-annular needle as well as a single-nozzle emulsion electrospinning set-up and the formation of core-shell fibers. Figure 6 displays a transmission electron microscopy (TEM) image of a core-shell fiber formed from a polyacrylonitrile (PAN) and PMMA emulsion solution and the corresponding fiber cross-section imaged using atomic force microscopy (AFM).

Hollow fibers can be formed by removing the core material via heating or solvent extraction. Sanders *et al.* used a two solvent single-nozzle electrospinning method to encapsulate the protein bovine serum albumin in poly(ethylene co-vinyl acetate) (EVA) fibers. Rather than a continuous core-shell structure as generated by certain emulsion solutions, the fibers contained microencapsulated aqueous domains containing the protein in the surrounding EVA fibers [115]. A wide variety of core-shell fibers have been generated, including PCL/gelatin, the conjugated polymer polyhexylthiophene with a polypyrrolidone sheath, fibers with various encapsulated live cells, and hollow ceramic fibers using oil as the core [98,102,103,116,117].

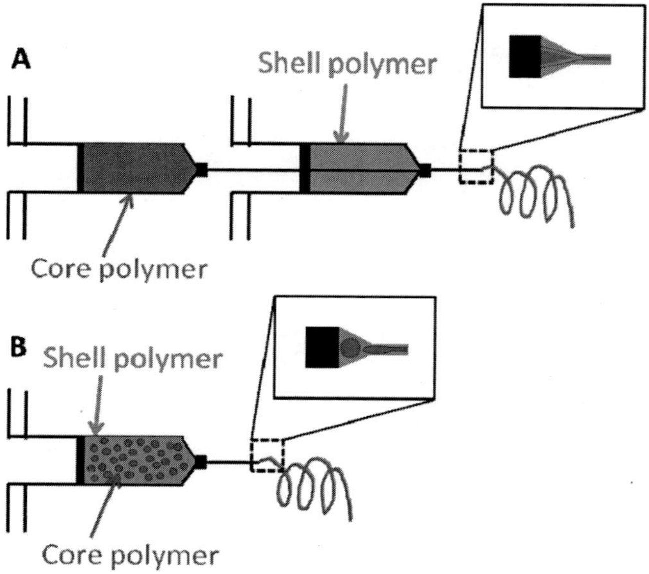


**Figure 5.** (**A**) Schematic of co-annular needle set-up to generate core-shell electrospun fibers; (**B**) single-nozzle set-up to generate core-shell fibers from an emulsion solution. Insets show a detailed schematic of the core-shell jet.

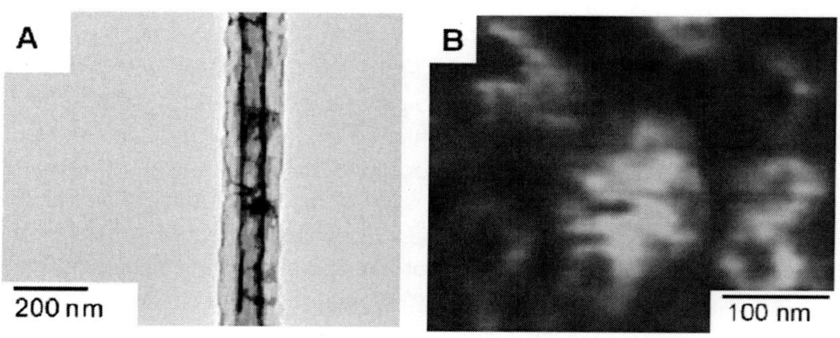


**Figure 6.** Core-shell PMMA-PAN electrospun nanofibers prepared from an emulsion solution. (**A**) TEM image of core-shell fibers; (**B**) AFM phase image of core-shell fiber cross-section. (Reprinted with permission from Zander *et al.* [114]. Copyright 2011 American Chemical Society).

### 3.2.2. Fibers Formed from Block Copolymers

Block copolymers are known to self-assemble into a variety of nanoscale morphologies including spheres, rods, micelles, lamella and cylinders [118]. Block copolymers are generally studied in the bulk or in thin films, but recently there have been some studies of block copolymers in electrospun fibers. The same domains as in the bulk have been observed, but they are generally not as well ordered or as regular in shape due to the fast solvent evaporation and mechanical deformation during the electrospinning process [118,119]. Ruotsalainen *et al.* observed lamellar-within-spherical structures for a poly(styrene-*b*-vinylpyridine) system [119]. Rutledge*et al.* prepared electrospun block copolymers using a co-annular needle system, with the copolymer confined within a shell with a high glass transition temperature [120,121]. The shell served as a process aid for the core material during electrospinning, poly(styrene-*b*-isoprene-*b*-styrene), and also confined the copolymer during the subsequent annealing step [120]. Gyroid-type structures were formed when the block copolymer poly(styrene-*b*-dimethyl siloxane) was used with a poly(methacrylic acid) shell [121]. Ma *et al.* reported the formation of superhydrophobic fibers by the electrospinning poly(styrene-b-dimethylsiloxane). The water contact angle of 163° was attributed to the combined effect of surface enrichment of siloxane and increased surface roughness of the nanometer diameter fibers [122]. These internally structured polymers have a wide variety of applications from advanced filters, optical materials, and templates for nanoscale objects and to direct nanoparticle assembly [120,123].

## 3.3. Helical

Inspired by the structure of proteins and DNA, nanofibers with nano- and micro-scale helical structures exhibit some unique properties, such as higher porosity, elasticity, and toughness, which make them good candidates for surgical fibers, textiles, filters, and components in microelectromechanical systems (MEMs). Helical nanofibers made from conjugated polymers have unique electronic and optical properties. They are expected to be used as building blocks in novel organic nanophotonic and electronic devices, as well as field-effect transistors, energy storage units and sensors [124,125,126,127,128]. Nanoscale helices have been made from the contraction of a manganese oxide sol gel upon solvent evaporation, as well as from zinc oxide and silicon dioxide using vapor processing methods [129,130,131]. Recently, researchers have found that electrospinning is an efficient and facile method to fabricate such helical nanofibers. Usually, two polymers with different properties, such as conductivity or elasticity, are electrospun, either as a blend or as a core-shell or side-by-side system, to generate bi-component fibers. Some examples of these type of bicomponent fibers are poly(*p*-phenylene vinylene) (PPV)/PVP fibers or PEO and the conductive polymer poly(aniline sulfonic acid). The authors suggest that the helical micro-coils observed were formed spontaneously upon contact with a conducting substrate due to the viscoelastic contraction of the fiber upon partial charge neutralization [132,133]. Lin *et al.* developed a microfluidic electrospinning nozzle for the side-by-side deposition of two polymers to make a bi-component fiber [134]. Bi-component fibers with helical morphology were fabricated using this nozzle with equal flow rates from an elastomeric (polyurethane) and thermoplastic (polyacrylonitrile) polymer solution. Adjusting the flow rate of the two polymer solutions influenced the fiber morphology. Higher flow rates of the thermoplastic polymer reduced the helical nature of the resulting fibers, whereas increased flow rates for the polyurethane resulted in curly and beaded fibers. Chen *et al.* electrospun core-shell fibers of a thermoplastic polyurethane and the stiff polymer Nomex®, and observed helices with the addition of LiCl to increase the solution conductivity [135]. The relative amount of the stiff polymer in the core was varied, and the most homogenous nanosprings were formed when the relative ratio of the core and shell polymers was near 50:50.

Other researchers have shown the formation of helical nanofibers from a pure single-component fiber. Shin *et al.* reported the formation of helices from poly(2-acrylamido-2-methyl-1-propanesulfonic acid) dissolved in a mixture of water and ethanol [136]. The helical fibers were able to be linearly oriented when collected on parallel electrodes. In the aforementioned work, the formation of helical structures was more heavily influenced by the physical forces caused by the bending instability of the jet compared to the effects of electrical charge or difference in polymer elastic modulus and differential shrinkage [134]. Canejo *et al.* observed the spontaneous twisting of cellulose fibers in the liquid crystalline phase. They attributed the helical formation to the native chirality of the polymer and the high solution concentration used in the study [137]. Several researchers have focused on using a Near Field Electrospinning approach in which the helical patterns are formed in the region before the jet experiences electrical instability. According to the Han *et al.* model, the straight electrospinning jet spontaneously forms a coil in response to the positive changes carried by the jet upon impingement on the collector [138]. A combination of solution concentration, voltage and frequency of the applied voltage (e.g., sinusoidal) varies the pitch length and radii of the coils. A wide variety of buckling/coiled patterns, including sinusoidal trajectories, coiled structures, and figure-of-eights, can be achieved. Figure 7 displays a selection of coiled fiber patterns generated using the electrospinning approach. Using this technique, researchers have fabricated PS,

PEO, polylactic acid (PLLA) and nylon fibers with helical morphology [138]. In many cases, the addition of salt to the polymer solution increased the conductivity of the solution, although the exact mechanism of how this increases coil formation remains unknown [135,139]. Both stationary collectors, such as water, and laterally moving collectors have been used to collect coiled fibers, with slower collector speeds generating more uniform coils [140,141].

Yu *et al.* fabricated helical PCL fibers, and controlled the coil size based on solution concentration. A tilted glass slide was used as the collector and it was found that tilt angles greater than 45° did not result in helical fibers [142]. Helical light-emitting conjugated polymers in which the polymer molecular weight affects the loop size and corresponding color shift have also been fabricated using the electrospinning technique [143]. The fibers with the most stretched helical geometries were formed from polymer solutions with the highest molecular weights. Xin *et al.* electrospun PPV blended with PEO. By varying the PEO composition, the range of fluorescence could be tuned from yellow-green to blue, making these fibers ideal for micro- and nano-optoelectronic devices [144].

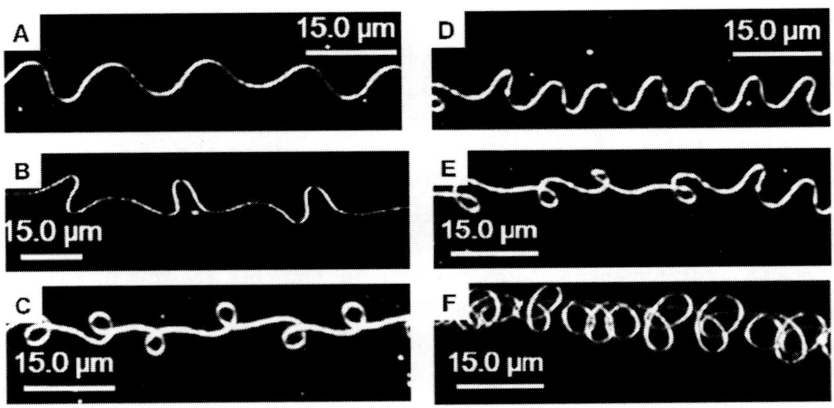


**Figure 7.** Buckled patterns created from electrified jets of polyethylene oxide collected on glass slides (Reprinted with permission from Han *et al.* [138]. Copyright 2007 Polymer).

Helical or curled fibers have also been observed on special collectors such as a nail tip, which concentrates the electric field [145]. The mechanism for the observed fiber morphology has been proposed to be contraction or uniform mechanical buckling of the fiber jet upon contact with the collector

[133,134,136,138,146,147]. The converging electrical field drags the front part of the electrospinning jet toward the collector tip by increasing the bending instabilities of the charged jet. Thus, first- and higher-order bending instabilities will be more likely to occur, increasing the proportion of helical fibers. An increase in the applied voltage increases the speed with which the jet moves toward the collector, making the jet buckling effect more dominant.

The cross-sectional shape of most electrospun fibers, including helical fibers, is circular. However, in some cases ribbon-like cross-sections are formed because of the formation of a skin on the polymeric jet that collapses into a ribbon upon solvent evaporation [141,148]. The electrospinning of nylon-6/clay composites, polycarbonates, polyvinyl alcohol, elastin and hemoglobin have generated ribbon-like fibers [149].

## 3.4. Beads-on-a-String

As discussed briefly in the introduction, beads-on-a-string structures can be fabricated using electrospinning by controlling the solution concentration/viscosity. In general, dilute solutions or solutions made from polymers of low molecular weights aid in the formation of beads-on-a-string-type fibers. However, if chain entanglements are too low, only beads or electrospray will form. The shape and morphology of the beads, from spherical to highly stretched ellipses, can also be tailored by changing the solvent compositions [47,48]. Jiang *et al.* fabricated PS nanofibers with beads in the 3 to 7 μm range in order to mimic lotus leaves [150]. The morphology of the fibers was controlled by adjusting the concentration of PS in the spinning solution. The resulting porous microbeads greatly enhanced the surface roughness and hydrophobicity of the fiber mats. The resulting fibers were superhydrophobic with a water contact angle up to 160°. Ma *et al.* formed superhydrophobic PCL fibers by varying electrospinning parameters to control bead size and density. Fiber mats were then coated with a thin layer of a polymerized perfluoroalkyl ethyl methacrylate to increase the hydrophobicity. Contact angle increased with increasing bead density and decreasing bead size, due to increased surface roughness. Water contact angles as high as 139° for native PCL and 175° for coated PCL fibers were observed [151].

## 3.5. Nanonets or Honeycomb Structure

Honeycomb-like fiber mats or "nanonets" with fiber diameters less than 50 nm were first reported by Deng *et al.* from nylon 6 and poly(acrylic acid) [152]. The mechanism for the net formation was reported to be due to the formation and distortion of micro-sized droplets by various forces such as the electrostatic force, gravity, Coulombic repulsion, surface tension and viscoelastic forces. Rapid phase separation of the distorted droplet and energy minimization may lead to the formation of the spider-web-like structure of the fiber mats [153,154,155,156]. These nanonets are expected to make excellent filters for the removal of nanometer-sized particles and viruses.

# 4. HIERARCHICAL SECONDARY STRUCTURES

## 4.1. Introduction

A variety of interesting secondary structures such as nanopores, nanopillars and nanorods can be prepared using the electrospinning process. These structures may require a two-step process, such as growing carbon nanotubes from nanoparticles doped in electrospun fibers, or single-step processes in which the humidity and solvent parameters are carefully controlled to generate nanometer sized pores along the fiber axis [157,158]. These hierarchical secondary structures, as opposed to smooth fibers traditionally generated by the electrospinning process, impart a variety of properties to the fibers, including superhydrophobicity or superhydrophilicity, improved catalytic, sensing, and ultra-filtration capabilities [159].

## 4.2. Nanopores

The formation of hierarchically-ordered arrays in polymer fibers is of interest because of potential applications in photonic band-gap materials, sensors, and patterned light-emitting diodes [160,161,162,163,164]. These patterned arrays also increase the roughness of the material, which, according to the Cassie-Baxter law, affects the wettability or water contact angle of the surface because it is governed by both the surface energy and roughness [165]. Thus, roughened hydrophobic surfaces are rendered more hydrophobic and roughened hydrophilic surfaces become more hydrophilic [166].

Nanopores on electrospun fibers have been formed using thermally induced phase separation, selective removal of polymer blends and the breath figure process [167,168,169,170,171]. Breath figures are nano- to micron-sized patterned arrays of defects formed when water droplets condense onto a polymer solution surface during drying. This process is driven by the evaporation of the solvent under humid conditions, which leads to a decrease in the temperature at the air-liquid interface. The resulting water condensation begins with the nucleation of droplets in random positions followed by coalescence and potentially organization to minimize thermodynamic energy [172]. Water droplets then sink into the film or cause indentations (depending on solution density) and evaporate, leaving behind a hexagonally-patterned array of pores [161,173].

Figure 8 displays pores formed on electrospun PS fibers. Casper *et al.* reported controlled pore formation on PS and PMMA fibers by varying the solvent composition and solution concentration [158]. Bognitzki *et al.* have reported the formation of nanoscale structures on electrospun PLLA, polycarbonate and poly(vinylcarbazole) electrospun from solutions in dichloromethane [174]. The pores were elliptical in shape and in the range of 200 nm, with the longer dimension oriented with the fiber axis. The cause of pore formation was postulated to be due to the rapid phase separation of the polymer and solvent during the electrospinning process. The use of solvents with lower vapor pressures reduced pore formation. Han *et al.* reported pore formation on cellulose triacetate fibers [175]. Pores in the range of 200 to 500 nm were observed for the 90/10 solvent ratio of dichloromethane to ethanol. However, little or no pore formation was observed when the ethanol content was increased. Miyauchi *et al.* mimicked silver ragwort leaves by preparing PS fibers with high water contact angles from solvent mixtures of tetrahydrofuran (THF) and DMF. Oval-shaped pores were created along the fiber axis with a 1:1 weight ratio of the solvents. When the composition of DMF was increased to 75 wt %, short groves appeared [89]. The change in surface morphology was attributed to rapid phase separation during the electrospinning process [176]. Ma reported the formation of superhydrophobic PMMA fibers formed in a humid environment. The increased surface roughness from the nanometer sized pores and a thin coating of poly(perfluoroalkyl ethyl methacrylate) resulted in water contact angles up to 163° [177]. Cao *et al.*fabricated hydrophobic PLLA nanofibers with nanopores by varying the solvent ratios of DMF and dichloromethane. The resulting water contact angle depended on the pore size, with the 150 nm diameter pores having contact angles of 146.6° [178]. Zhang *et al.* used oxygen etching of PVA/$SnCl_4$ to generate highly porous tin oxide fibers for ethanol sensing. Detection limits of less than 1 ppb were achieved for the porous fibers [179].

## 4.3. Nanoprotrusions

The addition of nanoparticles is another way to increase the surface area and roughness of a material and alter its properties. Lin *et al.* prepared superhydrophobic nanofibers containing both nanoprotrusions and nanogrooves by electrospinning PS solutions with silica nanoparticles [180]. Fiber mats containing 14.3 wt % silica exhibited the highest water contact angles (157.2°), approaching that of the lotus leaf (160°). Fang *et al.* applied a surface coating of silica nanoparticles and a fluorinated alkyl silane to electrospun PAN fibers. The loading of the nanoparticles and resulting surface roughness was found to be critical in generating a superhydrophobic surface [181]. Asmatulu *et al.* incorporated $TiO_2$ and graphene nanoparticles into PS and polyvinyl chloride nanofibers in order to generate self-cleaning photoelectrodes for dye-sensitized solar cells [182]. Wen *et al.* incorporated $SiO_2$ nanoparticles into $SiO_2$ fibers. The specific surface area increased substantially upon addition of the nanoparticles from 4.14 $m^2/g$ to 345 $m^2/g$ with 4.29% loading. These materials are expected to find important applications in dental composites [183].

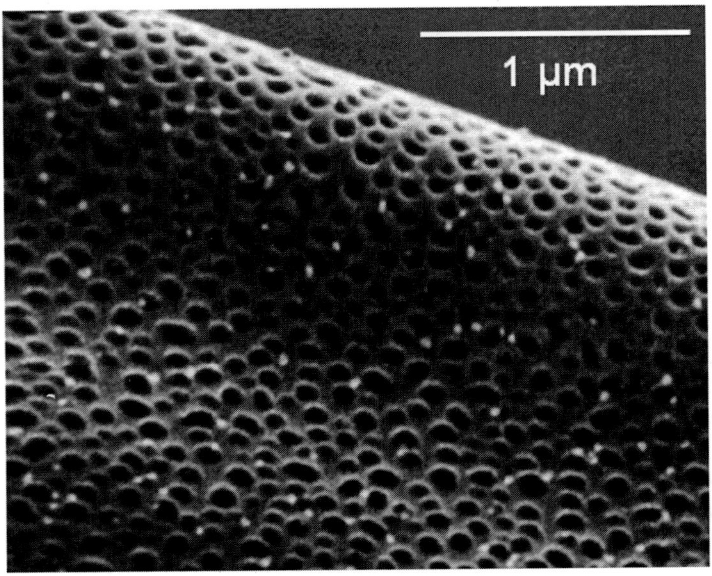


**Figure 8.** Scanning electron microscopy image of electrospun PS in THF at 50% relative humidity. (Reprinted with permission from Casper *et al.* [158]. Copyright 2004 American Chemical Society).

Several researchers have used two-step processes to generate nanoprotrusions on electrospun fibers. Chen *et al.* created nanofiber "shish kebabs" by electrospinning polycaprolactone followed by a solution crystallization of PCL or a copolymer of PCL and PEO [184]. They observed a series of nanoprotrusions from spikes to blocks oriented perpendicular to the fiber axis. The mechanism for this process was similar to the crystallization mechanism of carbon nanotube growth from a surface. Reneker *et al.* and Lai *et al.* grew carbon nanostructures from electrospun carbon nanofibers with iron and palladium nanoparticles, respectively [157,185]. In the latter work, the type of carbon dictated the morphology of the resulting nanostructure. Toluene as the carbon source yielded straight nanotubes; pyridine gave coiled and y-shaped nanotubes; and chlorobenzene formed nanoribbons. Figure 9 displays a variety of carbon nanostructures from rods to y-shaped protrusions.

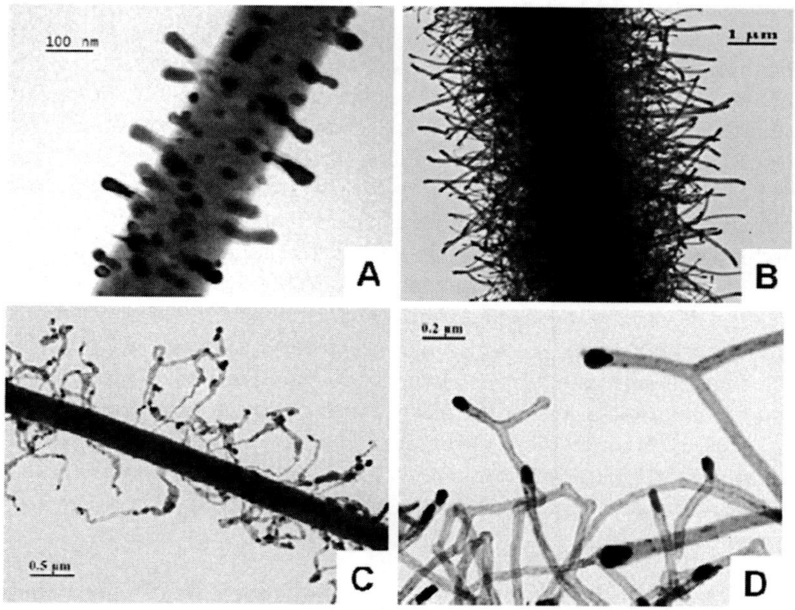


**Figure 9.** TEM images of carbon nanostructures grown on carbon nanofibers. (**A**) Short carbon nanotubes; (**B**) densely packed carbon nanotubes; (**C**) carbon nanoribbons; (**D**) Y-shaped carbon nanotubes. (Reprinted with permission from Lai *et al.* [157]. Copyright 2008 IOP Publishing).

He *et al.* grew polyaniline nanowires on electrospun carbon nanofibers *via in situ* polymerization of aniline. These materials had a high specific capacitance of 976.5 F $g^{-1}$ at a current density of 0.4 A $g^{-1}$ and are expected to serve as electrodes for supercapacitors [186]. Ostermann *et al.* grew $V_2O_5$ nanorods on $V_2O_5/TiO_2$ nanofibers. The size of the resulting nanorods was controlled by the calcination temperature and electrospinning conditions [187]. Wang *et al.* grew

ZnO nanostructures on $TiO_2$ fibers. Both ZnO nanorods and nanoplates were observed [188]. Liu *et al.* grew ZnO nanoneedles on ZnO nanofibers using an aqueous solution process. These needle structures were over 500 nm long and had quite high aspect ratios [189]. A possible application for these materials is in high-performance field emission devices. Chen *et al.* used anodic aluminum oxide templates to create nanopillars on PMMA and PS electrospun fibers. The formed fibers were annealed above their glass transition temperature on the templates, cooled and removed to create ordered arrays of nanorods [88].

# 5. CONCLUSIONS AND OUTLOOK

Although most of the research in the electrospinning field has focused on the use of smooth 1-D nanofibers and the control of their gross morphology (fiber diameter, alignment and quality, and mat porosity), there is an ever-increasing interest in unique primary and secondary structures of nanofibers. Through control of electrospinning processing parameters such as polymer chemistry, solvent, humidity, electric field strength and collector type, a variety of hierarchical structures (helices, beads-on-a-string, core-shell or hollow fibers, and nanoporous or fibers with nanoprotrusions) can be made. These hierarchical fibers greatly expand and enhance the possible applications of electrospun fibers in the fields of nanoelectronics, sensors, MEMs devices, tissue engineering, energy harvesting and storage, meta-materials and photonics, nano/micro-fluidics, catalysis and filtration. It is expected that the range of hierarchical structures and controlled nanofiber organization/patterning will continue to grow, along with the applications of such materials. Many future advances will also most likely be driven by the applications.

Electrospinning offers many advantages for the fabrication of nanomaterials, because it is generally an inexpensive, straightforward facile process to manipulate the alignment and morphology of electrospun fibers via changes in processing conditions. Yet many challenges still remain, including the formation of sub-100-nm-sized fibers, electrospinning of materials with limited solubility, such as conjugated organic polymers, and scale-up of the electrospinning process, especially in the generation of patterned nanofibers or those with hierarchical structures [190]. In addition, more studies need to be done to better correlate processing parameters and the resulting primary and secondary hierarchical structures. Finally, the diversity and scope of electrospun materials still needs to be expanded. More reproducible methods to generate fibers into well-defined arrays with hierarchical structures are needed [191]. As these electrospinning techniques are perfected, and through interdisciplinary collaborations, electrospinning is expected to become more significant in the fields of nanotechnology and bioengineering [143].

## REFERENCES

1. Gopal, R.; Kaur, S.; Feng, C.Y.; Chan, C.; Ramakrishna, S.; Tabe, S.; Matsuura, T. Electrospun nanofibrous polysulfone membranes as pre-filters: Particulate removal. J. Membr. Sci. 2007, 289, 210–219.

2. Boudriot, U.; Dersch, R.; Greiner, A.; Wendorff, J.H. Electrospinning approaches toward scaffold engineering—A brief overview. Artif. Organs 2006, 30, 785–792.

3. Li, W.J.; Laurencin, C.T.; Caterson, E.J.; Tuan, R.S.; Ko, F.K. Electrospun nanofibrous structure: A novel scaffold for tissue engineering. J. Biomed. Mater. Res. 2002, 60, 613–621.

4. Khil, M.S.; Bhattarai, S.R.; Kim, H.Y.; Kim, S.Z.; Lee, K.H. Novel fabricated matrix via electrospinning for tissue engineering.J. Biomed. Mater. Res. B 2005, 72, 117–124.

5. Yoshimoto, H.; Shin, Y.M.; Terai, H.; Vacanti, J.P. A biodegradable nanofiber scaffold by electrospinning and its potential for bone tissue engineering. Biomaterials 2003, 24, 2077–2082.

6. Patel, A.C.; Li, S.X.; Yuan, J.M.; Wei, Y. In situ encapsulation of horseradish peroxidase in electrospun porous silica fibers for potential biosensor applications. Nano Lett. 2006, 6, 1042–1046.

7. Ding, B.; Kim, J.H.; Miyazaki, Y.; Shiratori, S.M. Electrospun nanofibrous membranes coated quartz crystal microbalance as gas sensor for NH3 detection. Sens. Actuators B Chem. 2004, 101, 373–380.

8. Tepper, G.; Kessick, R. Electrospun polymer composite fiber arrays for the detection and identification of volatile organic compounds. Sens. Actuators B Chem. 2006, 117, 205–210.

9. Nakata, K.; Fujii, K.; Ohkoshi, Y.; Gotoh, Y.; Nagura, M.; Numata, M.; Kamiyama, M. Poly(ethylene terephthalate) nanofibers made by sea-island-type conjugated melt spinning and laser-heated flow drawing. Macromol. Rapid Commun. 2007, 28, 792–795.

10. Bansal, V.; Shambaugh, R.L. On-line determination of diameter and temperature during melt blowing of polypropylene. Ind. Eng. Chem. Res. 1998, 37, 1799–1806.

11. Deitzel, J.M.; Kleinmeyer, J.; Harris, D.; Tan, N.C.B. The effect of processing variables on the morphology of electrospun nanofibers and textiles. Polymer 2001, 42, 8163–8170.

12. Hohman, M.M.; Shin, M.; Rutledge, G.; Brenner, M.P. Electrospinning and electrically forced jets. I. Stability theory. Phys. Fluids 2001, 13, 2201–2220.

13. Hohman, M.M.; Shin, M.; Rutledge, G.; Brenner, M.P. Electrospinning and electrically forced jets. II. Applications. Phys. Fluids 2001, 13, 2221–2236.

14. Huang, Z.M.; Zhang, Y.Z.; Kotaki, M.; Ramakrishna, S. A review on polymer nanofibers by electrospinning and their applications in nanocomposites. Compos. Sci. Technol. 2003, 63, 2223–2253.

15. Theron, A.; Zussman, E.; Yarin, A.L. Electrostatic field-assisted alignment of electrospun nanofibres. Nanotechnology 2001, 12, 384–390.

16. Munir, M.M.; Widiyandari, H.; Iskandar, F.; Okuyama, K. Patterned indium tin oxide nanofiber films and their electrical and optical performance. Nanotechnology 2008, 19, 375601–375607.

17. Wang, Y.; Wang, G.; Chen, L.; Li, H.; Yin, T.; Wang, B.; Lee, J.C.M.; Yu, Q. Electrospun nanofiber meshes with tailored architectures and patterns as potential tissue-engineering scaffolds. Biofabrication 2009, 1, 015001:1–015001:9.

18. Zhang, D.; Chang, J. Electrospinning of three-dimensional nanofibrous tubes with controllable architectures. Nano Lett. 2008, 8, 3283–3287.

19. Zhang, K.; Wang, X.; Jing, D.; Yang, Y.; Zhu, M. Bionic electrospun ultrafine fibrous poly(L-lactic acid) scaffolds with a multi-scale structure. Biomed. Mater. 2009, 4, 035004:1–035004:6.

20. Li, D.; Wang, Y.L.; Xia, Y.N. Electrospinning of polymeric and ceramic nanofibers as uniaxially aligned arrays. Nano Lett.2003, 3, 1167–1171.

21. Beachley, V.; Wen, X. Effect of electrospinning parameters on the nanofiber diameter and length. Mater. Sci. Eng. C Mater. Biol. Appl. 2009, 29, 663–668.

22. Liu, L.; Dzenis, Y.A. Analysis of the effects of the residual charge and gap size on electrospun nanofiber alignment in a gap method. Nanotechnology 2008, 19, 355307:1–355307:7.

23. Ishii, Y.; Sakai, H.; Murata, H. A new electrospinning method to control the number and a diameter of uniaxially aligned polymer fibers. Mater. Lett. 2008, 62, 3370–3372.

24. Li, D.; Xia, Y.N. Electrospinning of nanofibers: Reinventing the wheel? Adv. Mater. 2004, 16, 1151–1170.

25. Li, M.Y.; Mondrinos, M.J.; Gandhi, M.R.; Ko, F.K.; Weiss, A.S.; Lelkes, P.I. Electrospun protein fibers as matrices for tissue engineering. Biomaterials 2005, 26, 5999–6008.

26. Li, J.; He, A.; Zheng, J.; Han, C.C. Gelatin and gelatin-hyaluronic acid nanofibrous membranes produced by electrospinning of their aqueous solutions. Biomacromolecules 2006, 7, 2243–2247.

27. Li, C.M.; Vepari, C.; Jin, H.J.; Kim, H.J.; Kaplan, D.L. Electrospun silk-BMP-2 scaffolds for bone tissue engineering.Biomaterials 2006, 27, 3115–3124.

28. Zhang, Y.Z.; Ouyang, H.W.; Lim, C.T.; Ramakrishna, S.; Huang, Z.M. Electrospinning of gelatin fibers and gelatin/PCL composite fibrous scaffolds. J. Biomed. Mater. Res. B 2005, 72, 156–165.

29. Zhang, Y.Z.; Venugopal, J.; Huang, Z.M.; Lim, C.T.; Ramakrishna, S. Crosslinking of the electrospun gelatin nanofibers.Polymer 2006, 47, 2911–2917.

30. Zhong, S.; Teo, W.E.; Zhu, X.; Beuerman, R.W.; Ramakrishna, S.; Yung, L.Y.L. An aligned nanofibrous collagen scaffold by electrospinning and its effects on in vitro fibroblast culture. J. Biomed. Mater. Res A 2006, 79, 456–463.

31. Matthews, J.A.; Wnek, G.E.; Simpson, D.G.; Bowlin, G.L. Electrospinning of collagen nanofibers. Biomacromolecules 2002, 3, 232–238.

32. Zhou, Z.P.; Lai, C.L.; Zhang, L.F.; Qian, Y.; Hou, H.Q.; Reneker, D.H.; Fong, H. Development of carbon nanofibers from aligned electrospun polyacrylonitrile nanofiber bundles and characterization of their microstructural, electrical, and mechanical properties. Polymer 2009, 50, 2999–3006.

33. Zhmayev, E.; Zhou, H.; Joo, Y.L. Modeling of non-isothermal polymer jets in melt electrospinning. J. Non-Newton. Fluid Mech.2008, 153, 95–108.

34. Zhang, X.; Thavasi, V.; Mhaisalkar, S.G.; Ramakrishna, S. Novel hollow mesoporous 1D TiO2 nanofibers as photovoltaic and photocatalytic materials. Nanoscale 2012, 4, 1707–1716.

35. Wu, H.; Hu, L.; Rowell, M.W.; Kong, D.; Cha, J.J.; McDonough, J.R.; Zhu, J.; Yang, Y.; McGehee, M.D.; Cui, Y. Electrospun metal nanofiber webs as high-performance transparent electrode. Nano Lett. 2010, 10, 4242–4248.

36. Sahay, R.; Kumar, P.S.; Sridhar, R.; Sundaramurthy, J.; Venugopal, J.; Mhaisalkar, S.G.; Ramakrishna, S. Electrospun composite nanofibers and their multifaceted applications. J. Mater. Chem. 2012, 22, 12953–12971.

37. Ramakrishna, S.; Jose, R.; Archana, P.S.; Nair, A.S.; Balamurugan, R.; Venugopal, J.; Teo, W.E. Science and engineering of electrospun nanofibers for advances in clean energy, water filtration, and regenerative medicine. J. Mater. Sci. 2010, 45, 6283–6312.

38. Greiner, A.; Wendorff, J.H. Electrospinning: A fascinating method for the preparation of ultrathin fibres. Angew. Chem. Int. Ed.2007, 46, 5670–5703.

39. Dersch, R.; Graeser, M.; Greiner, A.; Wendorff, J.H. Electrospinning of nanofibres: Towards new techniques, functions, and applications. Aust. J. Chem. 2007, 60, 719–728.

40. Theron, S.A.; Zussman, E.; Yarin, A.L. Experimental investigation of the governing parameters in the electrospinning of polymer solutions. Polymer 2004, 45, 2017–2030.

41. Shin, Y.M.; Hohman, M.M.; Brenner, M.P.; Rutledge, G.C. Experimental characterization of electrospinning: the electrically forced jet and instabilities. Polymer 2001, 42, 9955–9967.

42. Fridrikh, S.V.; Yu, J.H.; Brenner, M.P.; Rutledge, G.C. Controlling the fiber diameter during electrospinning. Phys. Rev. Lett.2003, 90, 144502:1–144502:4.

43. Demir, M.M.; Yilgor, I.; Yilgor, E.; Erman, B. Electrospinning of polyurethane fibers. Polymer 2002, 43, 3303–3309.

44. Kim, H.W.; Yu, H.S.; Lee, H.H. Nanofibrous matrices of poly(lactic acid) and gelatin polymeric blends for the improvement of cellular responses. J. Biomed. Mater. Res. A 2008, 87, 25–32.

45. Zong, X.H.; Ran, S.F.; Kim, K.S.; Fang, D.F.; Hsiao, B.S.; Chu, B. Structure and morphology changes during in vitrodegradation of electrospun

poly(glycolide-co-lactide) nanofiber membrane. Biomacromolecules 2003, 4, 416–423.

46. Reneker, D.H.; Kataphinan, W.; Theron, A.; Zussman, E.; Yarin, A.L. Nanofiber garlands of polycaprolactone by electrospinning. Polymer 2002, 43, 6785–6794.

47. Liu, J.; Rasheed, A.; Dong, H.; Carr, W.W.; Dadmun, M.D.; Kumar, S. Electrospun micro- and nanostructured polymer particles. Macromol. Chem. Phys. 2008, 209, 2390–2398.

48. Liu, J.; Kumar, S. Microscopic polymer cups by electrospinning. Polymer 2005, 46, 3211–3214.

49. Fong, H.; Chun, I.; Reneker, D.H. Beaded nanofibers formed during electrospinning. Polymer 1999, 40, 4585–4592.

50. Zong, X.H.; Kim, K.; Fang, D.F.; Ran, S.F.; Hsiao, B.S.; Chu, B. Structure and process relationship of electrospun bioabsorbable nanofiber membranes. Polymer 2002, 43, 4403–4412.

51. Baiguera, S.; Del Gaudio, C.; Fioravanzo, L.; Bianco, A.; Grigioni, M.; Folin, M. In vitro astrocyte and cerebral endothelial cell response to electrospun poly(epsilon-caprolactone) mats of different architecture. J. Mater. Sci. Mater. Med. 2010, 21, 1353–1362.

52. Balguid, A.; Mol, A.; van Marion, M.H.; Bank, R.A.; Bouten, C.V.C.; Baaijens, F.P.T. Tailoring fiber diameter in electrospun poly(epsilon-caprolactone) scaffolds for optimal cellular infiltration in cardiovascular tissue engineering. Tissue Eng. A 2009,15, 437–444.

53. Nam, J.; Huang, Y.; Agarwal, S.; Lannutti, J. Improved cellular infiltration in electrospun fiber via engineered porosity. Tissue Eng. 2007, 13, 2249–2257.

54. Leong, M.F.; Rasheed, M.Z.; Lim, T.C.; Chian, K.S. In vitro cell infiltration and in vivo cell infiltration and vascularization in a fibrous, highly porous poly(D,L-lactide) scaffold fabricated by cryogenic electrospinning technique. J. Biomed. Mater. Res. A2009, 91, 231–240.

55. Huang, Y.Y.; Wang, D.Y.; Chang, L.L.; Yang, Y.C. Fabricating microparticles/nanofibers composite and nanofiber scaffold with controllable pore size by rotating multichannel electrospinning. J. Biomater. Sci. Polym. E. 2010, 21, 1503–1514.

56. Kim, S.J.; Jang, D.H.; Park, W.H.; Min, B.M. Fabrication and characterization of 3-dimensional PLGA nanofiber/microfiber composite scaffolds. Polymer 2010, 51, 1320–1327.

57. Pham, Q.P.; Sharma, U.; Mikos, A.G. Electrospun poly(epsilon-caprolactone) microfiber and multilayer nanofiber/microfiber scaffolds: Characterization of scaffolds and measurement of cellular infiltration. Biomacromolecules 2006, 7, 2796–2805.

58. Ju, Y.M.; Choi, J.S.; Atala, A.; Yoo, J.J.; Lee, S.J. Bilayered scaffold for engineering cellularized blood vessels. Biomaterials2010, 31, 4313–4321.

59. Thorvaldsson, A.; Stenhamre, H.; Gatenholm, P.; Walkenstrom, P. Electrospinning of highly porous scaffolds for cartilage regeneration. Biomacromolecules 2008, 9, 1044–1049.

60. Baker, B.M.; Gee, A.O.; Metter, R.B.; Nathan, A.S.; Marklein, R.A.; Burdick, J.A.; Mauck, R.L. The potential to improve cell infiltration in composite fiber-aligned electrospun scaffolds by the selective removal of sacrificial fibers. Biomaterials 2008, 29, 2348–2358.

61. Katta, P.; Alessandro, M.; Ramsier, R.D.; Chase, G.G. Continuous electrospinning of aligned polymer nanofibers onto a wire drum collector. Nano Lett. 2004, 4, 2215–2218.

62. Li, D.; Ouyang, G.; McCann, J.T.; Xia, Y.N. Collecting electrospun nanofibers with patterned electrodes. Nano Lett. 2005, 5, 913–916.

63. Sundaray, B.; Subramanian, V.; Natarajan, T.S.; Xiang, R.Z.; Chang, C.C.; Fann, W.S. Electrospinning of continuous aligned polymer fibers. Appl. Phys. Lett. 2004, 84, 1222–1224.

64. Teo, W.E.; Ramakrishna, S. A review on electrospinning design and nanofibre assemblies. Nanotechnology 2006, 17, R89–R106.

65. Ner, Y.; Asemota, C.; Olson, J.R.; Sotzing, G.A. Nanofiber alignment on a flexible substrate: Hierarchical order from macro to nano. ACS Appl. Mater. Interfaces 2009, 1, 2093–2097.

66. Li, D.; Wang, Y.L.; Xia, Y.N. Electrospinning nanofibers as uniaxially aligned arrays and layer-by-layer stacked films. Adv. Mater. 2004, 16, 361–366.

67. Zussman, E.; Theron, A.; Yarin, A.L. Formation of nanofiber crossbars in electrospinning. Appl. Phys. Lett. 2003, 82, 973–975.

68. Lin, J.; Wang, X.; Ding, B.; Yu, J.; Sun, G.; Wang, M. Biomimicry via electrospinning. Crit. Rev. Solid State Mater. Sci. 2012,37, 94–114.

69. Ding, Z.; Salim, A.; Ziaie, B. Selective nanofiber deposition through field-enhanced electrospinning. Langmuir 2009, 25, 9648–9652.

70. Zucchelli, A.; Fabiani, D.; Gualandi, C.; Focarete, M.L. An innovative and versatile approach to design highly porous, patterned, nanofibrous polymeric materials. J. Mater. Sci. 2009, 44, 4969–4975.

71. Ye, X.Y.; Huang, X.J.; Xu, Z.K. Nanofibrous mats with bird's nest patterns by electrospinning. Chin. J. Polym. Sci. 2012, 30, 130–137.

72. Sun, D.H.; Chang, C.; Li, S.; Lin, L.W. Near-field electrospinning. Nano Lett. 2006, 6, 839–842.

73. Chang, C.; Limkrailassiri, K.; Lin, L. Continuous near-field electrospinning for large area deposition of orderly nanofiber patterns. Appl. Phys. Lett. 2008, 93, 123111:1–123111:3.

74. Zheng, G.; Li, W.; Wang, X.; Wu, D.; Sun, D.; Lin, L. Precision deposition of a nanofibre by near-field electrospinning. J. Phys. D Appl. Phys. 2010, 43, 415501:1–415501:6.

75. Kameoka, J.; Craighead, H.G. Fabrication of oriented polymeric nanofibers on planar surfaces by electrospinning. Appl. Phys. Lett. 2003, 83, 371–373.

76. Kameoka, J.; Orth, R.; Yang, Y.N.; Czaplewski, D.; Mathers, R.; Coates, G.W.; Craighead, H.G. A scanning tip electrospinning source for deposition of oriented nanofibres. Nanotechnology 2003, 14, 1124–1129.

77. Tao, S.; Li, G.; Yin, J. Fluorescent nanofibrous membranes for trace detection of TNT vapor. J. Mater. Chem. 2007, 17, 2730–2736.

78. Yang, D.; Lu, B.; Zhao, Y.; Jiang, X. Fabrication of aligned fibirous arrays by magnetic electrospinning. Adv. Mater. 2007, 19, 3702–3706.

79. Shi, J.; Wang, L.; Chen, Y. Microcontact printing and lithographic patterning of electrospun nanofibers. Langmuir 2009, 25, 6015–6018.

80. Carlberg, B.; Wang, T.; Liu, J. Direct photolithographic patterning of electrospun films for defined nanofibrillar microarchitectures. Langmuir 2010, 26, 2235–2239.

81. Sharma, C.S.; Sharma, A.; Madou, M. Multiscale carbon structures fabricated by direct micropatterning of electrospun mats of su-8 photoresist nanofibers. Langmuir 2010, 26, 2218–2222.

82. Ding, B.; Wang, M.; Wang, X.; Yu, J.; Sun, G. Electrospun nanomaterials for ultrasensitive sensors. Mater. Today 2010, 13, 16–27.

83. Charra, F.; Cousty, J. Surface-induced chirality in a self-assembled monolayer of discotic liquid crystal. Phys. Rev. Lett. 1998,80, 1682–1685.

84. Schull, G.; Douillard, L.; Fiorini-Debuisschert, C.; Charra, F.; Mathevet, F.; Kreher, D.; Attias, A.J. Single-molecule dynamics in a self-assembled 2D molecular sieve. Nano Lett. 2006, 6, 1360–1363.

85. Yang, W.Y.; Lee, E.; Lee, M. Tubular organization with coiled ribbon from amphiphilic rigid-flexible macrocycle. J. Am. Chem. Soc. 2006, 128, 3484–3485.

86. Akagi, K.; Piao, G.; Kaneko, S.; Sakamaki, K.; Shirakawa, H.; Kyotani, M. Helical polyacetylene synthesized with a chiral nematic reaction field. Science 1998, 282, 1683–1686.

87. Wang, J.; Zheng, Y.; Nie, F.Q.; Zhai, J.; Jiang, L. Air bubble bursting effect of lotus leaf. Langmuir 2009, 25, 14129–14134.

88. Chen, J.T.; Chen, W.L.; Fan, P.W. Hierarchical structures by wetting porous templates with electrospun polymer fibers. ACS Macro Lett. 2012, 1, 41–46.

89. Miyauchi, Y.; Ding, B.; Shiratori, S. Fabrication of a silver-ragwort-leaf-like super-hydrophobic micro/nanoporous fibrous mat surface by electrospinning. Nanotechnology 2006, 17, 5151–5156.

90. Li, X.; Ding, B.; Lin, J.; Yu, J.; Sun, G. Enhanced mechanical properties of superhydrophobic microfibrous polystyrene mats via polyamide 6 nanofibers. J. Phys. Chem. C 2009, 113, 20425–20457.

91. Huang, Z.M.; Zhang, Y.Z.; Ramakrishna, S. Double-layered composite nanofibers and their mechanical performance. J. Polym. Sci. B Polym. Phys. 2005, 43, 2852–2861.

92. Jiang, H.L.; Hu, Y.Q.; Li, Y.; Zhao, P.C.; Zhu, K.J.; Chen, W.L. A facile technique to prepare biodegradable coaxial electrospun nanofibers for controlled release of bioactive agents. J. Control. Release 2005, 108, 237–243.

93. Sun, B.; Duan, B.; Yuan, X.Y. Preparation of core/shell PVP/PLA ultrafine fibers by coaxial electrospinning. J. Appl. Polym. Sci. 2006, 102, 39–45.

94. Srikar, R.; Yarin, A.L.; Megaridis, C.M.; Bazilevsky, A.V.; Kelley, E. Desorption-limited mechanism of release from polymer nanofibers. Langmuir 2008, 24, 965–974.

95. Moghe, A.K.; Gupta, B.S. Co-axial electrospinning for nanofiber structures: Preparation and applications. Polym. Rev. 2008, 48, 353–377.

96. Zhang, Y.Z.; Venugopal, J.; Huang, Z.M.; Lim, C.T.; Ramakrishna, S. Characterization of the surface biocompatibility of the electrospun PCL-collagen nanofibers using fibroblasts. Biomacromolecules 2005, 6, 2583–2589.

97. Cao, G.; Brinker, C.J. Annual Review of Nano Research; World Scientific: Hackensack, NJ, USA, 2006.

98. Sun, Z.C.; Zussman, E.; Yarin, A.L.; Wendorff, J.H.; Greiner, A. Compound core-shell polymer nanofibers by co-electrospinning. Adv. Mater. 2003, 15, 1929–1932.

99. Yu, J.H.; Fridrikh, S.V.; Rutledge, G.C. Production of submicrometer diameter fibers by two-fluid electrospinning. Adv. Mater.2004, 16, 1562–1566.

100. Greiner, A.; Wendorff, J.H.; Yarin, A.L.; Zussman, E. Biohybrid nanosystems with polymer nanofibers and nanotubes. Appl. Microbiol. Biotechnol. 2006, 71, 387–393.

101. Zussman, E.; Yarin, A.L.; Bazilevsky, A.V.; Avrahami, R.; Feldman, M. Electrospun polyacrylonitrile/poly (methyl methacrylate)-derived turbostratic carbon micro-/nanotubes. Adv. Mater. 2006, 18, 348–353.

102. Loscertales, I.G.; Barrero, A.; Marquez, M.; Spretz, R.; Velarde-Ortiz, R.; Larsen, G. Electrically forced coaxial nanojets for one-step hollow nanofiber design. J. Am. Chem. Soc. 2004, 126, 5376–5377.

103. Li, D.; Xia, Y.N. Direct fabrication of composite and ceramic hollow nanofibers by electrospinning. Nano Lett. 2004, 4, 933–938.

104. Choi, S.H.; Ankonina, G.; Youn, D.Y.; Oh, S.G.; Hong, J.M.; Rothschild, A.; Kim, I.D. Hollow ZnO nanofibers fabricated using electrospun polymer templates and their electronic transport properties. Acs Nano 2009, 3, 2623–2631.

105. Gao, Y.; Li, X.; Gong, J.; Fan, B.; Su, Z.; Qu, L. Polyaniline nanotubes prepared using fiber mats membrane as the template and their gas-response behavior. J. Phys. Chem. C 2008, 112, 8215–8222.

106. Park, J.Y.; Choi, S.W.; Lee, J.W.; Lee, C.; Kim, S.S. Synthesis and gas sensing properties of TiO2-ZnO core-shell nanofibers.J. Am. Ceram. Soc. 2009, 92, 2551–2554.

107. Zhan, S.; Chen, D.; Jiao, X.; Tao, C. Long TiO2 hollow fibers with mesoporous walls: Sol-gel combined electrospun fabrication and photocatalytic properties. J. Phys. Chem. B 2006, 110, 11199–11214.

108. Zhang, Y.Z.; Wang, X.; Feng, Y.; Li, J.; Lim, C.T.; Ramakrishna, S. Coaxial electrospinning of (fluorescein isothiocyanate-conjugated bovine serum albumin)-encapsulated poly(epsilon-caprolactone)nanofibers for sustained release. Biomacromolecules2006, 7, 1049–1057.

109. Townsend-Nicholson, A.; Jayasinghe, S.N. Cell electrospinning: A unique biotechnique for encapsulating living organisms for generating active biological microthreads/scaffolds. Biomacromolecules 2006, 7, 3364–3369.

110. Klein, S.; Kuhn, J.; Avrahami, R.; Tarre, S.; Beliavski, M.; Green, M.; Zussman, E. Encapsulation of bacterial cells in electrospun microtubes. Biomacromolecules 2009, 10, 1751–1756.

111. Han, D.; Steckl, A.J. Superhydrophobic and oleophobic fibers by coaxial electrospinning. Langmuir 2009, 25, 9454–9462.

112. Bazilevsky, A.V.; Yarin, A.L.; Megaridis, C.M. Co-electrospinning of core-shell fibers using a single-nozzle technique.Langmuir 2007, 23, 2311–2314.

113. Kim, C.; Jeong, Y.I.; Ngoc, B.T.N.; Yang, K.S.; Kojima, M.; Kim, Y.A.; Endo, M.; Lee, J.W. Synthesis and characterization of porous carbon nanofibers with hollow cores through the thermal treatment of electrospun copolymeric nanofiber webs. Small2007, 3, 91–95.

114. Zander, N.; Strawhecker, K.; Orlicki, J.; Rawlett, A.; Beebe, T. Coaxial electrospun poly(methyl methacrylate)-polyacrylonitrile nanofibers: Atomic force microscopy and compositional characterization. J. Phys. Chem. B 2011, 115, 12441–12447.

115. Sanders, E.H.; Kloefkorn, R.; Bowlin, G.L.; Simpson, D.G.; Wnek, G.E. Two-phase electrospinning from a single electrified jet: Microencapsulation of aqueous reservoirs in poly(ethylene-co-vinyl acetate) fibers. Macromolecules 2003, 36, 3803–3805.

116. Zhang, Y.Z.; Huang, Z.M.; Xu, X.J.; Lim, C.T.; Ramakrishna, S. Preparation of core-shell structured PCL-r-gelatin Bi-component nanofibers by coaxial electrospinning. Chem. Mater. 2004, 16, 3406–3409.

117. Li, D.; Babel, A.; Jenekhe, S.A.; Xia, Y.N. Nanofibers of conjugated polymers prepared by electrospinning with a two-capillary spinneret. Adv. Mater. 2004, 16, 2062–2066.

118. Kalra, V.; Kakad, P.A.; Mendez, S.; Ivannikov, T.; Kamperman, M.; Joo, Y.L. Self-assembled structures in electrospun poly(styrene-block-isoprene) fibers. Macromolecules 2006, 39, 5453–5457.

119. Ruotsalainen, T.; Turku, J.; Heikkila, P.; Ruokolainen, J.; Nykanen, A.; Laitinen, T.; Torkkeli, M.; Serimaa, R.; ten Brinke, G.; Harlin, A.; Ikkala, O. Towards internal structuring of electrospun fibers by hierarchical self-assembly of polymeric comb-shaped supramolecules. Adv. Mater. 2005, 17, 1048–1052.

120. Ma, M.L.; Krikorian, V.; Yu, J.H.; Thomas, E.L.; Rutledge, G.C. Electrospun polymer nanofibers with internal periodic structure obtained by microphase separation of cylindrically confined block copolymers. Nano Lett. 2006, 6, 2969–2972.

121. Ma, M.L.; Thomas, E.L.; Rutledge, G.C.; Yu, B.; Li, B.H.; Jin, Q.H.; Ding, D.T.; Shi, A.C. Gyroid-forming diblock copolymers confined in cylindrical geometry: A case of extreme makeover for domain morphology. Macromolecules 2010, 43, 3061–3071.

122. Ma, M.L.; Hill, R.M.; Lowery, J.L.; Fridrikh, S.V.; Rutledge, G.C. Electrospun poly(styrene-block-dimethylsiloxane) block copolymer fibers exhibiting superhydrophobicity. Langmuir 2005, 21, 5549–5554.

123. Kalra, V.; Mendez, S.; Lee, J.H.; Nguyen, H.; Marquez, M.; Joo, Y.L. Confined assembly in coaxially electrospun block-copolymer fibers. Adv. Mater. 2006, 18, 3299–3203.

124. Grevin, B.; Rannou, P. Electrochemistry: Arrays of polymer nanowires. Nat. Mater. 2004, 3, 503–504.

125. di Benedetto, F.; Camposeo, A.; Pagliara, S.; Mele, E.; Persano, L.; Stabile, R.; Cingolani, R.; Pisignano, D. Patterning of light-emitting conjugated polymer nanofibres. Nat. Nanotechnol. 2008, 3, 614–619.

126. Babel, A.; Li, D.; Xia, Y.N.; Jenekhe, S.A. Electrospun nanofibers of blends of conjugated polymers: Morphology, optical properties, and field-effect transistors. Macromolecules 2005, 38, 4705–4711.

127. Gupta, V.; Miura, N. High performance electrochemical supercapacitor from electrochemically synthesized nanostructured polyaniline. Mater. Lett. 2006, 60, 1466–1469.

128. Li, W.G.; Wang, H.L. Oligomer-assisted synthesis of chiral polyaniline nanofibers. J. Am. Chem. Soc. 2004, 126, 2278–2279.

129. Veretennikov, I.; Indeikina, A.; Chang, H.C.; Marquez, M.; Suib, S.L.; Giraldo, O. Mechanism for helical gel formation from evaporation of colloidal solutions. Langmuir 2002, 18, 8792–8798.

130. Kong, X.Y.; Wang, Z.L. Spontaneous polarization-induced nanohelixes, nanosprings, and nanorings of piezoelectric nanobelts.Nano Lett. 2003, 3, 1625–1631.

131. Zhang, H.F.; Wang, C.M.; Buck, E.C.; Wang, L.S. Synthesis, characterization, and manipulation of helical SiO2 nanosprings.Nano Lett. 2003, 3, 577–580.

132. Xin, Y.; Huang, Z.H.; Yan, E.Y.; Zhang, W.; Zhao, Q. Controlling poly(p-phenylene vinylene)/poly(vinyl pyrrolidone) composite nanofibers in different morphologies by electrospinning. Appl. Phys. Lett. 2006, 89, 053101:1–053101:3.

133. Kessick, R.; Tepper, G. Microscale polymeric helical structures produced by electrospinning. Appl. Phys. Lett. 2004, 84, 4807–4809.

134. Lin, T.; Wang, H.X.; Wang, X.G. Self-crimping bicomponent nanofibers electrospun from polyacrylonitrile and elastomeric polyurethane. Adv. Mater. 2005, 17, 2699–2703.

135. Chen, S.L.; Hou, H.Q.; Hu, P.; Wendorff, J.H.; Greiner, A.; Agarwal, S. Polymeric nanosprings by bicomponent electrospinning.Macromol. Mater. Eng. 2009, 294, 265–271.

136. Shin, M.K.; Kim, S.I.; Kim, S.J. Controlled assembly of polymer nanofibers: From helical springs to fully extended. Appl. Phys. Lett. 2006, 88, 223109:1–223109:3.

137. Canejo, J.P.; Borges, J.P.; Godinho, M.H.; Brogueira, P.; Teixeira, P.I.C.; Terentjev, E.M. Helical twisting of electrospun liquid crystalline cellulose micro- and nanofibers. Adv. Mater. 2008, 20, 4821–4825.

138.Han, T.; Reneker, D.H.; Yarin, A.L. Buckling of jets in electrospinning. Polymer 2007, 48, 6064–6076.

139.Xin, Y.; Reneker, D.H. Garland formation process in electrospinning. Polymer 2012, 53, 3629–3635.

140.Xin, Y.; Reneker, D.H. Hierarchical polystyrene patterns produced by electrospinning. Polymer 2012, 53, 4254–4261.

141.Reneker, D.H.; Yarin, A.L. Electrospinning jets and polymer nanofibers. Polymer 2008, 49, 2387–2425.

142.Yu, J.; Qiu, Y.; Zha, X.; Yu, M.; Yu, J.; Rafique, J.; Yin, J. Production of aligned helical polymer nanofibers by electrospinning.Eur. Polym. J. 2008, 44, 2838–2844.

143.Pagliara, S.; Camposeo, A.; Cingolani, R.; Pisignano, D. Hierarchical assembly of light-emitting polymer nanofibers in helical morphologies. Appl. Phys. Lett. 2009, 95, 263301:1–263301:3.

144.Xin, Y.; Huang, Z.; Jiang, Z.; Che, L.; Sun, M.; Wang, C.; Liu, S. Fluorescent poly(p-phenylene vinylene)/poly(ethylene oxide) nanofibers obtained by electrospinning. J. Polym. Res. 2011, 18, 477–482.

145.Tang, C.C.; Chen, J.C.; Long, Y.Z.; Yin, H.X.; Sun, B.; Zhang, H.D. Preparation of curled microfibers by electrospinning with tip collector. Chin. Phys. Lett. 2011, 28, 056801:1–056801:3.

146.Reneker, D.H.; Yarin, A.L.; Fong, H.; Koombhongse, S. Bending instability of electrically charged liquid jets of polymer solutions in electrospinning. J. Appl. Phys. 2000, 87, 4531–4547.

147.Zhang, B.; Li, C.; Chang, M. Curled poly(ethylene glycol terephthalate)/poly(ethylene propanediol terephthalate) nanofibers produced by side-by-side electrospinning. Polym. J. 2009, 41, 252–253.

148.Koombhongse, S.; Liu, W.X.; Reneker, D.H. Flat polymer ribbons and other shapes by electrospinning. J. Polym. Sci. B Polym. Phys. 2001, 39, 2598–2606.

149.Barnes, C.P.; Smith, M.J.; Bowlin, G.L.; Sell, S.A.; Tang, T.; Matthews, J.A.; Simpson, D.G.; Nimtz, J.C. Feasibility of electrospinning the globular proteins hemoglobin and myoglobin. J. Eng. Fibers Fabr. 2006, 1, 16–29.

150. Jiang, L.; Zhao, Y.; Zhai, J. A lotus-leaf-like superhydrophobic surface: A porous microsphere/nanofiber composite film prepared by electrohydrodynamics. Angew. Chem. Int. Ed. 2004, 43, 4338–4341.

151. Ma, M.L.; Mao, Y.; Gupta, M.; Gleason, K.K.; Rutledge, G.C. Superhydrophobic fabrics produced by electrospinning and chemical vapor deposition. Macromolecules 2005, 38, 9742–9748.

152. Bin, D.; Chunrong, L.; Miyauchi, Y.; Kuwaki, O.; Shiratori, S. Formation of novel 2D polymer nanowebs via electrospinning.Nanotechnology 2006, 17, 3685–3691.

153. Wang, X.F.; Ding, B.; Yu, J.Y.; Wang, M.R.; Pan, F.K. A highly sensitive humidity sensor based on a nanofibrous membrane coated quartz crystal microbalance. Nanotechnology 2010, 21, 055502–055507.

154. Yang, S.B.; Wang, X.F.; Ding, B.; Yu, J.Y.; Qian, J.F.; Sun, G. Controllable fabrication of soap-bubble-like structured polyacrylic acid nano-nets via electro-netting. Nanoscale 2011, 3, 564–568.

155. Wang, X.F.; Ding, B.; Yu, J.Y.; Yang, J.M. Large-scale fabrication of two-dimensional spider-web-like gelatin nano-nets via electro-netting. Colloid Surf. B 2011, 86, 345–352.

156. Wang, X.F.; Ding, B.; Yu, J.Y.; Si, Y.; Yang, S.B.; Sun, G. Electro-netting: Fabrication of two-dimensional nano-nets for highly sensitive trimethylamine sensing. Nanoscale 2011, 3, 911–915.

157. Lai, C.; Guo, Q.H.; Wu, X.F.; Reneker, D.H.; Hou, H. Growth of carbon nanostructures on carbonized electrospun nanofibers with palladium nanoparticles. Nanotechnology 2008, 19, 195303–195309.

158. Casper, C.L.; Stephens, J.S.; Tassi, N.G.; Chase, D.B.; Rabolt, J.F. Controlling surface morphology of electrospun polystyrene fibers: Effect of humidity and molecular weight in the electrospinning process. Macromolecules 2004, 37, 573–578.

159. Zheng, J.F.; Zhang, H.Y.; Zhao, Z.G.; Han, C.C. Construction of hierarchical structures by electrospinning or electrospraying.Polymer 2012, 53, 546–554.

160. Boker, A.; Lin, Y.; Chiapperini, K.; Horowitz, R.; Thompson, M.; Carreon, V.; Xu, T.; Abetz, C.; Skaff, H.; Dinsmore, A.D.; Emrick, T.; Russell, T.P. Hierarchical nanoparticle assemblies formed by decorating breath figures. Nat. Mater. 2004, 3, 302–306.

161. Park, M.S.; Kim, J.K. Breath figure patterns prepared by spin coating in a dry environment. Langmuir 2004, 20, 5347–5352.

162. Yabu, H.; Tanaka, M.; Ijiro, K.; Shimomura, M. Preparation of honeycomb-patterned polyimide films by self-organization.Langmuir 2003, 19, 6297–6300.

163. Yabu, H.; Shimomura, M. Simple fabrication of micro lens arrays. Langmuir 2005, 21, 1709–1711.

164. Karikari, A.S.; Williams, S.R.; Heisey, C.L.; Rawlett, A.M.; Long, T.E. Porous thin films based on photo-cross-linked star-shaped poly(D,L-lactide)s. Langmuir 2006, 22, 9687–9693.

165. Cassie, A.B.D.; Baxter, S. Wettability of porous surfaces. Trans. Faraday Soc. 1944, 40, 546–551.

166. Wenzel, R.N. Resistance of solid surfaces to wetting by water. Ind. Eng. Chem. 1936, 28, 988–994.

167. Gibson, P.; Schreuder-Gibson, H.; Rivin, D. Transport properties of porous membranes based on electrospun nanofibers. Colloid Surf. A 2001, 187, 469–481.

168. Doshi, J.; Reneker, D.H. Electrospinning process and applications of electrospun fibers. J. Electrost. 1995, 35, 151–160.

169. Srinivasan, G.; Reneker, D.H. Structure and morphology of small-diameter electrospun aramid fibers. Polym. Int. 1995, 36, 195–201.

170. Peng, J.; Han, Y.C.; Yang, Y.M.; Li, B.Y. The influencing factors on the macroporous formation in polymer films by water droplet templating. Polymer 2004, 45, 447–452.

171. Barrow, M.S.; Jones, R.L.; Park, J.O.; Srinivasarao, M.; Williams, P.R.; Wright, C.J. Physical characterisation of microporous and nanoporous polymer films by atomic force microscopy, scanning electron microscopy and high speed video microphotography. Spectrosc. Int. J. 2004, 18, 577–585.

172. Marcosmartin, M.; Beysens, D.; Bouchaud, J.P.; Godreche, C.; Yekutieli, I. Self-diffusion and visited surface in the droplet condensation problem (breath figures). Physica A 1995, 214, 396–412.

173. Srinivasarao, M.; Collings, D.; Philips, A.; Patel, S. Three-dimensionally ordered array of air bubbles in a polymer film. Science2001, 292, 79–83.

174. Bognitzki, M.; Czado, W.; Frese, T.; Schaper, A.; Hellwig, M.; Steinhart, M.; Greiner, A.; Wendorff, J.H. Nanostructured fibers via electrospinning. Adv. Mater. 2001, 13, 70–72.

175. Han, S.O.; Son, W.K.; Youk, J.H.; Lee, T.S.; Park, W.H. Ultrafine porous fibers electrospun from cellulose triacetate. Mater. Lett. 2005, 59, 2998–3001.

176. Megelski, S.; Stephens, J.S.; Chase, D.B.; Rabolt, J.F. Micro- and nanostructured surface morphology on electrospun polymer fibers. Macromolecules 2002, 35, 8456–8466.

177. Ma, M.; Gupta, M.; Li, Z.; Zhai, L.; Gleason, K.K.; Cohen, R.E.; Rubner, M.F.; Rutledge, G.C. Decorated electrospun fibers exhibiting superhydrophobicity. Adv. Mater. 2007, 19, 255–259.

178. Cao, S.; Hu, B.; Liu, H. Fabrication of nano-porous structured polylactide (plla) fibers through electrospinning. Acta Polym. Sin.2010, 10, 1193–1198.

179. Zhang, Y.; Li, J.; An, G.; He, X. Highly porous SnO2 fibers by electrospinning and oxygen plasma etching and its ethanol-sensing properties. Sens. Actuators B 2010, 144, 43–48.

180. Lin, J.Y.; Cai, Y.; Wang, X.F.; Ding, B.; Yu, J.Y.; Wang, M.R. Fabrication of biomimetic superhydrophobic surfaces inspired by lotus leaf and silver ragwort leaf. Nanoscale 2011, 3, 1258–1262.

181. Fang, J.; Wang, H.; Wang, X.; Lin, T. Superhydrophobic nanofibre membranes: Effects of particulate coating on hydrophobicity and surface properties. J. Text. Inst. 2012, 103, 937–944.

182. Asmatulu, R.; Ceylan, M.; Nuraje, N. Study of superhydrophobic electrospun nanocomposite fibers for energy systems.Langmuir 2011, 27, 504–507.

183. Wen, S.P.; Liu, L.; Zhang, L.F.; Chen, Q.; Zhang, L.Q.; Fong, H. Hierarchical electrospun SiO2 nanofibers containing SiO2nanoparticles with controllable surface-roughness and/or porosity. Mater. Lett. 2010, 64, 1517–1520.

184. Chen, X.; Dong, B.; Wang, B.B.; Shah, R.; Li, C.Y. Crystalline block copolymer decorated, hierarchically ordered polymer nanofibers. Macromolecules 2010, 43, 9918–9927.

185. Hou, H.Q.; Reneker, D.H. Carbon nanotubes on carbon nanofibers: A novel structure based on electrospun polymer nanofibers.Adv. Mater. 2004, 16, 69–73.

186. He, S.J.; Hu, X.W.; Chen, S.L.; Hu, H.; Hanif, M.; Hou, H.Q. Needle-like polyaniline nanowires on graphite nanofibers: Hierarchical micro/nano-architecture for high performance supercapacitors. J. Mater. Chem. 2012, 22, 5114–5120.

187. Ostermann, R.; Li, D.; Yin, Y.D.; McCann, J.T.; Xia, Y.N. V2O5 nanorods on TiO2 nanofibers: A new class of hierarchical nanostructures enabled by electrospinning and calcination. Nano Lett. 2006, 6, 1297–1302.

188. Wang, N.; Sun, C.; Zhao, Y.; Zhou, S.; Chen, P.; Jiang, L. Fabrication of three-dimensional ZnO/TiO2 heteroarchitectures via a solution process. J. Mater. Chem. 2008, 18, 3909–3911.

189. Liu, Y.X.; Xie, Y.Z.; Chen, J.T.; Liu, J.; Gao, C.T.; Yun, C.; Lu, B.G.; Xie, E.Q. High field emission performance of needle-on-fiber hierarchical-structure ZnO. J. Am. Ceram. Soc. 2011, 94, 4387–4390.

190. Bellan, L.M.; Craighead, H.G. Applications of controlled electrospinning systems. Polym. Adv. Technol. 2011, 22, 304–309.

191. Yang, J.; Zhan, S.; Wang, N.; Wang, X.; Li, Y.; Li, Y.; Ma, W.; Yu, H. A mini review: Electrospun hierarchical nanofibers. J. Dispers. Sci. Technol. 2010, 31, 760–769.

# Chapter 3

# Preparation of Cellulose-based Nanofibers Using Electrospinning

**Youn-Mook Lim[1], Hui-Jeong Gwon, Joon Pyo Jeun and Young-Chang Nho**

[1] Advanced Radiation Technology Institute, Korea Atomic Energy Research Institute, Republic of Korea

## 1. INTRODUCTION

Eelctrospinning is a straightforward method to prepare fibers with diameters as small as several tens of nanometers (Doshi & Reneker, 1995). In eclectrospinning, a high electrostatic voltage is imposed on a drop of polymer solution held by its surface tension at the end of a capillary. The surface of the liquid is distorted into a conical shape known as the Taylor cone. Once the voltage exceeds a critical value, the electrostatic force overcomes the solution surface tension and a stable liquid jet is ejected from the cone tip. Solvent evaporates as the jet travels through the air, leaving behind ultrafine polymeric fibers collected on an electrically grounded target (Fong et al., 1999, 2002; Shin et al., 2001). Electrospun mats have a larger specific surface area and small pore size compared to commercial non-woven fabrics. They are of interest in a wide variety of applications including semi-permeable membranes, tissue engineering scaffolds and drug delivery systems (Tsai et al., 2002; Gibson et al., 2001; Kenawy et al., 2002; Luu et al., 2003).

Recently, electrospun nanofibers (NFs) based on cellulose and its derivatives have been studied as potential candidates for applications within the field of pharmaceuticals. For instance, several reports deal with the investigation of electrospun fiber mats as delivery vehicles, showing dosage forms with useful and controllable dissolution properties. This interest in cellulose-based NFs is primarily driven by its environmental value as a biomaterial. The cellulose is an abundant and renewable resource found in most parts of the world, which makes it a cheap raw material for various applications (Zeng et al., 2003; Jiang et al., 2004; Verreck et al., 2003; Liu & Hsieh, 2002). However, little research has been done on the use of cellulose and cellulose derivatives as a raw material within electrospinning. The complications involved in electrospinning of cellulose are mainly due to the many difficulties ascribed to the material, one being its reluctance to interact with conventional solvents. Therefore, the choice of solvent systems is very important.

Ethyl-cellulose (EC) is a kind of cellulose ether, and it shows a non-biodegradable and biocompatible polymer. EC is one of the extensively studied encapsulating materials for the controlled release of pharmaceuticals (Prasertmanakit et al., 2009). The film made from EC has quite good permeability, it has been widely used industrial air filter (Park et al., 2007).

Hydroxypropyl methylcellulose (HPMC) is frequently used as the basis for sustained release hydrophilic matrix tablets (Ford, 1999). HPMC backbone is composed of glucose units linked by β-1,4-glycosidic bonds (Zhao et al., 2009; Péreza et al., 2008). The hydrophobic (methyl) and hydrophilic (hydropropyl) groups are distributed along the cellulose backbone. It can be lower the surface tension when adsorbed at fluid interface (Zhao et al., 2009; Kulicke et al., 1998). The HPMC molecule is rod-like and has many hydroxyl groups to participate in intermolecular hydrogen bonds, which allows the cellulose chains to form bundles or fibrils (Zhao et al., 2009; Péreza et al., 2008).

However, this chapter deals with the electrospinning of cellulose derivatives such as EC and HPMC. The fiber morphology and influence of experimental parameters including the solution concentration, flow rate and electric voltage and pattern of the derivatives on the eclectrospinning process were discussed.

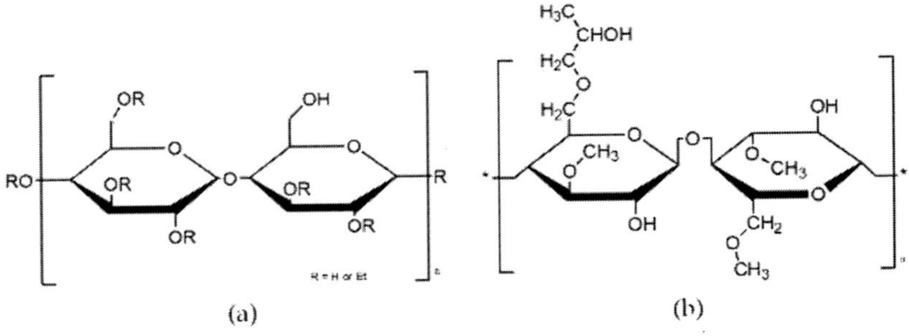


**Figure 1.** The structure of (a) ethylcellulose (EC) and (b) hydroxypropyl methylcellulose (HPMC).

# 2. EXPERIMENTAL

## 2.1. Materials

EC (ethoxy content, 48%; viscosity, 46 cps) and HPMC ($M_n$, 120000; D.S. (methoxy), 1.10-1.60; M.S. (propylene oxide), 0.10-0.30; viscosity, 100000cps 2wt% in $H_2O$ at 20°C) were obtained from Aldrich Chemical and used without further purification. 2,2,2-Trifluoroethanol (Aldrich, USA) was used as a solvent and used as received from commercial supplier.

## 2.2. Preparation and Characterization Of Cellulose Derivatives Solutions

EC and HPMC solutions were prepared by dissolving a measured amount of EC and HPMC at room temperature. Four concentrations were prepared about EC: 6, 8, 10 and 12 wt%. Two concentrations were prepared about HPMC: 0.5, 1.0 wt%. These solutions were characterized for their viscosity using a Brookfield DV-II viscometer.

## 2.3. Electrospinning Of Cellulose Derivatives Solutions

In the electrospinning process, a high electric potential was applied to a droplet of EC solution at the tip (ID 0.15 mm) of a syringe needle. The electrospun EC NFs were collected on a target drum which was placed at a distance of 5 cm from the syringe tip. A voltage of 5-28 keV was applied to the collecting target by a high voltage power supply. The flow rate of polymer solution was 0.001-0.01 ml/min. The HPMC NFs were collected on a target plate which was placed at a distance of 5-15 cm from the syringe tip. A voltage of 10, 15 keV was applied to the collecting target by a high voltage power supply. The flow rate of polymer solution was 0.05 ml/min. Fig. 2 shows the photograph of electrospinning apparatus.

## 2.4. Characterization Of Cellulose Nfs

The morphological changes of the samples were observed using a scanning electron microscope (SEM, JMS6390 JEOL Company). Specimen surfaces were coated with a thin layer of gold palladium alloy by sputtering to provide a conductive surface. The fiber diameter of the electrospun fibers was measured by Adobe Photoshop 5.0 software from the SEM pictures in original magnification of 10k X.

**Figure 2.** The photograph of electrospinning apparatus.

**Table 1.** Electrospinning condition of the cellulose solutions.

| Polymers | Concentration [wt%] | Electric voltage [keV] | Flow rate [ml/ min] | Tip-target distance [cm] | Note |
|---|---|---|---|---|---|
| EC | 6 | 10 | 0.001 | 5 | |
| | 8 | 15 | 0.003 | 7 | |
| | 10 | 20 | 0.005 | 9 | |
| | 12 | 25 | 0.01 | 11 | |
| HPMC | 0.5 | 10 | 0.05 | 5 | |
| | 1.0 | 15 | | 10 | |
| | - | - | | 15 | |

# 3. RESULT AND DISCUSSION

## 3.1. Effect of The Solution Concentration

The viscosity of the polymer solutions are summarized in Table 1. Increasing the concentration of the cellulose solution increased. Viscosity plays an important role in determining the rage of concentrations from which continuous fibers can be obtained by electrospinning. At low viscosity, surface tension is the dominant effect on fiber morphology. Below a certain concentration, drops will form instead of fibers. At high concentrations, processing will be prohibited by an inability to control and maintain the flow of a polymer solution to the tip of the needle and by the cohesive nature of the high viscosity. As shown in Fig. 3, with increasing concentration, the morphology was changed from beaded fiber to uniform fiber structure and the fiber diameter was also increased gradually. At 6wt%, a large amount of smooth fibers and beaded fibers were obtained. (With increasing the concentration of the solution to 8 wt%, very small amount of beads was present.) With further increasing concentration of the solution to 10 wt%, only smooth, bead-less fibers were obtained. At 12 wt%, the average fiber diameter was much larger than those of fibers spun at lower concentrations.

**Table 2.** Viscosity of polymer solutions as a function of concentration.

| Polymers | Concentration [wt%] | Viscosity [cP] | note |
|---|---|---|---|
| EC | 6 | 461 | |
| | 8 | 1245 | |
| | 10 | 2918 | |
| | 12 | 6347 | |
| HPMC | 0.5 | 1284 | |
| | 1.0 | 8614 | |

The morphologies of HPMC NFs in Fig. 4. (??) No big change in the morphology of fibers as a function of concentration. At 0.5wt%, a large amount of smooth fibers and thick fibers was obtained. Increasing the concentration of the solution to 1.0 wt%, only smooth, bead-less fibers were obtained. The average fiber diameter was investigated 155 and 162nm, respectively.

In electrospinning, the coiled macromolecules in solution were transformed by the elongational flow of the jet into oriented entangled networks that persist with fiber solidification. Below this concentration, chain entanglements were insufficient to stabilize the jet and the contraction of the diameter of the jet driven by surface tension caused the solution to form beads or beaded fibers. At high concentration, viscoelastic force which resisted rapid changes in fiber shape resulted in uniform fiber formation. However, it was impossible to electrospin if the solution concentration or the corresponding viscosity was too high due to the difficulty in liquid jet formation.

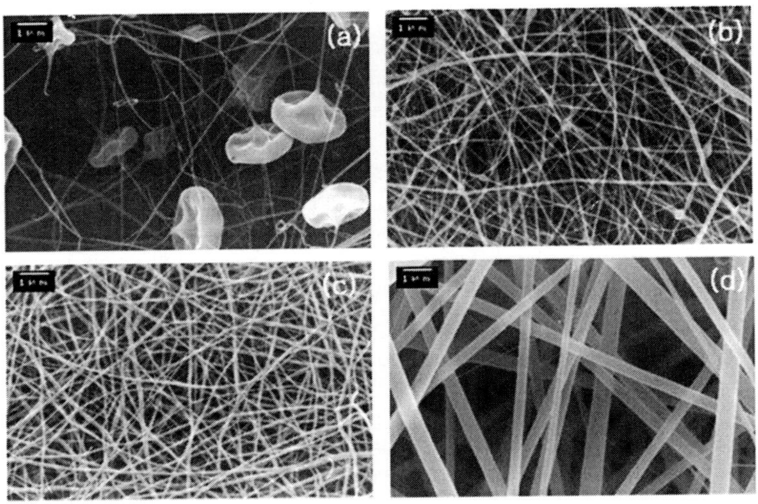


**Figure 3.** SEM micrographs of electrospun fibers from EC solution with different solution concentration (voltage=10 keV, flow rate=0.005 ml/min). EC concentration; (a) 6 wt%; (b) 8 wt%; (c) 10 wt%; (d) 12 wt%.

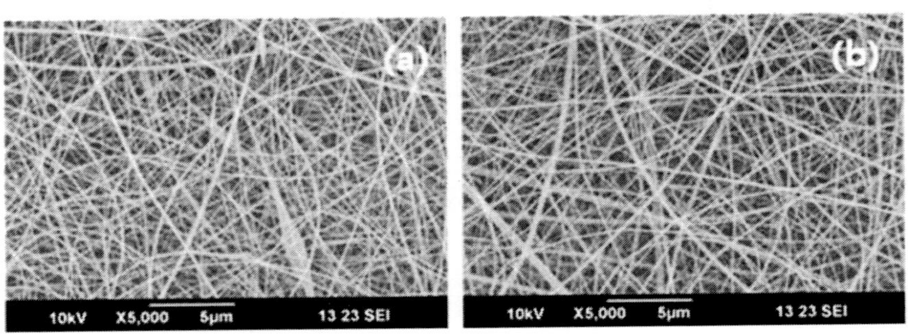


**Figure 4.** SEM micrographs of electrospun fibers from HPMC solution with different solution concentration (voltage=15 keV, flow rate=0.05 ml/min). HPMC concentration; (a) 0.5 wt%; (b) 1.0 wt%.

## 3.2. Effect of The Electrospinning Flow Rate

The morphological structure can be changed by changing the solution flow rate as shown in Fig. 5. At the flow rate of 0.001 ml/min, uniform fiber structure was observed. But increasing flow rate, the average fiber diameter was much larger than those of fibers spun at lower flow rate. At the flow rate of 0.01 ml/min, a considerable amount of thick fibers with diameters above 1mm were found. When the flow rate exceeded a critical value, the delivery rate of the solution jet to the capillary tip exceeded the rate at which the solution was removed from the tip by the electric forces. This shift in the mass-balance resulted in sustained but unstable jet and fibers with broad distribution in the fiber diameter were formed.

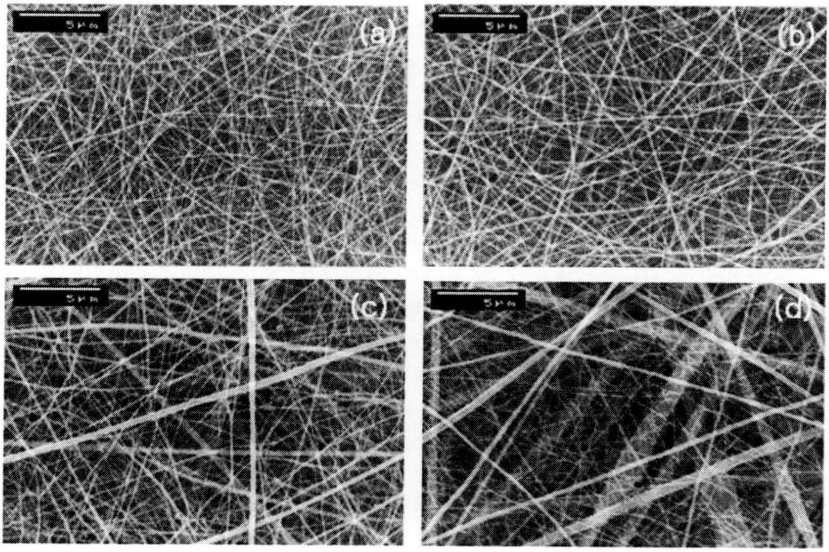


**Figure 5.** Effect of flow rate of 10 wt% EC solution on fiber morphology (voltage=15 kV). Flow rate: (a) 0.001 ml/min; (b) 0.003 ml/min; (c) 0.005 ml/min; (d) 0.01 ml/min.

## 3.3. Effect of The Electric Voltage

A series of experiments were carried out when the applied voltage was varied from 5 to 28 keV. The results are shown in Fig. 6 and Fig. 7. Weaker voltage, such as 5 keV, was not strong enough to overcome the surface tension and viscoelastic forces of the polymer solution, but at higher voltages at this distance, electrical discharge would occur. There was a slight decrease in average fiber diameter with increasing applied electric field. A considerable amount of thin fibers with diameters below 100 nm were found when the applied voltage above 20 keV. Increasing the applied voltage, i.e., increasing field strength will increase the electrostatic repulsion force on the fluid jet which favors the thinner fiber formation. On the other hand, the solution will be removed from the capillary tip more quickly as jet is ejected from Taylor cone. This results in the increase of the fiber diameter. Corona discharge was observed at voltage above 28 keV, making electrospinning impossible.

The diameter of HPMC NFs was measured (??) a narrow distribution when the electric voltage decreased. It was found that the fibers had an average diameter of 129nm at a voltage of 10keV and an average diameter of 140nm at a voltage of 15keV. Electrospun NFs of HPMC showed similar results to EC NFs

## 3.4. Effect of The Needle Tip-Target Distance

Fig. 8 shows that NFs formed with decreasing tip-target distance of EC. As tip-target distance of NFs was increased, the diameter of NFs. The NFs had non-uniform diameters. It may be assumed that the lower electric strength solvent from evaporating. However, (?? we were observed that formation of HPMC NFs regardless of tip-target distance. ??) The most uniform NFs derived with our experimental conditions were afforded at 5cm of tip-target distance. Increasing diameter of NFs at 10cm and decreasing diameter of NFs at 15cm (??).

# 4. CONCLUSIONS

Submicron EC and HPMC fibers have been successfully prepared by electrospinning of polymer solutions. The morphology of the fibers was strongly affected by the parameters such as polymer concentration, tip-target distance, solution flow rate, and applied voltage. At below 6 wt% EC solutions, electrospinning was not enhanced due to the low viscosity. (the concentration of EC solution,) the morphology was changed from beaded fiber to uniform fiber structure and the fiber diameter was also increased.

There was an increase in fiber diameter and fiber distribution with increasing solution flow rate, while there was a slightly decrease in average fiber diameter with increasing applied electric field. The HPMC NFs are feasible for use as wound healing and drug delivery systems for biomedical applications.

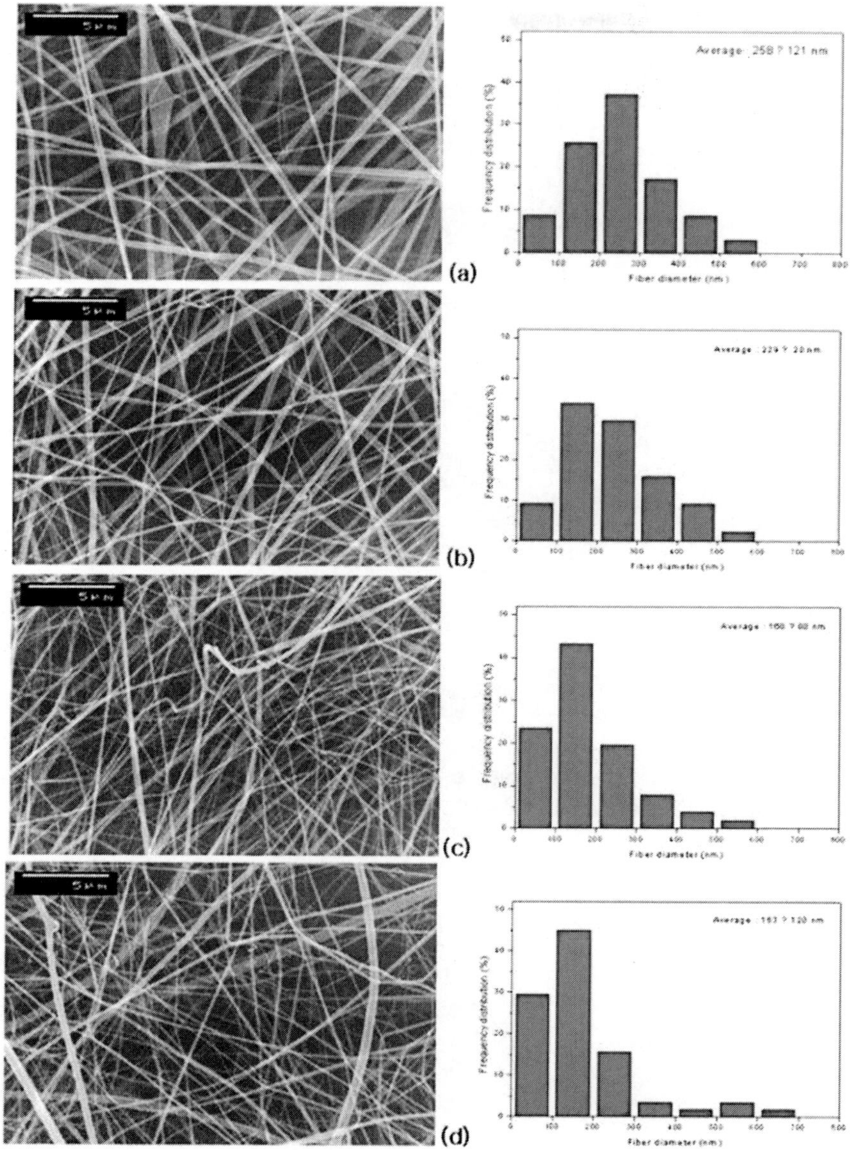


**Figure 6.** Effect of electric voltage on EC fiber morphology (solution concentration=10 wt%, flow rate=0.005 ml/min). Voltage: (a) 10 keV; (b) 15 keV; (c) 20 keV; (d) 25 keV.

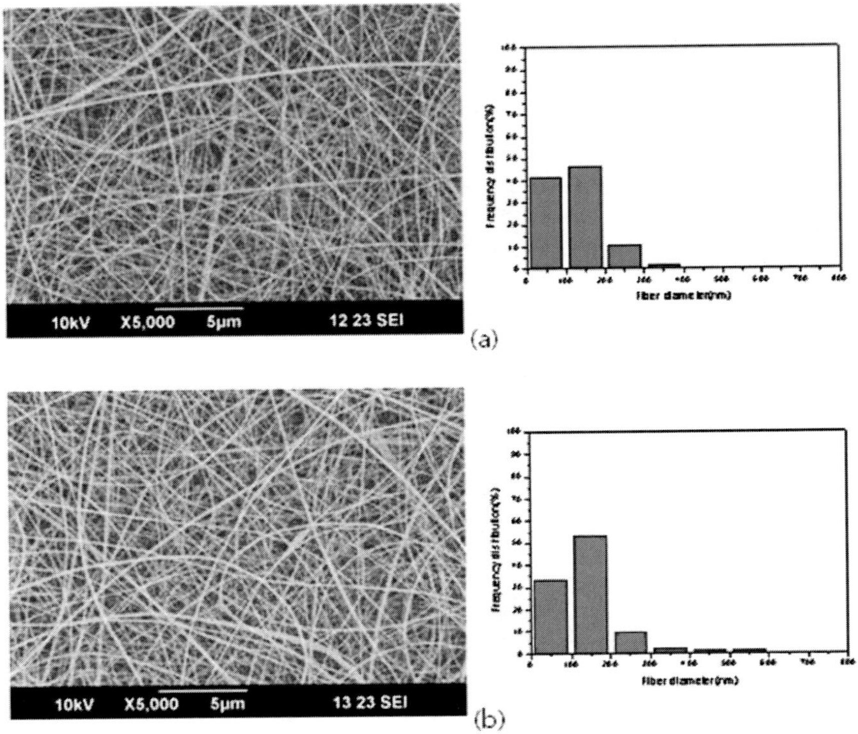


**Figure 7.** Effect of electric voltage on HPMC fiber morphology (solution concentration=1.0 wt%, flow rate=0.05 ml/min). Voltage: (a) 10 keV; (b) 15 keV.

**Figure 8.** Effect of needle tip-target distance of EC NFs (solution concentration, 10wt%; electric voltage, 15 keV; flow rate, 0.005 ml/min). (a) 5 cm, (b) 7 cm, (c) 9 cm, (d) 11 cm.

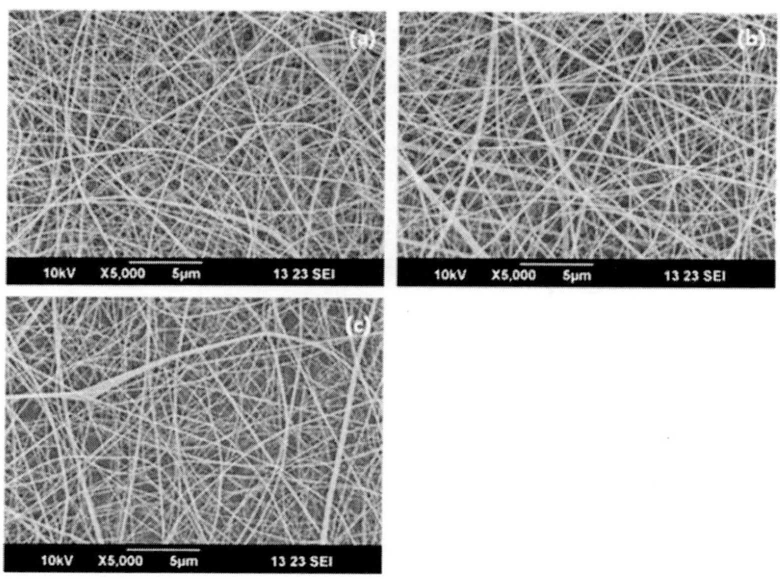


**Figure 9.** Effect of needle tip-target distance of HPMC NFs (solution concentration, 1.0wt%; electric voltage, 15 keV; flow rate, 0.05 ml/min). (a) 5 cm, (b) 10 cm, (c) 15 cm.

# REFERENCES


1. J. Doshi, D. H. Reneker, 1995 Electrospinning Process and applications of Electrospun Fibers. J. Electrostat., 35, 151-160, 0304-3886

2. H. Fong, W. Liu, C. S. Wang, R. A. Vaia, 2002 Generation of electrospun fibers of nylon 6 and nylon 6-montmorillonite nanocomposite. Polym., 43, 775-780, 0032-3861

3. H. Fong, I. Chun, D. H. Reneker, 1999 Beaded nanofibers formed during electrospinning. Polym., 40, 4585 4592, 0032-3861

4. J. L. Ford, 1999 Thermal analysis of hydoxypropylmethylcellulose and methylcellulose: powders, gels and matrix tablets. Int. J. Pharm., 179, 209-228, 0378-5173

5. P. Gibson, H. S-Gibson, D. Rivin, 2001 Transport properties of porous membranes based on electrospun nanofibers. Colloid. Surface A., 187-188, 469 481 , 0927-7757

6. H. Jiang, D. Fang, B. Hsiao, B. Chu, W. Chen, 2004 Preparation and characterization of ibuprofen-loaded poly(lactide-co-glycolide)/ poly (ethylene glycol)-g-chitosan electrospun membranes. J. Biomater. Sci. Polym. Ed. 15, 279-, 0920-5063

7. El-R. Kenawy, G. L. Bowlin, K. Mansfield, J. Layman, D. G. Simpson, E. H. Sanders, G. E. Wnek, 2002 Release of tetracycline hydrochloride from electrospun poly(ethylene-co-vinylacetate), poly(lactic acid), and a blend. J. Control. Release, 81, 57-64, 0168-3659

8. M. W. Kulicke, O. Aredt, M. Berger, 1998 Characterization of hydroxypropyl methylcellulose-stabilized emulsion. Part I. Particle size distribution. Colloid polym. Sci., 276, 1019 1023, 0303-402X

9. H. Q. Liu, Y. L. Hsieh, 2002 Ultrafine Fibrous Cellulose Membranes from Electrospinning of Cellulose Acetate. J. Polym. Sci. Part B: Polym. Phys., 40, 2119 2129 , 0887-6266

10. Y. K. Luu, K. Kim, B. S. Hsiao, B. Chu, M. Hadjiargyou, 2003 Development of a nanostructured DNA delivery scaffold via electrospinning of PLGA and PLA-PEG block copolymers. J. Control. Release, 89, 341 353 , 016-3659

11. J. Y. Park, S. W. Han, I. H. Lee, 2007 Preparation of Electrospun Porous Ethyl Celluolose Fiber by THF/DMAc Binary Solvent System. J. Ind. Eng. Chem., 13, 6, 1002-1008, 1226-086X

12. O. E. Péreza, C. C. Sánchezb, A. M. R. Pilosofa, J. M. R. Patino, 2008 Dynamics of adsorption of hydroxypropyl methylcellulose at the air-water interface. Food Hydrocolloids, 22, 397-402, 0268-005X

13. S. Prasertmanakit, N. Praphairaksit, W. Chiangthong, N. Muangsin, 2009 Ethyl Cellulose Microcapsules for Protecting and Controlled Release of Folic Acid. AAPS PharmSciTech., article in press, 1530-9932

14. Y. M. Shin, M. M. Hohman, M. P. Brenner, G. C. Rutledge, 2001 Experimental characterization of electrospinning: the electrically forced jet and instabilities. Polym., 42, 9955-9967, 0032-3861

15. P. P. Tsai, H. S-Gibson, P. Gibson, 2002 Different electroststic methods for making electrets filters. J. Electrostat., 54, 333-341, 0304-3886

16. G. Verreck, I. Chun, J. Peeters, J. Rosenblatt, M. E. Brewster, 2003 Preparation and Characterization of Nanofibers containing Amorphous Drug Dispersions Generated by Electrotatic Spinning. Pharmcol. Res., 20, 810-817, 1043-6618

17. J. Zeng, X. Xu, X. Chen, Q. Liang, X. Bian, L. Yang, X. Jing, 2003 Biodegradable electrospun fibers for drug delivery. J. Control. Release, 92, 227 231, 016-3659

18. Q. Zhao, M. Zhao, J. Li, B. Yang, G. Su, C. Cui, Y. Jiang, 2009 Effect of hydroxypropyl methylcellulose on the textural and whipping properties of whipped cream. Food Hydrocolloids, 23, 2168 2173, 0268-005X

# Chapter 4

# Physical Properties of Polypeptide Electrospun Nanofiber Cell Culture Scaffolds on a Wettable Substrate

**Donald T. Haynie [1,2,*], Dhan B. Khadka [1] and Michael C. Cross [1]**

[1]Nanomedicine and Nanobiotechnology Laboratory, Department of Physics, University of South Florida, 4202 East Fowler Avenue, Tampa, FL 33620, USA; Email:

[2]Center for Integrated Functional Materials, Department of Physics, University of South Florida, 4202 East Fowler Avenue, Tampa, FL 33620, USA

## ABSTRACT

Physical properties of poly(L-ornithine) (PLO), a polycation, poly(L-glutamic $acid_4$-*co*-L-tyrosine) (PLEY), a polyanion, and electrospun fibers made of these polymers have been determined and compared. The polymers adopted random coil-like conformations in aqueous feedstocks at neutral pH and in dehydrated cast films and fibers on glass, and the fibers comprised numerous counterions, according to spectral analysis. Adsorption of model proteins and serum proteins onto hydrated and crosslinked fibers depended on the electrical charge of the proteins and the fibers. The surface charge density of the fibers will be comparable to, but less than, the charge density on the outer leaflet of the plasma membrane of usual eukaryotic cells. The present analysis thus advances understanding of cell behavior on electrospun fiber scaffolds, a topic of considerable current interest.

## KEYWORDS

cell; electrospinning; fiber; physics; polypeptide; scaffold

## 1. INTRODUCTION

Electrospun fiber mats hold out promise for realizing perceived advantages of nanostructured materials in many areas, including medicine and biotechnology. Fibers can be made of biocompatible and biodegradable synthetic organic polymers. Such polymers include poly(lactide-*co*-glycolide) and poly (caprolactone). Interest in electrospun materials for *in vitro* tissue culture coatings, *in vivo*medical implant device coatings and other applications is now global and likely to continue growing for the foreseeable future [1,2,3,4].

Investigators have taken a variety of approaches to increasing the desirability of electrospun materials for biomedical applications and reducing the environmental footprint of materials production [2,3,4,5,6,7]. In some cases, the polymers were polypeptides (see reference [7] for a recent review). Whether natural or biomimetic, whether pure or blended with synthetic organic polymers, these polymers can provide further advantages for compatibility, degradability, absorbability and, crucially, functionality. Most polypeptide electrospinning studies have involved purified proteins that require an organic solvent for chain entanglement [7]. Lately, interest has grown in polymer electrospinning from aqueous feedstocks [5,7,8]. Model polypeptides that have a simple amino acid sequence or composition, display little persistent structure in aqueous solution under usual conditions and therefore tend to be non-immunogenic are also receiving increasing attention (reviewed in reference [7]). Such peptides can represent different degrees of biomimicry in terms of amino acid composition and sequence and functional properties. In any case, the diameter of polypeptide fibers typically ranges from 100 nm to 2 μm, though thinner and thicker fibers are spun.

It is generally held that quantitative analysis of the physical properties of feedstock solutions, the electrospinning process and the resulting fibers will lead to advances in electrospinning technology. Deitzel *et al.* [9], Yarin *et al.* [10] and Minato *et al.* [11], for example, have studied the relationship of process variables to chain conformation and fiber spinnability, bending instability, diameter and morphology. Hohman *et al.* [12], Fridrikh *et al.* [13] and He *et al.* [14] have developed theoretical models of electrospun fibers, fiber diameter and fiber stability. The primary aim of the present study was to produce a quantitative description of physical properties of polypeptide fibers on a wettable solid support.

The synthetic model polypeptides PLO and PLEY have now been analyzed in greater depth than in our previous work [4,6]. Circular dichroism spectroscopy (CD) and Fourier transform infrared spectroscopy (FTIR) have been utilized to obtain structural information on these polymers in aqueous solution, cast films and fibers. Fiber composition has been analyzed by electron-dispersive X-ray spectroscopy (EDX). Protein adsorption has been analyzed by quantitative fluorescence microscopy. A model of key electrical properties of fibers has been described and compared to literature values for the corresponding properties of glass and cells. The surface energy of the fibers has been analyzed and compared to values for various cell culture materials. The results of this work are relevant to applications of electrospun materials in medicine and biotechnology, for example, tissue engineering scaffolds and wound healing biomaterials.

## 2. EXPERIMENTAL SECTION

All processes were carried out at ambient temperature and humidity unless indicated otherwise. All fiber mat properties were measured *in situ* on a glass substrate.

## 2.1. Polymers

PLO (153.5 kDa by viscometry and 196.8 kDa by multiple-angle laser light scattering) and PLEY (20–50 kDa by viscometry) were synthesized in solution and obtained from Sigma (USA) as lyophilized polydisperse salts. The polydispersity index, defined as the ratio of the weight-average molar mass to the number-average molar mass, or $M_W/M_N$, is a way of quantifying the heterogeneity of polymer length. Light scattering measures $M_N$; viscometry measures the viscosity-average, $M_V$, not $M_W$. Because $M_N < M_V < M_W$, $M_V/M_N$ can be taken as a lower bound on $M_W/M_N$. For PLO, $M_V/M_N \approx 1.3$ and $M_V/M_N$ was probably less than 20% higher. For PLEY, if the number density of polymers was uniform, $M_N = 35.0$ kDa, $M_W = 37.3$ kDa, $M_V = 37.1$ kDa (for the Mark-Houwink parameter $a = 0.8$, as for semi-flexible polymers [15]) and $M_W/M_N \approx 1.1$. Typical synthetic materials have $M_W/M_N \approx 4$, a value that is often improved by size-exclusion chromatography and polymer fractionation. The polydispersity of PLO or PLY is rather unlikely to have influenced the present results. The counterions were $Br^-$ for PLO and $Na^+$ for PLEY. Aqueous feedstocks for electrospinning were prepared by dissolving as-received peptide in deionized water (40% w/v or 2 mM polymer, and 50% w/v or 14 mM polymer, respectively). The feedstock solutions will therefore have consisted of three main components: water (and its ionized forms), ionized polypeptide and counterions.

## 2.2. Electrospinning

Fibers were prepared as described previously [4,6]. Briefly, polymer feedstock was taken up into a 1 mL syringe outfitted with a 0.5 mm blunt-end needle (Jensen Global, USA). A positive potential in the 5–20 kV range was applied to the needle with a PS/FX20P15.0-11 power supply (Glassman, USA). Fibers were collected 5–15 cm from the spinneret on a glass plate on top of a grounded sheet of metal or on a sheet of 60 $\Omega/in^2$ indium tin oxide-coated poly(ethyleneterephthalate) (ITO-PET; Sigma-Aldrich). The feedstock flow rate was not regulated.

## 2.3. Crosslinking

Two methods were utilized for *in situ* polymer crosslinking at ambient temperature: Exposure to glutaraldehyde vapor (25% w/v in water; Sigma) for PLO [4], and immersion in 50 mM 1-ethyl-3-(3-dimethylaminopropyl) carbodiimide (EDC; Thermo Scientific, USA) in 90% ethanol/10% water for PLEY [6]. Relevant details of the chemistry now follow. The symmetrical glutaraldehyde molecule will have reacted with amino groups in two ornithine side chains or, with much lower probability, *N*-termini, yielding no change in the number of amide groups present. As to EDC, a symmetrical diimide, the most likely reaction in this study will have involved the carboxylate moieties of two glutamic acid side chains. The resulting acid anhydrides, however, will have been unstable in aqueous solution. Formation of a stable crosslink, a peptide bond, will have required an EDC molecule to react with a glutamate side chain and the resulting *O*-acylisourea intermediate with an amino terminus; *N*-termini are the only amino groups in PLEY. The byproduct of this reaction will have been a soluble urea derivative. Additional stable crosslinks may have formed between the phenolic oxygen atoms of tyrosinate residues and *N*-termini, albeit with lower probability. Following crosslinking, PLO and PLEY fiber mesh samples were rinsed extensively with aqueous solution at neutral pH. Residual *O*-acylisourea intermediates, whether formed with glutamic acid or tyrosine, will have been unstable in aqueous solution; *N*-hydroxysulfosuccinimide (sulfo-NHS) was not utilized for intermediate stabilization; failure to react with an amine will have resulted in hydrolysis, regenerating glutamate or tyrosinate and releasing an *N*-unsubstituted urea molecule to solution.

## 2.4. Dye Labeling and Protein Adsorption

Model proteins were assayed for adsorption onto fibers versus glass. Visualization of adsorbed material was enabled by labeling with fluorescein isothiocyanate (FITC), which is selective for free amino groups; no sulfhydryl groups were present. The test species were serum proteins, FITC-labeled hen egg white lysozyme (FITC-HEWL; Sigma) in phosphate-buffered saline (PBS) and FITC-labeled bovine serum albumin (FITC-BSA; Sigma) in PBS. For PLEY, fibers and serum proteins were labeled *in situ*; HEWL and BSA,*ex situ*. For PLO, serum proteins, HEWL and BSA were labeled *ex situ*.

All labeling reactions were carried out overnight at 4 °C, the dye concentration was 1 mg/mL in dimethylsulfoxide, and the buffer was 20 mM sodium carbonate, pH 9. For HEWL and BSA labeling, the protein concentration was 2 mg/mL. Carbonate buffer was diluted 1:100 in PBS after labeling, and the final protein concentration was adjusted to 2 mg/mL, the approximate concentration of BSA in cell culture medium; 10% fetal bovine serum (FBS) has a protein concentration of 3–4.5 mg/mL, and the concentration of BSA is 0.69–1.52 mg/mL [16]. For direct labeling of PLEY fibers, FITC will have reacted with free amino termini. Non-specific adsorption of FITC onto fibers or glass may also have been strong enough to remain bound after rinsing. For PLO fibers, too many free amino groups were present in side chains for direct labeling. Some PLEY fiber samples were incubated overnight with fibroblast culture medium (Lonza, USA) supplemented with 10% FBS. These samples were then rinsed with PBS and labeled with FITC as described above. FITC will have reacted with free amino groups on bound serum proteins and amino termini in fibers. All fiber samples were rinsed with deionized water for visualization and image capture by fluorescence microscopy.

## 2.5. Spectroscopy and Microscopy

### 2.5.1. CD

Far-UV spectra of PLO and PLEY were collected with an Aviv Biomedical 215 instrument (USA). 5 scans were averaged for measurement over a range of 190–270 nm at a rate of 1 nm/s, a step size of 1 nm and a bandwidth of 1 nm. The peptide concentration was 1.3 μM for PLO and 5.7 μM for PLEY. The path length was 1.0 cm. The temperature was 25 °C.

### 2.5.2. FTIR

A Jasco 4100 instrument (Japan) was utilized to obtain FTIR spectra of cast films, fibers and crosslinked fibers of PLO and PLEY on ITO-PET over the 1000 to 4000 $cm^{-1}$ range in attenuated total reflectance mode. The ATR accessory (Harrick Scientific, USA) was equipped with a ZnSe crystal. The bandwidth was 4 $cm^{-1}$; 256 spectra were averaged in each case.

### 2.5.3. EDX

Spectra of glass, PLO fibers on glass and PLEY fibers glass were collected with an INCA X-sight 7582M energy-dispersive spectrometer (Oxford Instruments, UK) mounted on a JSM-6390LV scanning electron microscope (SEM; JEOL, Japan). The fiber samples were not crosslinked. The working distance was 10 mm, and the accelerating voltage was 15 kV. Copper tape was utilized as a calibration standard without changing focus, spot size or accelerating voltage.

### 2.5.4. Microscopy

Bright-field and fluorescence micrographs of fibers after protein adsorption were obtained with a Zeiss Axiovert 200 M inverted microscope (Germany) equipped with an incandescent source, a mercury vapor source, a filter set, a 10× objective lens and a Roper Scientific MicroMAX System CCD camera (USA). The excitation wavelength was 488 nm. A JEOL JSM-6390LV scanning electron microscope (Japan) was utilized for fiber morphometry after metalizing samples with 10 nm of gold. The accelerating potential was 30 kV.

## 3. RESULTS

## 3.1. Polymer Conformation

PLO and PLEY chains will have been extensively ionized in water at 25 °C and neutral pH [17,18]. Both polymers adopted a mostly random coil conformation under these conditions, according to analysis by CD (Figure 1). Evidence is provided by the large narrow absorption band near 205 nm, which is a general indicator of a disordered polypeptide backbone, and by the absence of a negative band in the 210–220 nm range, which would be expected for $\alpha$ helix or $\beta$ sheet [19]. The main difference between the PLO and PLEY spectra is the contribution of the tyrosine rings in PLEY. The molar absorptivity of the tyrosine side chain is *ca.* 8000 $M^{-1}$ $cm^{-1}$ at 225 nm [20], and this can influence the CD signal to within about 25 nm of the center of the band [21]. What is important here is that both polymers were mostly unstructured in the aqueous feedstock.

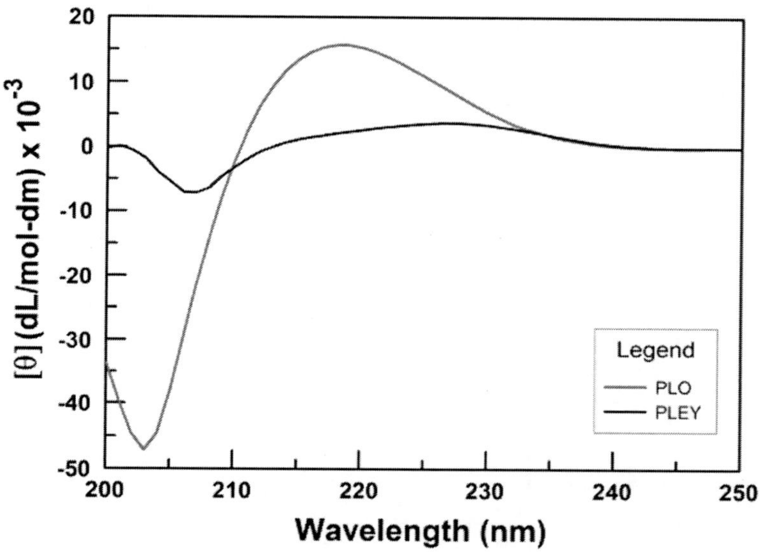


**Figure 1.** Far-UV dichroic spectra of polymers in water. Poly(ornithine) (red), poly(glutamic acid, tyrosine) (black). The polymer concentration, 0.02 mg/mL in each case, was *ca.* 2000× lower than for electrospinning.

Electrospinning the polymer solutions represented by Figure 1 produced fibers represented by the scanning electron micrographs shown in Figure 2. Panels a and b show PLEY fibers before and after crosslinking; panels c and d, PLO fibers. Fiber diameter ranged from ~0.1 to several microns, based on analysis of digital micrographs. Mat thickness, a function of the fiber production rate and time, was always under 50 μm here. There was no obvious change in fiber morphology on crosslinking.

Analysis of the amide I region of the IR spectra of fibers prepared as in Figure 2 showed that most polymers were in a random coil conformation prior to crosslinking, in PLO and in PLEY. Key differences were evident, however, on closer comparison. Resonances in the amide I region are primarily due to stretching vibrations of main-chain carbonyl groups. For PLEY, the close similarity of the fiber spectrum to the cast film spectrum suggests that molecules lacked a preferred orientation in fibers and, therefore, that polymer chains in the fiber interior had little regular structure (Figure 3a). Some residues evidently adopted a *β* sheet conformation prior to crosslinking, however, based on the amplitude, resolvability and width of absorbance bands in the amide I region, especially near 1651 $cm^{-1}$ (coil), 1617 $cm^{-1}$ (*β* sheet) and 1681 $cm^{-1}$ (turn) [22,23]. *β* sheets are characterized by hydrogen bonds formed between polymer backbone amide groups and carbonyl groups. Here, *β* sheet formation will presumably have made a favorable contribution to the enthalpy of the polymer system during solvent evaporation; it may also have influenced the entropy by releasing to bulk solution and then to the vapor phase water molecules initially hydrogen bonded to donors and acceptors in the polymer backbone.

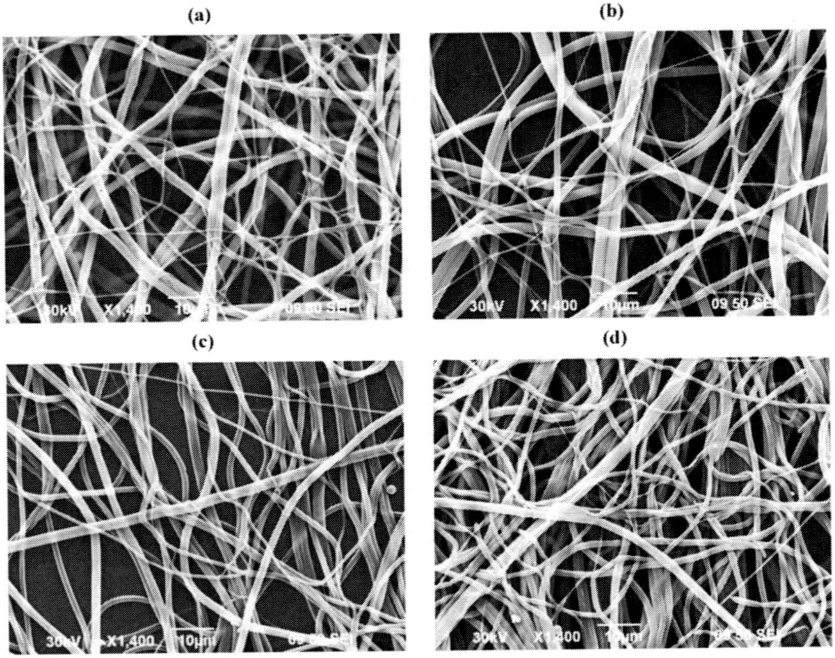


**Figure 2.** Fiber mat morphology. PLEY: (**a**) Non-crosslinked; (**b**) Crosslinked. PLO: (**c**) Non-crosslinked; (**d**) Crosslinked. All micrographs were obtained at 1400× magnification and 30 kV accelerating potential.

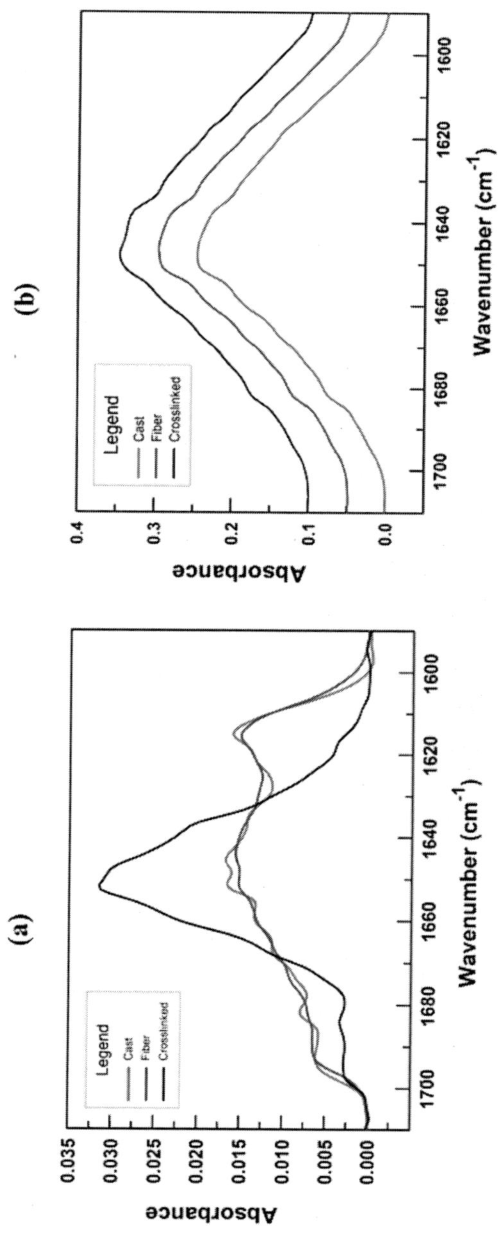


**Figure 3.** Infrared analysis of cast films and fibers. Cast film (red), fibers (blue) and crosslinked fibers (black). All but the amide I envelope was subtracted from the spectra during baseline treatment. (**a**) PLEY; (**b**) PLO. There is a vertical offset of 0.05 between the spectra in this panel.

Crosslinking is necessary to make PLEY fibers and PLO fibers water-insoluble [4,6]. For PLEY, the EDC crosslinker will have reacted with the carboxylate group in a glutamate side chain and an *N*-terminal amino group. This will have increased the number of peptide bonds present, but only by a maximum of 1 per polymer chain; a mere 0.5% rise in the number of carbonyl groups contributing to the amide I spectrum for the PLEY molecules of this study. The *β*-sheet bands around 1617 cm$^{-1}$ and 1682 cm$^{-1}$ decreased and the random-coil band around 1651 cm$^{-1}$ increased on crosslinking. The large change in shape of the amide I envelope (Figure 3a) must therefore be due to a large change in the average conformation of polymers in the fibers—despite a lack of change in fiber morphology (Figure 2a,b). The crosslinking reaction evidently decreased the ability of PLEY molecules to form *β* sheets in fibers. Rinsing crosslinked fibers in aqueous solution will presumably have reversed any changes in the average backbone conformation of PLEY due to ethanol, the solvent for EDC. The spectral changes on crosslinking were *internal* to the fibers.

As to PLO, crosslinking depended on the reactivity of side chains, just as for PLEY. Each end of a glutaraldehyde molecule will have reacted with a side-chain amino group, forming an imine. Unlike PLEY, PLO showed almost no difference between the cast film, fiber and crosslinked fiber spectra (Figure 3b). The largest peak in the spectra, at 1651 cm$^{-1}$, can be assigned to random coil. A second large peak, at *ca*. 1640 cm$^{-1}$, may be due to *α* helix. Both peaks are very broad, consistent with both a broad range of dihedral angles in any secondary structure present and limited thermostability [22]. In any case, most residues had random backbone dihedral angles in PLO fibers and in PLEY fibers after crosslinking.

## 3.2. Fiber Stability and Composition

Both PLO and PLEY fibers display considerable post-spinning stability in dry air in the absence of crosslinking [4,6]. This alone implies that counterions will be present inside the fibers, which would otherwise be very unstable due to charge repulsion. EDX analysis has confirmed this view (Figure 4). For PLEY, the relative abundance of carbon, oxygen and sodium atoms was respectively 62.3%, 56.7% and 15.4% for the fiber sample on glass and 6.4%, 44.4% and 3.9% for the glass control.

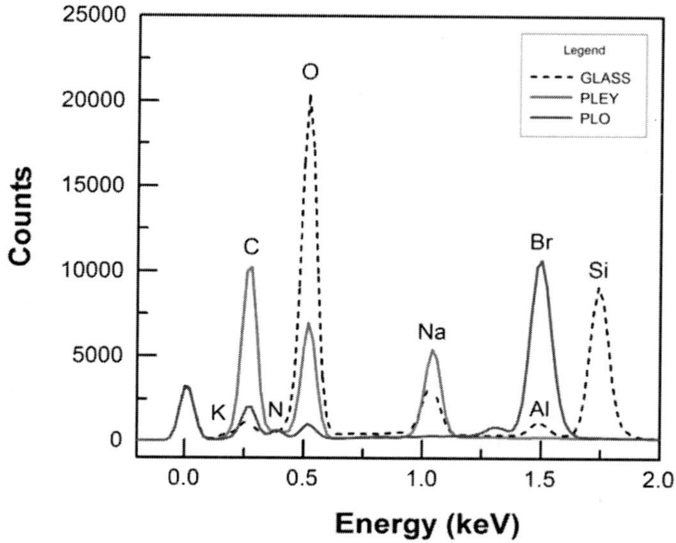


**Figure 4.** Energy-dispersive X-ray spectroscopic analysis of fibers. Spectra of glass, PLEY fibers on glass and PLO fibers on glass. Note the relative sizes of the oxygen, sodium and silicon peaks. The silicon peak occurs at *ca*. 1.8 keV. In the fibers samples, the silicon peak is lost in the noise; bare glass contributes relatively little to the measured values; the most prominent peaks are for carbon and oxygen (PLEY and PLO), sodium (PLEY) and bromide (PLO).

The abundance of silicon was 9.1% and <1%, respectively. These measurements translate into atomic ratios of 16.3 for C:N, 1.1 for C:O and 4.1 for O:Na for PLEY fibers on glass. The corresponding data for PLO are shown in spectral form in Figure 4.

The EDX results can be compared with predictions based on polymer structure. For PLEY, for instance, there were 30 C atoms for every 5 of N, 14 of O and 4 of Na, assuming complete ionization of glutamic acid side chains, no ionization of tyrosine side chains and complete charge compensation by counterions. The corresponding ratios are 6 for C:N, 2.1 for C:O and 7.5 for C:Na. The difference from the corresponding measured value may be attributed to a deviation of Glu:Tyr from 4.0:1.0, incomplete side chain ionization, incomplete removal of excess $Na^+$ from PLEY prior to lyophilization or substitution of $H_3O^+$ for $Na^+$ in PLEY fibers. PLO could not be analyzed in this way because the extent of crosslinking was not determined. The main point here is that $Na^+$ was abundant in PLEY fibers and $Br^-$ was in PLO fibers.

## 3.3. Protein Adsorption

Serum protein adsorption onto the surface of a biomaterial influences the foreign body response and therefore clinical complications arising from implantation. Protein binding affinity to a material is determined by Coulomb forces and van der Waals interactions. Adsorption of the model proteins HEWL and BSA and serum proteins onto PLEY fibers and PLO fibers has been studied here. Unmodified HEWL has an isoelectric point of 11.5 and therefore a positive net charge at neutral pH [24]; BSA has an isoelectric point of 4.7 and therefore a negative net charge at neutral pH [25]. Limited FITC labeling will not have altered either isoelectric point by a significant margin.

Figure 5a shows PLEY fibers after direct labeling with FITC. Dye molecules will have become conjugated to *N*-termini still available after EDC crosslinking or bound non-specifically to fibers or glass. PLEY fibers were scarcely more densely coated with FITC-labeled serum proteins than was the glass substrate, according to Figure 5b. Many of the proteins adsorbed onto fibers will have been positively charged by virtue of ionized lysine residues. This amino acid, which has a primary amino group in the side chain, is labeled by FITC. The high background staining evident in panels a and b can be attributed to *in situ* labeling and non-specific FITC adsorption. Contrast in the FITC-BSA micrograph (Figure 5c) is comparatively high, and the overall signal intensity is low. This suggests that BSA had greater affinity for fibers than glass, despite the negative charge on PLEY. BSA is well-known to be "sticky", to bind non-specifically to many materials; the hydrophobic component of binding may have been greater for fibers than glass. Panel d shows the result of incubating FITC-HEWL with PLEY fibers. This cationic protein was expected to bind both PLEY fibers and glass. The contrast between fibers and glass was correspondingly low. A hydrophobic contribution to HEWL binding cannot be excluded. Similar results were obtained for PLO binding to fibers.

Figure 5h,i summarizes the fluorescence data for PLEY fibers (panels a–d) and PLO fibers (panels e–g) and quantify fluorescence as distributions of pixel intensities. BSA binding to PLEY fibers resulted in a high frequency of low grayscale values (light intensities). This makes sense because both BSA and PLEY were anionic in the experiments. The maximum in the BSA distribution was at a much higher grayscale value for PLO fibers. Just the opposite was found for HEWL, which was cationic in the experiments. The inward spikes in the BSA profile in Figure 5i are image processing artifacts that arose from increasing the contrast between the fiber signal and the glass signal; the shape of the envelope was not altered. The protein adsorption data are substantially explained by the sign of charge of the glass, the sign of charge and relative charge density of the fibers and the sign of charge of the model proteins.

Evidently, serum contains proteins that will bind to either positive fibers or negative fibers having the surface density of charge of the fibers studied here.

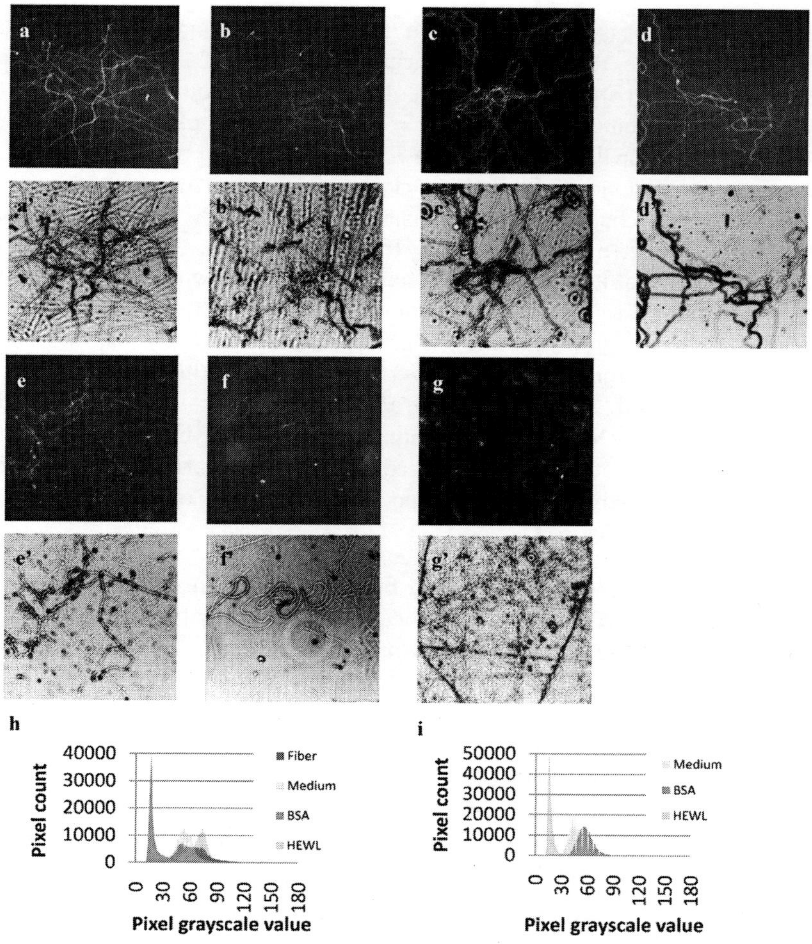


**Figure 5.** Protein adsorption onto fibers or glass. Proteins became deposited during overnight incubation at 37 °C. Bright-field images have been inverted and equalized to increase contrast; there was no processing of the fluorescence micrographs. (**a**, **a'**) Fluorescein (FITC)-labeled PLEY fibers. FITC-labeled proteins on (**b**, **b'**) PLEY fibers or (**e**, **e'**) PLO fibers after deposition from culture medium. FITC-bovine serum albumin on (**c**, **c'**) PLEY fibers or (**f**, **f'**) PLO fibers. FITC-hen egg white lysozyme on (**d**, **d'**) PLEY fibers or (**g**, **g'**) PLO fibers; (**h**) Grayscale histogram of pixel intensities for fluorescence micrographs in (**a–d**); (**i**) Grayscale histogram of pixel intensities for fluorescence micrographs in (**e–g**).

# 4. DISCUSSION

## 4.1. Fiber Formation, Composition and Structure

Chain entanglement is a major requirement for fiber spinning from a polymer feedstock [26]. Here, the feedstock concentration was ~40% (w/v) for PLO and ~50% for PLEY. The corresponding concentration of small counterions was ~1 M in each case, assuming the charge per residue at neutral pH was *ca.* +0.8 for PLO and *ca.* −0.8 for PLEY [27]. Charge repulsion between polymer molecules will have been significant, though at close range only. Such considerations suggest that PLO chains and PLEY chains will behave as statistical coils in aqueous solution at neutral pH below the solubility limit. The spectral data in Figure 1 show that PLO formed little helical structure in water at neutral pH. Even at strongly basic pH, PLO will adopt only a modest amount of helical structure, even though poly(lysine), which differs from PLO by just one extra methylene group in the side chain, converts to helix when the side-chain amino groups are deionized [17]. Poly(glutamic acid) forms $\alpha$-helical structure at acidic pH, where the side-chain carboxyl groups are deionized, but the polymer is mostly unstructured at neutral pH, where the side chains are protonated [18]. PLEY, like poly(glutamic acid), displayed a random coil-like conformation in water at neutral pH (Figure 1).

Polymer conformation in solution may be compared with conformation in a cast film or in fibers. The IR data in Figure 3 indicate that a large proportion of residues in PLEY molecules participated in *β*-sheet structure in the cast film and in fibers prior to crosslinking. Bands at 1617 cm$^{-1}$ and 1681 cm$^{-1}$ support the claim. Similarities and differences with respect to globular proteins and amyloid fibrils are worth mentioning here for the sake of interpreting the IR data. In a recent study by Zandomeneghi *et al.*, for example, amide I′ data were reported for transthyretin, a *β*-sheet protein [28]. (The measurements were made in $D_2O$ rather than water; the change in isotope will have had little effect on the shape of the distribution of stretching vibrations of main-chain carbonyl groups.) The native protein in solution displayed a broad amide I′ envelope with a maximum at 1630 cm$^{-1}$. Fibrils of this protein, by contrast, showed a large and narrow band at 1615 cm$^{-1}$ and a corresponding peak of lower intensity at 1684 cm$^{-1}$, similar to amyloid fibrils. A close resemblance of the transthyretin fibril and amyloid fibril IR bands to the PLEY data in Figure 3 is evident. Zandomeneghi *et al.* further interpreted their results as follows. *β*-sheet residues will have greater structural homogeneity in the core of amyloid fibrils than in native proteins. Amyloid fibrils correspond to very low free energy states, much deeper than the potential wells of native proteins, even if the polymers are kinetically trapped in amyloid material. A large right-handed *β*-sheet twist angle correlated with a high wavenumber maximum for the amide I′ band, and the amide I′ band wavenumber maximum correlated with the average number of strands per β sheet in transthyretin. Protein structure analysis by Richardson concluded that β sheets with fewer strands tend to have larger average twist angles [29]. No significant shift in the amide I band of the β-sheet protein lithostathine occurred on oligomerization into non-amyloid fibrils [30]. Zandomeneghi *et al.* proposed on these grounds that IR analysis, and more specifically a shift of the amide I band from about 1630 $cm^{-1}$ to about 1615 $cm^{-1}$ and the simultaneous appearance of a band at about 1684 $cm^{-1}$, could be a diagnostic for amyloid fibril formation [28]. The data in Figure 3 show that bands at 1615 $cm^{-1}$ and 1684 $cm^{-1}$ are not unambiguous indicators of amyloid fibril formation.

The band shift from about 1630 $cm^{-1}$ to about 1615 $cm^{-1}$ merits further attention for reasons that will become evident presently. Dimerization of a carboxylic acid, for example, propanoic acid in $CCl_4$, can decrease the vibrational frequency of the carbonyl oxygen by as much as 45 $cm^{-1}$ [31]. A low wavenumber maximum for a β sheet, as in transthyretin fibrils and PLEY cast films or fibers, may therefore represent an average structure in which backbone hydrogen bonds deepen the free energy well more than in native proteins. The energetic unfavorability of an increased average twist angle in a β strand in a native protein can apparently be more than offset by other kinds of interaction, for example, hydrophobic contacts between side chains. Such interactions are less probable in a cast film or electrospun fiber of PLO or PLEY, because the amino acid sequence diversity of these molecules is vastly lower than in a protein-encoding polypeptide and there are therefore vastly fewer distinguishable ways in which side chains can pack together, especially under the constraint of forming a large number of energetically significant interactions with counterions, as in the present study (Figure 4). PLEY chain oligomerization was presumably uncoordinated and in any case rapid during the dehydration of polymer jets or cast films. Both processes were nevertheless sufficient to give rise to resonances near 1615 $cm^{-1}$ and 1680 $cm^{-1}$ (Figure 3). Therefore, these bands cannot be considered unambiguous markers of amyloid fibrils. Further, PLEY residues in a β sheet conformation are likely to have but modest average right-handed twist angles. It is less clear whether the available fiber data also provide evidence for a relatively large average number of strands per β sheet.

## 4.2. Crosslinking

The data in Figure 3 also show that the EDC crosslinking reaction disrupted the β sheets formed between PLEY molecules during fiber production or further dehydration on the collector. The enthalpy of hydrogen bond formation will be relatively small, ~4 kJ/mol [32], close to thermal energy at 25 °C. In PLEY fibers, perhaps 1 in 4 of all residues was in a β sheet conformation prior to crosslinking, judging by the area of the resonance at 1617 $cm^{-1}$ *vis-à-vis* the total area of the amide I envelope. The distribution of β-strand lengths was probably broad, some chains of ideal backbone geometry will not have formed a β sheet, some β sheets will have comprised non-ideal backbone angles, and the composition of individual β sheets will probably have fluctuated more rapidly than in amyloid fibrils. It is probable that the mean number of consecutive residues in a β strand was modest for the same reason that many relatively small α helices are more probable than one long helix in a long chain of poly(lysine) or poly(glutamic acid): Entropy maximization [33]. It is likely, therefore, that little energy was required to disrupt most β sheets in PLEY fibers during EDC crosslinking. This energy will have been provided by thermal fluctuations and, possibly, the interaction of ethanol with the fibers. It was not determined whether crosslinking involved polymers throughout the fiber or only on the fiber surface.

## 4.3. Electrical Properties

Elementary electrical properties of PLO and PLEY fibers are usefully discussed here for the sake of interpreting the data in Figure 5. Such properties are of interest not only for their own sake but also for a variety of potential applications of electrospun materials. The surface charge density and the sign of the net charge of a fiber mat could determine its utility, for instance, for ion exchange in a filter application or for protein adsorption in a tissue culture scaffold application or an implantable biomaterial application. Cell-substrate interactions in serum-supplemented tissue culture *in vitro* are relevant to cell adhesion, proliferation, morphology, activation, differentiation and senescence [34]. *In vivo*, serum protein adsorption onto a biomaterial plays a crucial role in the foreign body response, leading in turn to the broad range of clinical complications associated with device implantation [35].

The surface charge density for glass and for fibers can be compared as follows. The properties of glass are relevant here because it was the material on which fibers were collected in the present study. For glass in contact with deionized water, the effective charge density is ~$2 \times 10^3$ electronic charges $\mu m^{-2}$ [36,37]. This value will be lower in a physiological buffer, for example cell culture medium, which has a high ionic strength. As to fibers, assuming a density matching that of small globular proteins (~1.45 $g/cm^3$; reference [38]), the outermost 5 Å of a 1 μm radius fiber (confer Figure 2) with a length of 1 μm will have a mass of ~4.4 fg. (The thickness of the fiber surface in which side chains will contribute to the net charge of the fiber must be estimated in the absence of additional data.) The mass fraction of Glu in PLEY is 0.79 (assuming complete charge compensation with $Na^+$), giving ~14 × $10^6$ Glu side chains on the surface per 1 μm of fiber. The surface density of charge will then be ~2.2 × $10^6$ electronic charges $\mu m^{-2}$. This value should probably be taken as an upper bound, as the fibers are not likely to be as dense as small globular proteins, even after crosslinking, a small proportion of glutamic acid side chains will be lost on EDC crosslinking, and some unreacted glutamate side chains may be protonated, even at pH 7.4. The surface charge density will be of similar magnitude for PLO, but of course the sign of charge must be just the opposite. The point here is that the surface charge density on a PLO fiber or a PLEY fiber will be greater than the surface charge density on glass.

Now, $\sigma_G \approx \sigma_F/1100$, where the subscripts signify glass and fiber. Taking the nominal diameter of a well-spread adhesive cell to be 50 μm, the charge on the corresponding surface area of glass will be $Q_G \approx$ (2 × $10^3$ electronic charges/$\mu m^2$)/[π (25 μm)$^2$] ≈ 1 electronic charge. $Q_F \approx$ 1100 electronic charges for the same surface area. The field strength in air due to glass will be $E_G = \sigma/\varepsilon \approx 3.6 \times 10^{-5}$N/C at any distance from the surface (the field lines will be perpendicular to the surface and parallel to each other to the extent that glass is planar) and the field strength due to a fiber with a radius of 1 μm will be $E_F \approx 4000 \times 10^{-5}$ N/C at the fiber surface. The fields will be about 80-fold smaller due to water, which has a high relative permittivity, and smaller still and strongly distance-dependent with salt present, as in cell culture medium, due to Debye-Hückel screening; the Debye length, the distance over which the electrical potential falls by a factor of 1/*e*, is ~1 nm for a 1:1 electrolyte solution at 100 mM. In summary, thicker fibers will influence the local electric field more than thinner fibers, but only close to the fiber surface; charge screening will ensure that the field is negligible farther away.

Five points of contact between the foregoing electrostatic calculations, the experimental results of this work and cell culture applications of electrospun fibers will now be discussed in turn. One, the calculated upper bound on the surface charge density of PLEY or PLO fibers was ~2.2 × $10^6$ electronic charges $\mu m^{-2}$ (~3.5 × $10^{-13}$ C/$\mu m^2$). Surface charge is known to influence different aspects of adhesive cell behavior. The molecular composition of the membrane (sialic acid is a major contributor) and the pH and composition of the surrounding medium together give cells a negative surface charge. Curtis *et al.* [39] have observed that, whereas surfaces bearing cations are very adhesive for cells [40], adhesion of red blood cells on anionic surfaces decreases as charge density decreases [41]. This suggests that nucleated cell adhesion could be greater on fibers than glass after topographical differences were accounted for. Separate work has found that fibronectin, a prominent component of the extracellular matrix and serum, undergoes a transition from monolayer to multilayer adsorption on sulfonated poly(styrene) at a surface charge density above 3 × $10^{10}$ C/$\mu m^2$ and forms fibrillar networks beyond 8 × $10^{10}$ C/$\mu m^2$ [42]. These charge density values are many orders of magnitude greater than the calculated charge density of PLEY fibers or PLO fibers, so monolayer adsorption of fibronectin and other serum proteins is probable.

Two, if a cell can sense the electric field of glass, the cell may also be able to distinguish between PLEY or PLO fibers of different diameter. Surface charge density will be largely independent of fiber diameter, because the typical fiber radius is much greater than the length scale of an amino acid side chain. The net surface charge on a given length of fiber will therefore be higher for a thick fiber than a thin one, and the electric field will be higher for the thick fiber at the same perpendicular distance from the fiber axis. The mechanisms whereby a cell might detect differences in surface charge density are not well understood (see below).

Three, anionic fibers will attract not only $Na^+$ and other small counterions but also cationic serum proteins (Figure 5). The serum proteins that became bound to fibers or glass during overnight incubation with culture medium therefore depended on the composition of serum. The binding affinity and therefore the equilibrium ratio of bound versus free proteins will have varied with the net charge on the proteins at neutral pH. A cluster of acidic or basic residues on a protein may also have influenced binding, enabling an anionic protein, say, to bind an anionic fiber. Many adhesive cells, including fibroblasts, attach more readily to poly(lysine)-coated glass than glass, and PLO fibers will presumably attract anionic serum proteins as readily as PLEY fibers bind cationic serum proteins. The data in Figure 5are consistent with this view.

Four, application of a static external electric field to fibroblasts in aqueous cell culture medium reportedly makes them elongate, orienting their bodies, actin stress fibers and microtubules perpendicular to the direction of the field [43]. Within the aqueous medium, the field will be small, owing to the polarizability of water, and there will be a current of mobile ions. The presumed cellular mechanisms involved in responding to the applied field are varied and complex: Interactions with sensors located at the cell membrane, the iontophoresis of receptors, channels and other molecules, and the separation of charged membrane components are possible explanations [44]. Cell reorientation on application of an external field of 5 V/cm is essentially complete after 3 h. This field, it should be noted, is ~3 orders of magnitude larger than that calculated for PLEY fibers or PLO fibers in air.

Five, many cells, including fibroblasts, display net migration toward the cathode, the negative pole of an applied external field. Directed fibroblast migration occurs at applied fields as low as 0.1 V/cm in three-dimensional collagen gels but not conventional two-dimensional culture [45]. This field magnitude is thought to be relevant to embryonic development [46], healthy epithelial tissue function and wound healing in response to epithelial cell disruption [47]. For comparison, cell electroporation for drug delivery is achieved with external fields ~$10^5$ larger [48], and the electric field on the surface of a PLEY or PLO fiber in air is ~$10^3$ smaller. PLEY is negatively charged.

## 4.4. Maximum Charge Density on an Unmodified Protein Fiber

The surface charge density calculation for PLEY or PLO fibers may be utilized to estimate the maximum surface charge density on a protein or peptide fiber in the absence of post-translational modification. The qualification is needed because phosphorylation of side chains can significantly change the net charge on a protein. The mass fraction of glutamic acid in a PLEY fiber is ~0.8. What if fibers were made of polypeptides that were completely ionized? Taking 5 Å within the fiber surface as the limit for uncompensated side chain ionization, the maximum charge density would be ~$3 \times 10^6$ electronic charges $\mu m^{-2}$ for a mass density like that of a small globular protein. Essentially the same value is obtained by different reasoning. Let all polymer molecules on a polypeptide fiber surface have anti-parallel *β*-sheet geometry, neglect backbone twist, and let the surface formed by the *β* sheet coincide with the surface of the fiber. Successive side chains will point above and below the fiber surface. There will be ~1 electronic charge for every 30 Å of surface, or $3.3 \times 10^6$ electronic charges $\mu m^{-2}$. The value found above for the surface charge density of PLEY or PLO fibers, ~$2.2 \times 10^6$ electronic charges $\mu m^{-2}$, is therefore of the same order of magnitude as the theoretical maximum on a protein- or peptide-based material. These values may be compared with experimental data for cells. The surface charge density on red blood cells is ~$2.8 \times 10^5$ electronic charges $\mu m^{-2}$ [49], and for other eukaryotic cells it is as high as ~$1.3 \times 10^8$ electronic charges $\mu m^{-2}$ [50]. The maximum protein fiber surface charge density approaches but probably does not exceed the charge density of a membrane in a living cell. The maximum value is unlikely to be realized in a living organism. For a strongly anionic polypeptide could compete with DNA for DNA-binding proteins and thus interfere with regulation of gene expression; and a strongly cationic polypeptide could form an interpolyelectrolyte complex with DNA and thus interfere with gene expression and possibly DNA replication, and it could disrupt membrane structure.

Finally, if the surface charge density of a PLEY or a PLO fiber does depend on fiber radius, there are only two possibilities. A lower charge density for a thick fiber would imply greater charge repulsion on a thin fiber closer to the structure-imposed limit of surface charge density. This must be considered implausible. If the surface density of ionizable groups is essentially independent of fiber radius, there will be a larger driving force for side chain neutralization on a thin fiber, lowering the effective surface density of charge. Therefore, the surface charge density can only be higher on a thick fiber. The field of a thick fiber will in any case exceed that of a thin fiber at a given distance from the fiber axis.

## 5. CONCLUSIONS

Key physical properties of polypeptides in solution and peptide-based electrospun fibers on glass have been analyzed. PLEY and PLO were random coil-like in aqueous solution, whereas in fibers and cast films, a large fraction of residues in PLEY but not PLO adopted a *β*-sheet conformation prior to crosslinking. Significant IR absorption bands near 1615 $cm^{-1}$ and 1680 $cm^{-1}$ were displayed by cast films of PLEY, so the bands cannot be taken as unambiguous indicators of amyloid fibril formation. Crosslinking PLEY fibers with a diimide reagent resulted in a large increase in irregular backbone structure. The backbones of PLO molecules were irregular before and after crosslinking. EDX analysis has confirmed that PLEY fibers contained $Na^+$ and PLO fibers contained $Br^-$. The adsorption of model proteins onto PLEY fibers or PLO fibers on glass was consistent with the net charge on the proteins and the relative surface charge density of the fibers and the substrate at neutral pH. The electric field of a thick fiber is larger than that of a thin fiber for a given surface density of charge and a given distance from the fiber axis. Calculations showed that the maximum surface charge density on a peptide-based material will be less than the charge density on the outer leaflet of the plasma membrane of typical eukaryotic cells. The results, taken together, provide a foundation for analyzing the behavior of adhesive cells on electrospun fiber mats on a wettable substrate in the presence of serum. Such behavior is relevant to fiber mat applications in biotechnology and medicine, for instance, *in vitro* tissue engineering, *ex vivo* stem cell therapy, wound healing and biomaterial implantation.

## ACKNOWLEDGMENTS

We thank Ivar Giaever, Myung Kim and David Rabson for helpful discussions and the USF Mass Spectrometry/Peptide Facility and Analysis Center and the Laboratory for Advanced Materials Science and Technology for technical assistance. M.C.C. is the recipient of a fellowship from the Louis Stokes Alliance for Minority Participation-Bridge to the Doctorate program, which is supported by the National Science Foundation. This work was supported in part by Bankhead-Coley Research Foundation grant 1BC01-43279 to D.T.H.

## REFERENCES


1. Rutledge, G.C.; Fridrikh, S.V. Formation of fibers by electrospinning. Adv. Drug Deliv. Rev. 2007, 59, 1384–1391.

2. Sill, T.J.; von Recum, H.A. Electrospinning: Applications in drug delivery and tissue engineering. Biomaterials 2008, 29, 1989–2006.

3. Yoo, H.S.; Kim, T.G.; Park, T.G. Surface-functionalized electrospun nanofibers for tissue engineering and drug delivery. Adv. Drug Deliv. Rev. 2009, 61, 1033–1042.

4. Khadka, D.B.; Haynie, D.T. Insoluble synthetic polypeptide mats from aqueous solution by electrospinning. ACS Appl. Mater. Interfaces 2010, 2, 2728–2732.

5. Agarwal, S.; Greiner, A. On the way to clean and safe electrospinning-green electrospinning: Emulsion and suspension electrospinning. Polym. Adv. Technol. 2011, 22, 372–378.

6. Khadka, D.B.; Cross, M.C.; Haynie, D.T. A synthetic polypeptide electrospun biomaterial. ACS Appl. Mater. Interfaces 2011, 3, 2994–3001.

7. Khadka, D.B.; Haynie, D.T. Protein- and peptide-based electrospun nanofibers in medical biomaterials. Nanomedicine 2012. in press.

8. Ner, Y.; Stuart, J.A.; Whited, G.; Sotzing, G.A. Electrospinning nanoribbons of a bioengineered silk-elastin-like protein (SELP) from water. Polymer 2009, 50, 5828–5836.

9. Deitzel, J.M.; Kleinmeyer, J.; Harris, D.; Beck Tan, N.C. The effect of processing variables on the morphology of electrospun nanofibers and textiles. Polymer 2001, 42, 261–272.

10. Yarin, A.L.; Koombhongse, S.; Reneker, D.H. Bending instability in electrospinning of nanofibers. J. Appl. Phys. 2001, 89, 3018–3026.

11. Minato, K.I.; Ohkawa, K.; Yamamoto, H. Chain conformations of poly(γ-benzyl-L-glutamate) pre and post an electrospinning process. Macromol. Biosci. 2006, 6, 487–495.

12. Hohman, M.M.; Shin, M.; Rutledge, G.; Brenner, M.P. Electrospinning and electricity forced jets. I. Stability theory. Phys. Fluids 2001, 13, 2201–2220.

13. Fridrikh, S.V.; Yu, J.H.; Brenner, M.P.; Rutledge, G.C. Controlling the fiber diameter during electrospinning. Phys. Rev. Lett.2003, 90, 114502:1–114502:4.

14. He, J.H.; Xu, L.; Wu, Y.; Liu, Y. Mathematical models for continuous electrospun nanofibers and electrospun nanoporous microspheres. Polym. Int. 2007, 56, 1323–1329.

15. Strobl, G. The Physics of Polymers; Springer: Berlin, Germany, 2007.

16. Zheng, X.; Baker, H.; Hancock, W.S.; Fawaz, F.; McCaman, M.; Pungor, E. Proteomic analysis for the assessment of different lots of fetal bovine serum as a raw material for cell culture. Part IV. Application of proteomics to the manufacture of biological drugs. Biotechnol. Prog. 2006, 22, 1294–1300.

17. Chaudhuri, S.R.; Yang, J.T. Helix-coil transition of poly-L-ornithine in solution. Biochemistry 1968, 7, 1379–1383.

18. Wada, A. Helix-coil transformation and titration curve of poly-L-glutamic acid. Mol. Phys. 1960, 3, 409–416.

19. Greenfield, N.; Fasman, G.D. Computed circular dichroism spectra for the evaluation of protein conformation. Biochemistry1969, 8, 4108–4116.

20. Wetlaufer, D. Ultraviolet spectra of proteins and amino acids. Adv. Protein. Chem. 1962, 17, 303–390.

21. Chakrabartty, A.; Kortemme, T.; Padmahabhan, S.; Baldwin, R.L. Aromatic side-chain contribution to far-ultraviolet circular dichroism of helical peptides and its effect on measurement of helix propensities. Biochemistry 1993, 32, 5560–5565.

22. Chirgadze, Y.N.; Nevskaya, N.A. Infrared spectra and resonance interaction of amide-I vibration of the antiparallel-chain pleated sheet. Biopolymers 1976, 15, 637–648.

23. Jackson, M.; Mantsch, H.H. The use and misuse of FTIR spectroscopy in the determination of protein structure. Crit. Rev. Biochem. Mol. Biol. 1995, 30, 95–120.

24. Imoto, T.; Johnson, L.N.; North, A.T.C.; Phillips, D.C.; Rupley, J.A. Vertebrate lysozymes. In The Enzymes; Boyer, P.D., Lardy, H., Myrback, K., Eds.; Academic: New York, NY, USA, 1972; pp. 665–868.

25. Malamud, D.; Drysdale, J.W. Isoelectric points of proteins: A table. Anal. Biochem. 1978, 86, 620–647.

26. Shenoy, S.L.; Bates, W.D.; Frisch, H.L.; Wnek, G.E. Role of chain entanglements on fiber formation during electrospinning of polymer solutions: Good solvent, non-specific polymer-polymer interaction limit. Polymer 2005, 46, 3372–3384.

27. Doty, P.; Wada, A.; Yang, J.T.; Blout, E.R. Polypeptides. VII. Molecular configurations of poly-L-glutamic acid in water-dioxane solution. J. Polym. Sci. 1957, 23, 851–857.

28. Zandomeneghi, G.; Krebs, M.R.H.; McCammon, M.G.; Fändrich, M. FTIR reveals structural differences between native β-sheet proteins and amyloid fibrils. Protein Sci. 2004, 13, 3314–3321.

29. Richardson, J.S. The anatomy and taxonomy of protein structure. Adv. Protein Chem. 1981, 34, 167–339.

30. Laurine, E.; Gregoire, E.; Fändrich, M.; Engemann, S.; Marchal, S.; Thion, L.; Mohr, M.; Monsarrat, B.; Michel, B.; Dobson, C.M.; et al. Lithostathine quadruple-helical filaments from proteinase K-resistant deposits in Creutzfeldt-Jakob disease. J. Biol. Chem. 2003, 278, 51770–51778.

31. Nyquist, R.A.; Clark, T.D.; Streck, R. Infrared study of alkyl carboxylic acids in CCl4 and/or CHCl3 solutions. Vib. Spectrosc.1994, 7, 275–286.

32. Baldwin, R.L. In search of the energetic role of peptide hydrogen bonds. J. Biol. Chem. 2003, 278, 17581–17588.

33. Poland, D.; Scheraga, H.A. Theory of Helix-Coil Transition Theory in Biopolymers; Academic: New York, NY, USA, 1970.

34. Wozniak, M.A.; Modzelewska, K.; Kwong, L.; Keely, P.J. Focal adhesion regulation of cell behavior. Biochim. Biophys. Acta1692, 103–119.

35. Junge, K.; Binnebösel, M.; von Trotha, K.T.; Rosch, R.; Kling, U.; Neumann, U.P.; Jansen, P.L. Mesh biocompatibility: Effects of cellular inflammation and tissue remodelling. Langenbecks Arch. Surg. 2012, 397, 255–270.

36. Squires, T.M.; Brenner, M.P. Like-charge attraction and hydrodynamic interaction. Phys. Rev. Lett. 2000, 85, 4976–4979.

37. Behrens, S.H.; Grier, D.G. The charge of glass and silica surfaces. J. Chem. Phys. 2001, 115, 6716–6721.

38. Fischer, H.; Polikarpov, I.; Craievich, A.F. Average protein density is a molecular-weight-dependent function. Protein Sci. 2004,13, 2825–2828.

39. Curtis, A.S.G.; Forrester, J.V.; McInnes, C.; Lawrie, F. Adhesion of cells to polystyrene surfaces. J. Cell Biol. 1983, 97, 1500–1506.

40. Curtis, A.S.G. Cell adhesion. Prog. Biophys. Mol. Biol. 1973, 27, 317–375.

41. Gingell, D.; Todd, I. Red blood cell adhesion. II. Interferometric examination of the interaction with hydrocarbon oil and glass.J. Cell Sci. 1980, 41, 135–149.

42. Pernodet, N.; Rafailovich, M.; Sokolov, J.; Xu, D.; Yang, N.-L.; McLeod, K.J. Fibronectin fibrillogenesis on sulfonated polystyrene surfaces. J. Biomed. Mater. Res. A 2003, 64, 684–692.

43. Harris, A.K.; Pryer, N.K.; Paydarfar, D. Effects of electric fields on fibroblast contractility and cytoskeleton. J. Exp. Zool.1990, 253, 163–176.

44. Poo, M. In situ electrophoresis of membrane components. Annu. Rev. Biophys. Bioeng. 1981, 10, 245–276.

45. Sun, S.; Wise, J.; Cho, M. Human fibroblast migration in three-dimensional collagen gel in response to noninvasive electrical stimulus. I. Characterization of induced three-dimensional cell movement. Tissue. Eng. 2004, 10, 1548–1557.

46. Nuccitelli, R. A role for endogenous electric fields in wound healing. Curr. Top. Dev. Biol. 2003, 58, 1–26.

47. Borgens, R.B.; Vanable, J.W., Jr.; Jaffe, L.F. Bioelectricity and regeneration: Large currents leave the stumps of regenerating newt limbs. Proc. Nat. Acad. Sci. USA 1977, 74, 4528–4532.

48. Mir, L.M.; Bureau, M.F.; Gehl, J.; Rangara, R.; Rouy, D.; Caillaud, J.M.; Delaere, P.; Branellec, D.; Schwartz, B.; Scherman, D. High-efficiency gene transfer into skeletal muscle mediated by electric pulses. Proc. Nat. Acad. Sci. USA 1999, 96, 4262–4267.

49. Hochmuth, R.M. Micropipette aspiration of living cells. J. Biomech. 2000, 33, 15–22.

50. Kinraide, T.B.; Wang, P. The surface charge density of plant cell membranes (σ): An attempt to resolve conflicting values for intrinsic σ. J. Exp. Bot. 2010, 61, 2507–2518.

# Chapter 5

# Keratin-based Nanofibres

**Claudio Tonin[1], Annalisa Aluigi[1], Alessio Varesano[1] and Claudia Vineis[1]**

[1] ISMAC, Institute for Macromolecular Studies, CNR,
National Research Council, Italy

## 1. INTRODUCTION

The interest in biopolymers from renewable resources as alternatives to polymers made from oil and other fossil resources has been increasing over the years. Biopolymers are also considered environmentally friendly over their entire live-cycle. There is much recent literature on carbohydrates and proteins derived from plants and animals and polyesters made from the fermentation of plant material. As to whether it is ethically justifiable to convert valuable foodstuffs into commodities is open to question. The focus here is on keratin, one of the most abundant and mostly unexploited non-food proteins, being the major component of hair, feathers, nails and horns of mammals and birds. In spite of their important and interesting characteristics keratin wastes represent a rather complicated disposal challenge because burning for fuel is inefficient and polluting due to the high sulphur content (3-4% wt). The total amount of keratin (including fibre by-products from the wool textile industry, poor quality raw wools from farms and butchery waste) has been estimated worldwide at more than 5 million tonnes per year (Barone et al., 2005). Ground horn and nail is used as a nitrogenous fertilizer for gardening and, more recently, has been processed by caustic hydrolysis to produce biodegradable surfactants for fire extinguisher foams. However, most keratin wastes made from unserviceable wools and feathers from poultry are not valorised and are simply disposed of (Martínez-Hernández et al., 2007; Schmidt, 1998). Pooling and processing into biopolymers might be a better way of exploiting such a large quantity of protein biomass. Keratin-based materials can be used in biotechnological and biomedical fields for tissue engineering and the production of affinity membranes, due to their biocompatibility, their ability to support fibroblast growth and absorb heavy metal ions and volatile organic compounds (VOCs). Transforming keratin into nanofibres by electrospinning combines the aforementioned properties of keratin with the high surface to volume ratio and the high porosity of nano-structured textiles. This may be an original and promising approach for the fabrication of scaffolds for tissue engineering and filtration devices.

Keratin distinguishes itself from other structural proteins by the quantity of cysteine residues in the protein molecules (7-20% of the total amino acid residues). In particular, cysteine amino acids form inter and intra molecular disulphide bonds (cysteine residues) giving rise to a compact three-dimensional structure that confers a high stability to the protein (Dowling et al., 1986).

Because of their low molecular weight (9–60 kDa), keratin-based materials have poor mechanical properties. Moreover, like most natural polymers, keratin is not thermoplastic. For electrospinning keratin should be blended with suitable polymers using common, volatile and easy to handle solvents. This is also recommended when mechanical performance is needed (Aluigi et al., 2008; Katoh et al., 2004; Zoccola et al., 2008). This chapter describes the extraction of keratin from wool and electrospinning of keratin-based blends with high molecular weight polymers. Blends of keratin with poly(ethylene-oxide) (PEO) and fibroin are suitable for biomedical application (tissue engineering and medical textiles), while keratin/polyamide 6 (PA6) nanofibres can be used for active filtration of air and water.

## 2. KERATIN AND REGENERATED KERATIN

Keratins represent a group of fibrous proteins with high sulphur content produced in some epithelial cells of vertebrate such as reptiles, birds and mammals. In particular, the cysteine amino acid residues form inter and intra molecular disulphide bonds (cystine residues) that give rise to a compact three-dimensional structure that confers to keratin proteins a high resistance to chemical and enzymatic attacks (Dowling et al., 1986). There are two kinds of keratins: the “hard-keratin” and the “soft-keratin” according to the physical and chemical properties, particularly the sulphur content. Soft keratins, with a sulphur content <3% wt, are found in the *stratum corneum* of the skin whereas the hard keratins found in hair, wool, feather, nails and horns and have a sulphur content > 3% wt (Fraser et al., 1972). A further classification is based on the X-ray diffraction pattern obtained from different keratin proteins. The α-helix appears to be the basic fibrillar element in all soft keratins and in the hard keratins from mammals (Crick, 1953). Studies carried out on the structure of feather keratin have shown that about 28% of the protein molecule has a β-conformation; the remainder does not possess a geometrically regular secondary structure (Fraser et al., 1971).

A key problem in blending keratin with other polymers is to find a suitable solvent capable of dissolving the complicated structure of the protein. Generally a mixture of solvent systems in which different components have different functions is used, one component breaks the hydrogen bonds while the reducing or oxidising agent breaks the disulphide bonds. Some multiple solvent systems that have been utilized are: urea/reducing agent (Yamauchi et al., 1996), carbamide/$H_2O_2$/$H_2O$ and carbamide/2-mercaptoethanol (Garret & Grisham, 2002), or Cu-oxam metal complex system ([Cu($NH_3$)4($OH)_2$]) (Aluigi et al., 2004). In 2005, Xie and co-workers tested ionic liquids, particularly 1-butyl-3-methyllimidazolium-chloride ($BMIM^+Cl^-$), as a solvent for keratin proteins (Xie et al., 2005).

Recently, two chemical-free processes using just water and heat had been explored as methods of dissolving and converting keratinous materials (i.e. wool and feathers, respectively) : steam explosion (Tonin et al., 2006) and superheated water (Yin et al., 2007). Steam explosion is based on short time steam cooking of biomasses at a high temperature for several minutes, followed by explosive decompression. Using saturated steam at 220 C (~22 bar) for 10 min followed by a rapid decompression, wool fibres were disrupted into solid and liquid phases consisting of oligo-peptides, water-soluble peptides and free amino acids (Tonin et al., 2006). The second process consisted of treating feather barbs with superheated water at ~220 C and ~22 bar for 2 h followed by cooling. Most of the keratin was converted into oligo-peptides with a molecular weight of about 1.0 and 1.8 kDa (Yin et al., 2007).

However, for the electrospinning process, volatile and easy to handle solvents are needed and for this reason we have studied the solvation properties of water and formic acid for keratin.

## 2.1. Wool Keratin

Wool fibre is a complex multi-cell system composed of inanimate cells which differ in composition, morphology and properties. The principal component of wool fibres is keratin and there is also a small amount of lipids (0.1%) and mineral salts (0.5%). Wool cleaning for scientific purposes is carried out by means of extraction with petroleum ether, ethanol and water (von Bergen, 1963).

The disulphide bonds of cystine form both inter and intra chain cross-links and are responsible for the greater stability and lower solubility of keratin compared with most proteins. Therefore, the presence of cystine plays the most important role in the chemical, thermal and mechanical properties of wool. Although disulphide bonds are responsible for keratin reticulation in wool, there are other different kinds of cross-links: hydrophobic bonds occur between non polar groups; hydrogen bonds have a very important influence on the mechanical properties of the wool fibre; ionic bonds occur between carboxylic anions and ammonium cations. These ionic bonds depend on the pH of the environment. In fact, at the isoelectric point (pH=4.9), the quantity of ionic bonds is greatest because the protein exists as zwitterions ($^{+}$H3N—CHR—COO$^{-}$). When the pH is very acid or very basic, there is a reduction in the number of ionic bonds. At low pH values the $COO^-$ groups are protonated, while at high levels the $NH3^+$ groups are deprotonated. (von Bergen, 1963).

Proteins extracted from wool can be separated into four different groups: the low sulphur content keratins (LS) that have a molecular weight between 45 and 60 kDa; the high sulphur content keratins (HS) with a molecular weight of between 11 and 28 kDa and the high-glycine and tyrosine content proteins with a molecular weight of between 9 and 12 kDa (Jeffrey, P. D., 1972). The low sulphur content proteins have an α-helical crystalline structure (α-keratins), while the other proteins do not have a helicoidal structure.

X-ray studies showed that the secondary structure of α-keratins is dextrorotatory helicoidal. This α-helix is stabilized by the presence of hydrogen bonds between the N-H group of a peptide bond and the carbonyl group of a peptide bond placed over four residues. Although the strength of a single hydrogen bond (~5 kcal/mole) is only 5% of that of C-C and C-H covalent bonds the large number of hydrogen bonds confers a remarkable stability to the α-helix crystallites. The conformation of the α-helix is characteristic for the wool fibre in its natural state, therefore it is found in the fibre which is not stretched along its axis (Pauling & Corey, 1953). During stretching, the α-helix declines and β-keratin appears (Tonin et al., 2004).

## 2.2. Preparation and Characterization Of Keratin Solutions

Keratin extraction from wool may take place only after the reduction or oxidation of disulphide bonds. The reducing agents often used are thiols (thioglycolic acid, dithiothreitol and 2-mercaptoethanol) while oxidizing agents are peracetic or performic acids (Yamauchi et al., 1996; Thompson & O'Donnel, 1959 ). All these agents are always used in combination with protein denaturing agent (e.g. urea) that breaks hydrogen bonds and extraction yields range from 50 to 70% wt. However, for further scaled-up processes, we preferred to use sulphitolysis reaction, although the extraction yield is lower (37% wt), instead of harmful reducing and oxidizing agents. The sulphitolysis is a reaction of wool with sulphite ions that break disulphide bonds in thiols and S-sulphonate anions, known as Bunte salts (Cecil, 1963).

Five grams of cleaned fibres were cut into snippets and put in 100 ml of aqueous solution containing urea (8 M), m-bisulphite 0.5 M, adjusted to pH 6.5 with NaOH 5 N and treated by shaking for 2 h at 65 C. Successively, the solution was filtered through a 5 m pore-size filter using a vacuum pump and then dialyzed against distilled water using cellulose tubes (molecular mass cut-off 12000-14000) for three days, changing the outer solution four times a day. The dialyzed part was first concentrated with a rotary vacuum evaporator in order to prepare solutions at different keratin concentrations (0.5, 1.5, 3.5 and 5% wt). The solution at 5%wt was cast onto a polyester plate and dried at 50 C overnight in order to prepare keratin films regenerated from water (samples labelled as KW). A part of keratin regenerated from water was dissolved in formic acid for 4h, so as to have solutions at different keratin concentrations (0.5, 1.5, 3.5 and 5% wt). The solution of keratin in formic acid at 5% wt was cast onto polyester plates and dried at 50 C overnight in order to prepare keratin films regenerated from water (samples labelled as KF).

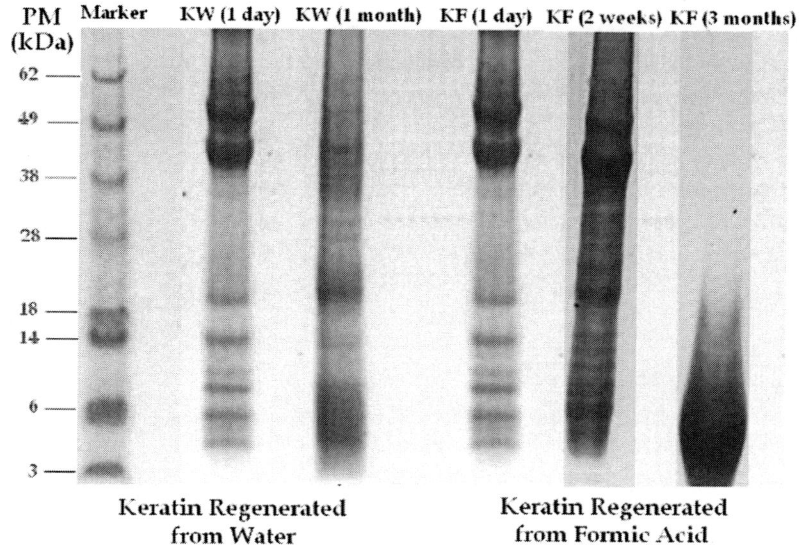


**Figure 1.** SDS-PAGE of KW and KF.

The stability of keratin dissolved in water and in formic acid solutions was studied both following the protein degradation during time and the flocculation process by turbidity measurements. The molecular weight distribution analysis conducted by electrophoresis SDS-PAGE (reported in Fig. 1), revealed that the keratin regenerated from a one month old aqueous solution was degraded, in fact the high molecular weight proteins (at 45 and 60 kDa) disappear, while a series of proteins having molecular weights below 38 kDa appear. Regarding keratin in a formic acid solution, the keratin regenerated from a fresh solution (1 day) is not degraded while a slight degradation appears after two weeks and after three months there is a complete digestion of the protein.

Turbidity *()* of solutions was calculated by measuring the transmittance T at 540 nm using the following Eq. 1:

$$\tau = -(\ln T) / c \tag{1}$$

where c is the cell length (1cm).

Fig. 2 shows that turbidity of the keratin aqueous solution increases with the increase in the keratin concentration, suggesting that higher concentrations promote protein chain aggregation (flocculation). On the other hand the turbidity of keratin dissolved in formic acid remains unchanged with increasing keratin concentration, indicating that flocculation does not occur, at least over a certain range of keratin concentration.

Viscosities of the keratin solutions at 5% wt, were plotted against the shear rate (Fig. 3). In the whole shear range investigated, the viscosity of keratin dissolved in formic acid was higher than that of keratin in water. This behaviour indicates that formic acid has better solvation properties for keratin than water. This is because formic acid is more polar than water and forms strong interactions with the polar side chain groups of keratin such as CO, OH, COOH and $NH_3^+$. As a consequence, the molecular chains become closer in formic acid than in water and this is confirmed by the lower turbidity.

## 3. ELECTROSPINNING

Several methods are dedicated to producing small diameter fibres for high-volume production such as fibrillation (Homonoff, 2008; Perez et al., 2000), island-in-sea (Pourdeyhimi et al., 2006), and a novel melt-blowing system (Bryner & Armantrout, 2006; Ellison et al., 2007); or highly accurate methods such as nanolithography (Tseng et al., 2005; Xie et al., 2006) and self-assembly (Zhang, 2002). However, their usefulness is restricted by combinations of narrow material ranges, high costs and low production rates. In comparison, electrospinning is a simple and low cost process, and has an intermediate production rate (Ramakrishna et al., 2006).

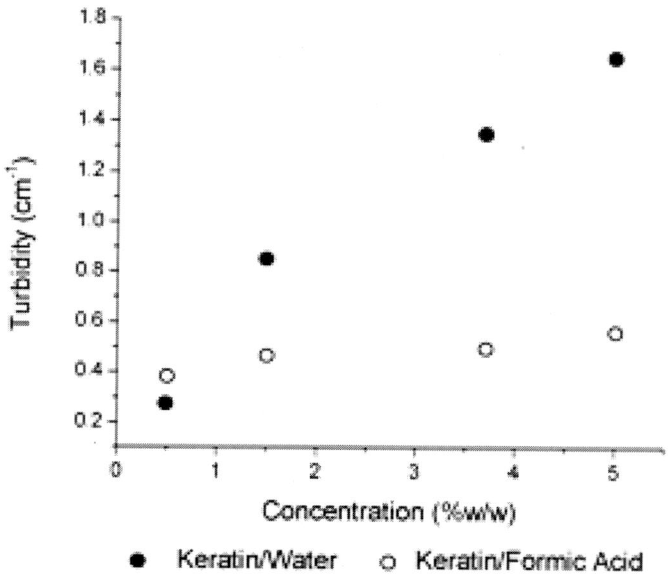


**Figure 2.** Turbidity of keratin dissolved in water and in formic acid.

Electrospinning is a physical phenomenon classified as a branch of electro-fluid-dynamics. The process is based on the electrostatic repulsion of charges ruled by the well-known Coulomb law. By means of an electrostatic field, an electrospinning apparatus generates electrically-driven polymer jets that solidify into sub-micron scale polymer-based filaments (usually called nanofibres). The idea was patented in the early 20th century (Cooley, 1902; Morton, 1902; Formhals, 1934); however, the invention was disregarded until the 1990s (Huang et al., 2003; Li & Xia, 2004), probably due to lack of knowledge and interest of both industries and scientists about tiny fibres. Electrospinning works with polymer melts and polymer solutions, but few research papers deal with electrospinning of molten polymer (so-called "melt-electrospinning") compared with the huge literature regarding electrospinning of polymer solutions (usually simply referred to as "electrospinning").

Electrospinning can be usefully employed in a wide range of applications requiring sub-micron scale fibre diameter such as: filtration (Tsai et al., 2002; Qin & Wang, 2006; Li et al., 2006; Barhate & Ramakrishna, 2007), membrane separation (Shin et al., 2005; Gopal et al., 2006), protective clothing and breathable garments (Gibson et al., 2001), wound dressings and scaffolds for tissue engineering (Pham et al., 2006; Lee et al., 2008; Sill & von Recum, 2008), and drug delivery (Sill & von Recum, 2008).

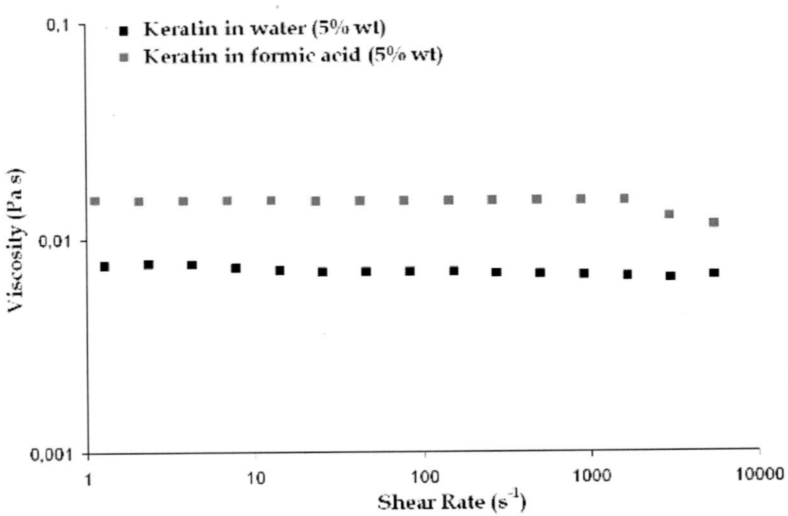


**Figure 3.** Rheological behaviour of the keratin in water and keratin in formic acid solutions.

A basic electrospinning setup consists of three elements: an electrical generator (high voltage supply), a capillary (jet source) and a metal collector (target). The solution is usually electrically charged by the generator and the collector is grounded, but it is also possible to invert the system by electrically charging the collector and grounding the solution (Kilic et al., 2008). A scheme of an electrospinning apparatus is shown in Fig. 4.

A pump pushes the polymer solution through the capillary with a fixed flow rate by means of a syringe. The generator, connected by a wire to the metal tip of the capillary, supplies the voltage to the solution. A stream of solution (jet) is produced from the tip toward the grounded metal collector on which electrospun nanofibres were collected. Usually, the production rate of a single jet is less than 10 g $h^{-1}$ of nanofibres (Tsai et al., 2002), depending on polymer concentration and process conditions, mainly the flow rate of solution.

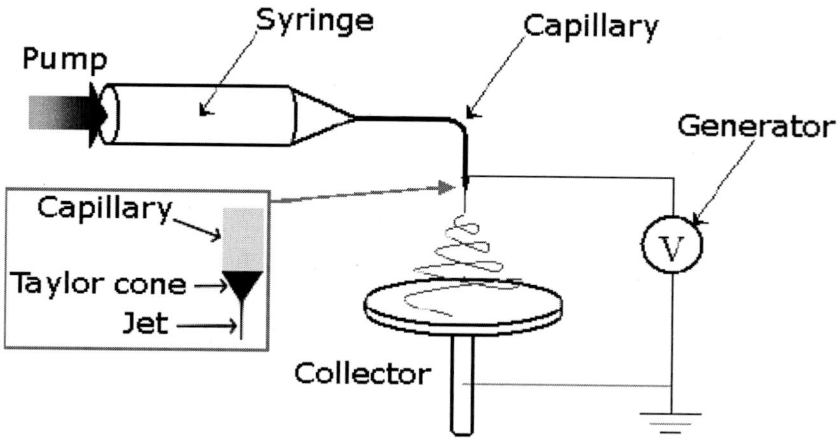


**Figure 4.** Scheme of an electrospinning setup. In the box: enlargement of the capillary tip.

The electrospinning process starts when the voltage generator is turned on. The electrical potential of the droplet surface at the capillary tip is increased to a sufficiently high value that causes the droplet to assume a steady conical shape, known as the Taylor cone (see box in Fig. 4), instead of a spherical shape. The electrostatic forces act in opposition to the surface tension of the fluid. The Coulomb repulsion between the charges promotes the formation of the cone; on the opposite the surface tension of the fluid favours sphere-like shapes. As the electrostatic forces exceed the surface tension, a charged jet of fluid is ejected from the vertex of the cone (Yarin et al., 2001; Shin et al., 2001).

The trajectory of the jet solution begins with a straight segment. The diameter of the jet is one or two orders of magnitude smaller that the inner diameter of the capillary from which the jet is generated, and progressively decreases during its journey towards the collector. The diameter reduction occurs thank to two mechanisms: (a) the electrostatic repulsion forces of the charges in the jet tend to extend the jet itself, and (b) the solvent evaporation leaves the solid materials in the jet (e.g. polymer), this effect is self-induced because as the diameter decreases the evaporation further increases. Moreover, it worth noting that as the solvent evaporates the concentration and the viscoelastic properties of the liquid jet change.

After the straight segment, the jet is subjected to a bending instability and its trajectory becomes disordered; the oscillation eventually evolves into a spiral, similar to the movement of a whip (so-called whipping motion), as illustrated in Fig. 4. Finally, the jet completely solidifies into a nanofibre and deposits on the collector as a disordered continuous filament (non-woven structure). In spite of the ease of the electrospinning, the onset of bending instability is quite a complex problem; several theories and mechanisms have been recently proposed to explain this phenomenon (Shin et al., 2001;Deitzel et al., 2001; Reneker & Yarin, 2008).

The fibre-forming process depends on operating parameters (applied voltage, working distance between tip and collector, flow-rate of solution, tip diameter), properties of the polymer solution (polymer molecular weight, concentration, relaxation time, viscosity, electrical conductivity, surface tension, and vapour tension of the solvent), and environmental parameters such as temperature, humidity and ventilation (Theron et al., 2004). Depending on working conditions and polymer solution properties, the electrospun jet might be subjected to many phenomena such as branching, splitting, buckling, jet break-up (spraying), production of flat ribbon or beaded nanofibres that engenders different morphologies (Reneker & Yarin, 2008).

# 4. KERATIN/PEO BLEND NANOFIBRES

Recently in a leading opinion paper, Furth and co-workers (Furth et al., 2007) stated that they had been investigating the utility of keratin-based biomaterials for regenerative medicine applications. Keratin contains intrinsic sites of cellular recognition that mimic the extra-cellular matrix and keratin-based biomaterials demonstrated cell instructive capabilities. Certain keratin materials have been shown to be mitogenic and chemotactic for a variety of cell types, and to mediate changes in gene expression consistent with the promotion of wound healing. Moreover, other recent works (Tachinaba et al., 2002;Tachinaba et al., 2005; Hamasaki et al., 2008) have highlighted the excellent cell adhesion onto wool keratin sponges produced by lyophilisation from aqueous solution. These results show that keratin is a useful biomaterial for scaffolds for cell cultivation.

The poor mechanical properties of regenerated keratin hinder its processability and restrict its practical applications to blends with appropriate polymers with better structural properties. Our attempts to obtain filaments of pure regenerated keratin from water were indeed unsuccessful; moreover, literature reports on the fabrication of regenerated keratin films (Tanabe et al., 2004) using cross linking agents, and the fabrication of composite nanofibres of regenerated silk fibroin blended with synthetic polymers such as poly(ethylene oxide) (PEO) (Jin et al., 2002). PEO is an amphiphilic water soluble and non-degradable polymer, with good biocompatibility (Desai & Hubbel, 1991) and low toxicity (Bergsma et al., 1995). This polymer is often used as an ideal model for the electrospinning process (Theron et al., 2004; Son et al., 2004) because it can be electrospun without defects from aqueous solutions in a rather narrow range of conditions.

We studied the production and the characterization of nanofibres produced by electrospinning of pure PEO and 50/50 wt/wt keratin/PEO blends, from aqueous solution of the polymers, in different operating conditions. The nanofibres were studied by Scanning Electron Microscopy (SEM) and Differential Scanning Calorimetry (DSC). The results were compared with those obtained by thin films produced by casting from the same solutions, with the aim of investigating the influence of the production processes on the structural arrangement of these materials.

## 4.1. Materials and Methods

Keratin was obtained from wool by means of a sulphitolysis extraction method. PEO powder with an average molecular weight of $4\ 10^5$ g mol$^{-1}$ was dissolved in distilled water at ambient temperature for about 12 h. The concentrations used were 5, 7 and 10 % wt.

The keratin/PEO blend solutions were prepared at room temperature in about 12 h by simply adding PEO powder to the keratin aqueous solution. The solutions of keratin/PEO blend had total polymer concentrations of 5, 7 and 10 % wt with a keratin/PEO weight ratio of 50/50. These solutions were electrospun at 20 cm working distance to ensure that the nanofibres were dried. The applied voltages were between 10 and 30 kV. The process was stopped after about 10 minutes. During the electrospinning process, environmental conditions were kept in check; in particular, the temperature was from 20 to 25 C and the relative humidity was in the range 55-65 %.

## 4.2. Solution Characterization

Viscosity is an important factor for complete fibre formation in the electrospinning of polymer solutions. In particular, fibres without beads were produced when polymer chain entanglements are present at sufficient concentration (Shenoy et al., 2005). Moreover, fibre formation is promoted at low polymer concentrations, increasing solution conductivity (Son et al., 2004).

A great increase in conductivity was measured from solutions containing the keratin because of SDS that remains associated with keratin through ionic interaction conferring negative charges to the protein (Schrooyen et al., 2001). In fact the conductivity of pure PEO was about 0.120 mS cm$^{-1}$ and the conductivity of keratin/PEO was 1.282 mS cm$^{-1}$ (at polymer concentration 7% wt).

Fig. 5 shows the flow curves of PEO and keratin/PEO solutions at 5, 7 and 10% wt of polymers in water. All solutions behave like shear-thinning fluids. At low shear rates, disentanglement is balanced by formation of new entanglements, thus the fluid has a Newtonian behaviour which corresponds to a constant viscosity value (zero-shear viscosity). At higher values of shear rate, the disentanglement rate exceeds the rate of entanglement formation, therefore the viscosity decreases and shear-thinning behaviour is observed. The onset of shear thinning shifts to lower values of shear rate with the increase of polymer concentrations for both pure PEO and keratin/PEO solutions. The zero-shear viscosity increases when the polymer concentration increases. It is worth nothing that the keratin/PEO solutions at concentrations of 7 and 10% wt show flow curves Comparable with 5 and 7% PEO solutions, respectively. Thus, the keratin with a relatively low molecular weight slightly increases the viscosities of the keratin/PEO solutions, but its contribution is not negligible (Varesano et al., 2008).

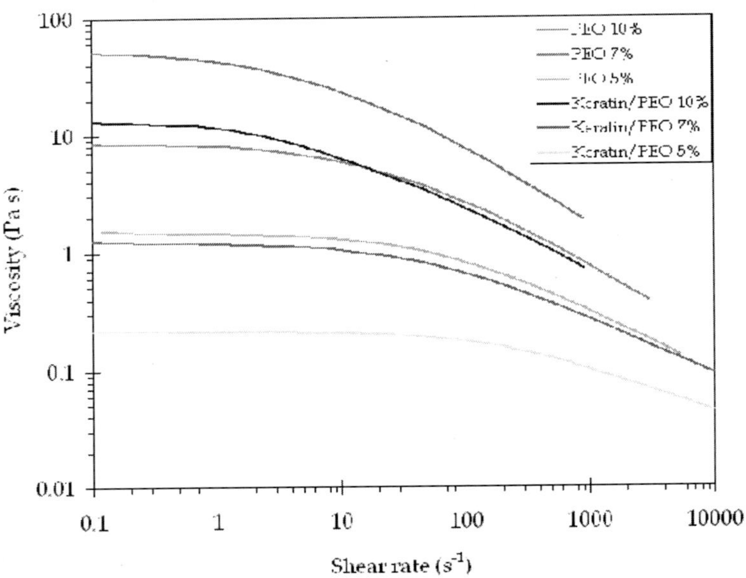


**Figure 5.** Viscosity flow curves of PEO and keratin/PEO solutions.

## 4.3. Electrospinning and Morphological Study

Morphologies of the electrospun materials from PEO and keratin/PEO solutions were investigated for comparison. The samples were produced by varying flow-rate, voltage and polymer concentrations.

PEO nanofibres without defects were produced from the 7% wt solution at 0.01 ml min$^{-1}$ flow-rate from 20 to 30 kV of voltage. For higher flow-rate (0.05 ml min$^{-1}$) macroscopic drops fall from the capillary also at the highest voltage (30 kV). The electrospinning of 5% wt solution produced nanofibres without defects using a flow-rate of 0.01 ml min$^{-1}$ with the voltage range from 13 to 30 kV. Increasing the flow-rate to 0.03 ml min$^{-1}$ the nanofibres became more irregular with some beads due to the insufficient stretching of the jet at voltages below 25 kV. For higher flow-rate (0.05 ml min$^{-1}$), the process started only at 30 kV; whereas at lower voltages the solution dripped from the capillary.

Nanofibres of keratin/PEO blend with regular diameter distribution and few defects were electrospun from the solutions at 7 and 10 % wt of polymers concentration applying a voltage from 20 to 30 kV and a solution flow-rate of 0.01 ml min$^{-1}$ (Fig. 6 a). The keratin/PEO solutions at high concentrations (7 and 10 % wt) produce fibres with few defects, like 5 and 7 % wt PEO solutions, probably because they have a similar flow behaviour. At low polymers concentration (5 % wt) the nanofibres were electrospun with many defects also at high voltage (25-30 kV), as Fig. 6 b shows, and the solution dripped from the capillary during the process at voltages below 20 kV because of the low viscosity.

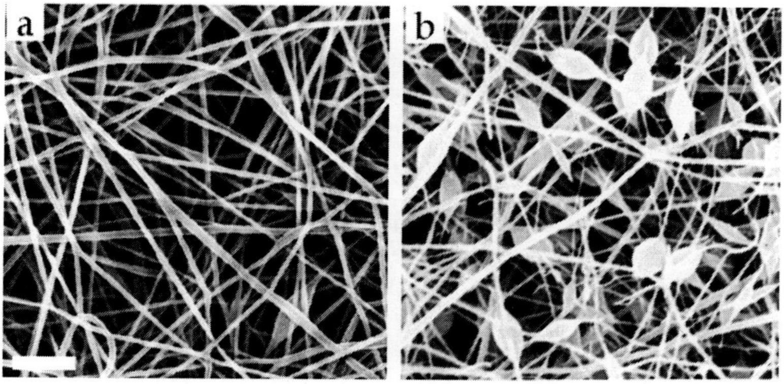


**Figure 6.** Electrospun nanofibres of 50/50 keratin/PEO blend at 25 kV, 0.01 ml min$^{-1}$ from (a) 7 % wt and (b) 5 % wt solutions. Scale bar = 2 μm.

The average diameters of the electrospun nanofibres, measured at 150 different points from SEM pictures for each sample produced, as a function of the applied voltage. The diameters of the nanofibres produced at the same voltage from 5 and 7 % wt of pure PEO solutions are generally comparable. The keratin/PEO solution produces nanofibres with small diameter at higher voltage (30 kV), whereas at 20 kV electrospun nanofibres have a diameter much higher than the pure PEO nanofibres at the same voltage. Moreover, the slope of the diameter/voltage trend-line for the keratin/PEO blend is the highest obtained in our experiment (Aluigi et al., 2007b). Thus, it seems that the presence of keratin strengthens the influence of the voltage on the size of the filaments. Since the keratin has many different functional groups, it is possible that inter- and intra-molecular bonds increase the jet rigidity when the solvent evaporates. Therefore, a higher voltage is required to stretch the solidifying keratin/PEO solution jet.

## 4.4. Thermal Analysis

The DSC curves of electrospun keratin/PEO samples, compared with the films produced by casting from the same solution, are shown in Fig. 7 with the aim of highlighting the structural changes induced by the electrospinning process.

The endothermic overlapped peaks at around 60 C are due to the fusion of PEO crystalline phase and to the evaporation of water, especially absorbed by keratin. It is worth noting that in the keratin/PEO film, the water evaporation occurs at lower temperature with respect to keratin/PEO nanofibres (50 C in the nanofibres and 80 C in the film). This is probably due to the high surface/volume ratio of the nanofibres which promotes water evaporation even at lower temperatures.

The DSC analysis the electrospun PEO exhibits a slight increase of the melting point. It is believed that the high stretching due to the electrospinning process promotes the orientation of the long polymer chains of PEO. This high degree of order shifts the melting point to a higher temperature.

The endothermic events observed in the range of 200-350 C are attributed to protein denaturation followed by protein degradation (Spei & Holzem, 1990). The thermograms show that the peaks related to the protein denaturation, which falls at 233 C in the film, shifts to lower temperature (213 C) in the electrospun sample. It could be presumed that the high draw, given by electrospinning process, and the quick water evaporation hinders the keratin self-assembly leading the protein chains to assume a less complex supermolecular organization which denatures at lower temperatures. This thermal behaviour is in agreement with the FT-IR observations (Aluigi et al.,

2007b); in fact, the keratin in the electrospun fibres shows a molecular conformation characterized by weaker hydrogen bonds that make the protein less thermally stable.

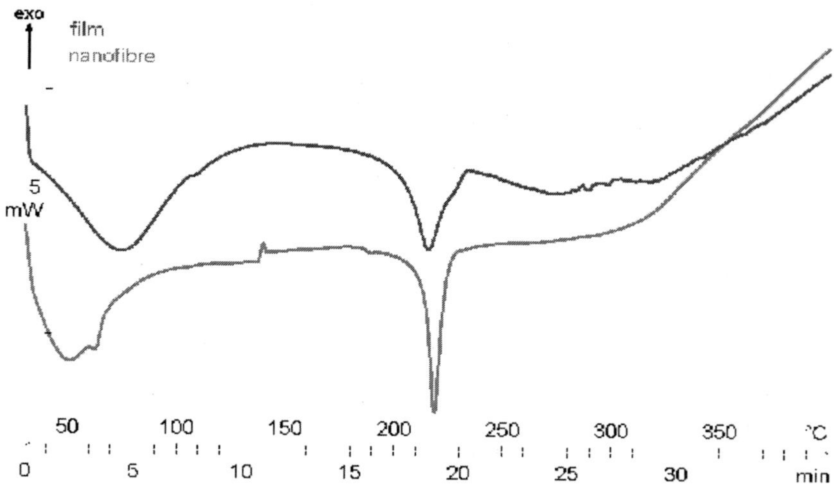


**Figure 7.** DSC analysis of keratin/PEO film from casting and keratin/PEO nanofibres from electrospinning.

## 5. KERATIN/PA6 BLEND NANOFIBRES

Since mid-1990s, there has been an increasing interest in the application of nano-sized fibres for high efficiency filtration (Montefusco, 2005; Gopal et al., 2006; Dotti et al., 2007). Indeed, high surface area associated to great surface cohesion and small pore dimensions of nanofibrous mats allow the capture of submicron particles improving filtration efficiency, cleaning operations and filter life. As far as "active" filtration is concerned, keratin is a bio-polymer that can absorb and remove toxic, sensitizing and suspected carcinogenic agents substances such as heavy-metal ions (copper, chromium, lead, mercury, etc.) (Kar & Misra, 2004; Ki et al., 2007; Aluigi et al., 2009), formaldehyde (Wortmann et al., 1999) and other hazardous VOCs.

Keratin is rich in amino acids having polar and ionisable side chains able to bind charged species such as metal ions. In particular, having a $pK_a$ about 4.5, free carboxyl groups of aspartyl and glutamyl residues are considered the most likely binding sites over a wide pH range (Maclaren & Milligan, 1981; Taddei et al., 2003). Recent studies showed that formic acid is a good media to dissolve keratin having strong solvation properties for keratin regenerated from wool (Alemdar et al., 2005; Aluigi et al., 2007a). Formic acid is also a solvent for polyamides so that it can be used to prepare keratin-polyamide 6 (PA6) blends in different proportions.

## 5.1. Solution Characterizations

Keratin/PA6 blend solutions with different blending ratios (100/0, 90/10, 70/30, 50/50, 30/70, 10/90 and 0/100), prepared by mixing solutions at 15% wt of pure keratin and pure PA6 separately dissolved in concentrated formic acid, can be processed into nanofibres by electrospinning, with different results according to viscosity and conductivity of the solutions. Viscosities measured at the shear rate 84 $s^{-1}$show deviation from linearity when compared with the theoretical values calculated by the additive rule (Eq. 2):

$$\ln \eta_T = \sum_i w_i \ln \eta_i \qquad (2)$$

where: w*i* is the weight fraction of the i$^{th}$ component, *i* is the solution viscosity of the i$^{th}$ component and *T* is the theoretical viscosity of the polymer blend. The fitting of the experimental measurements displays a negative deviation from the additive rule indicating a negligible interaction between the protein and the synthetic polymer. Moreover, viscosity of the blend solutions decreased with increasing the keratin content, while electrical conductivity (about of 2 mS $cm^{-1}$) did not change significantly.

## 5.2. Electrospinning and Morphological Study

Electrospinning is successful at room temperature, 25 kV applied voltage, 0.01 ml $min^{-1}$ flow rate and 10 cm working distance, producing very thin nanofibres in the range from 70 to 300 nm, with mean diameter of about 150 nm.

SEM images of the nanofibres with different compositions randomly deposited on the collecting screen are shown in Fig. 8. Nanofibres obtained by electrospinning keratin/PA6 blends are thin and free from defects, but pure keratin nanofibres show many bead defects since the jet breaks up into droplets as a result of the lower viscosity of the solution. The mean diameter of the nanofibres does not change significantly with the blend composition;

however, nanofibres rich in PA6 appear more homogeneous than keratin rich ones.

## 5.3. Testing and Characterization

Chemical interactions between the polymers can be excluded, according to the viscosity measurements and by the study of FT-IR spectra in the region 4000–2600 $cm^{-1}$ and 2000-900 $cm^{-1}$. Indeed, the nanofibre blend spectra appear as weighed overlapped spectra of pure keratin and pure PA6; a broadening of the Amide A and Amide I bands can be observed with increasing the keratin content and also the intensity of the 1025 $cm^{-1}$ peak (Bunte salt) increases with increasing the keratin content. But the absence of both band shifts and new peaks exclude chemical interaction.

The nanofibres containing more than 50% wt of keratin, are totally destroyed due to swelling of regenerated keratin when it contacts with water. On the other hand, morphology doesn't change for pure PA6 and keratin/PA6 10/90 nanofibres while those of keratin/PA6 30/70 start swelling and become flat. The adsorption of $Cr^{3+}$ ions by keratin can be studied only at low pH values in order to avoid the precipitation of chromium hydroxide above pH 4 (Maclaren & Milligan, 1981). $Cr^{3+}$ uptake of pure PA6 and blend nanofibres was calculated using the following Eq. 3:

$$q\ (\mu g/mg) = \frac{q_0 - q_1}{m} \tag{3}$$

where $q_0$ and $q_1$ are the $Cr^{3+}$ amount (g) in the 10 ml of standard solution before and after adsorption, respectively, and m is the mass of the nanofibres. The $Cr^{3+}$ adsorption of keratin/PA6 0/100, 10/90 and 30/70 nanofibres were compared with that of blend films obtained by casting from the same solutions. The chromium uptake of the nanofibres slightly increases with the increasing the keratin content. Moreover, the adsorption capacity of the nanofibrous mats is much higher than that of films with the same composition, since the higher surface area of the nanofibres increases the number of available binding sites. Both films and nanofibrous membranes were cut into square shapes of 4 $cm^2$ and placed in a standard solution (10 ml) of chromium ions, with an initial concentration of 50 g/L at pH 4, for 24 hours. The standard solution was prepared diluting with deionised water a stock solution of chromium (III) nitrate nonahydrate in nitric acid 1000 g/ml. The $Cr^{3+}$ concentration of the solution after immersion of the nanofibre mats was analysed by a Atomic Adsorption Spectrometer. Before the measurements, the calibration curve was made using standard solution of $Cr^{3+}$ at 25, 50, 75 and 100 g $l^{-1}$.

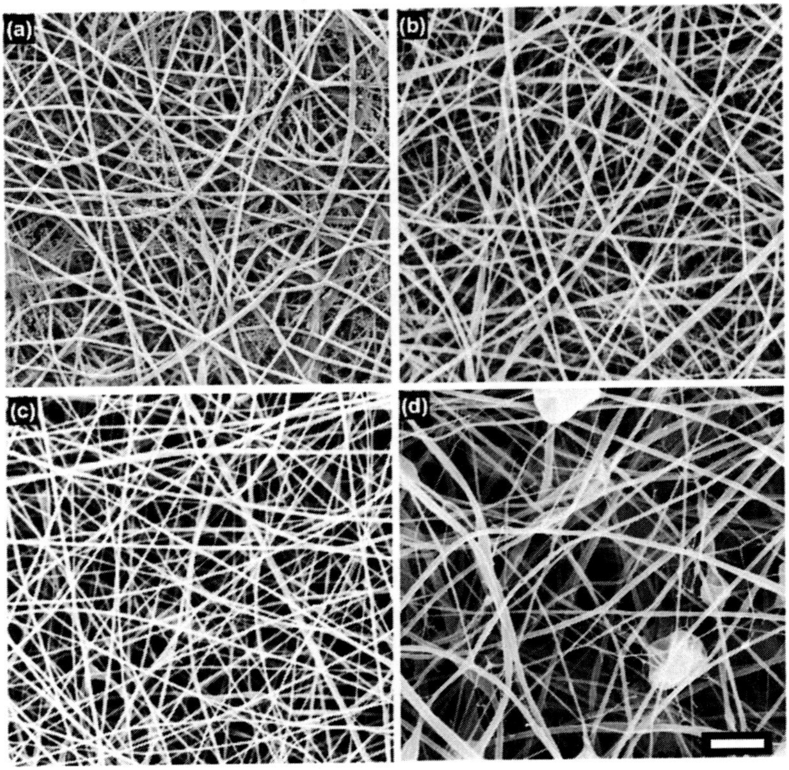


**Figure 8.** Keratin/PA6 blend nanofibres: (a) 30% keratin, (b) 70% keratin, (c) 90% keratin and (d) pure keratin. Scale bar = 2 m.

For the formaldehyde adsorption tests, multi-component filters of polypropylene (PP) and keratin/PA6 blend nanofibres were prepared by electrospinning the blend solutions directly onto the surface of the polypropylene filters. Adsorption tests were carried out using an apparatus composed as follows: a closed chamber having a volume of 3.3 L, a fan forcing air to pass through the filter, a filter holder and a Formaldemeter™ htV (PPM Technology Ltd.). About 0.6 ppm of formaldehyde was introduced in the chamber by using formaldehyde-releasing silica. When a stable concentration was reached in the chamber, the solid was removed, the multi-component filter was introduced in the chamber and the measurement started. The adsorption tests were performed at 20 C and 65% R.H. measuring the formaldehyde concentration versus time. The adsorption of gases on solid surfaces includes physiosorption and chemisorption. The physiosorption occurs when the interaction forces between adsorbent and adsorbate are intermolecular attractions (Van der Waals forces), while the chemisorption is a type of adsorption whereby the adsorbent adheres to the adsorbate through the formation of chemical bonds. Therefore, chemisorption is more selective than physiosoprtion.

It is known that formaldehyde has high reactivity towards proteins; in fact it is able to react with different side chains of amino acids such as the amino groups of lysine, arginine, glutamine and asparagine. The reaction scheme is reported in Fig. 9. The first reaction step involves amino groups and formaldehyde and produces an unstable product (aminomethylol derivative). The second reaction leads to a stable methylene bridge (Wortmann et al., 2005).

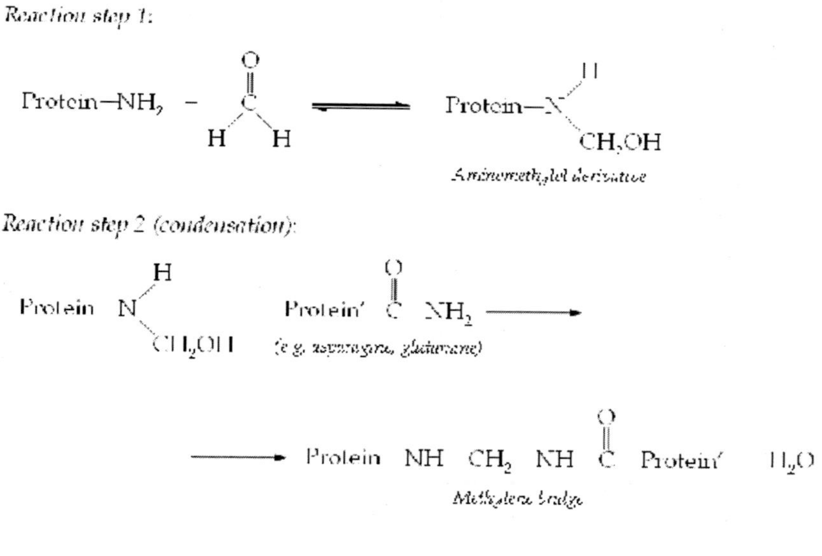


**Figure 9.** Reaction scheme of formaldehyde with keratin proteins.

The adsorption performances of keratin-based nanofibres were compared with that of PP filters and pure PA6 nanofibres/PP filters. Because of the selectivity of keratin towards formaldehyde, only the nanofibres containing the 90% wt of keratin were studied. It was found that PP filters and multi-component filters of PP sheets and PA6 nanofibres can reduce the formaldehyde concentration of about 30% and 40% respectively, but the reduction reaches 70% in the presence of keratin based nanofibres. The main reason is that PP and PA6 nanofibres/PP filters adsorb formaldehyde only by physiosorption. On the other hand keratin based nanofibres have stronger capability of adsorbing formaldehyde because in the adsorbing process both physiosorption and chemisorption occur simultaneously.

# 6. RESEARCH TRENDS

In the biomedical field great efforts have recently been made to mimic nature by producing nanofibrous materials solely composed of natural polymers in the assumption that such materials would be less subject to rejection by the host. Moreover, electrospinning of natural proteins offers a promising method to produce nanofibres with a similar nanofibrillar structure to that of a native extra-cellular matrix. Unfortunately, the electrospinning of natural macromolecules has demonstrated to be a challenge because they frequently do not possess the viscoelastic properties needed for good electrospinning. One exception is silk fibroin, a natural high molecular weight polypeptide. Silk fibroin is spun by Bombix mori and can be extracted in two molecular weights: 25 kDa (light chains) and 320–390 kDa (heavy chains). Silk fibroin is rich in glycine, alanine and serine amino acids. In the last years, regenerated silk fibroin has been proposed for the fabrication of a variety of biomaterials such as films, sponges, gels and fibres for biomedical applications.

Recently, keratin/fibroin blend has received much attention due to their potential applications in tissue engineering (Vasconcelos et al., 2008), and also for removal of metal ions and water purification (Ki et al., 2007; Baek et al., 2007).

Keratin/fibroin blended materials are a totally protein material (biodegradable and natural) that could combine the interesting properties of keratin (VOCs and metal ion adsorption) with the mechanical properties and electrospinnability of silk fibroin. Moreover, synergic effects could enhance some properties: for instance, it was found that films cast from silk fibroin and keratin decreased blood coagulation compared with silk fibroin or keratin alone (Lee et al., 1998).

Concerning the production of keratin/fibroin nanofibres by electrospinning, both silk fibroin and wool keratin can be dissolved in formic acid and mixed in several blending ratios (Ki et al., 2007; Baek et al., 2007; Zoccola et al., 2008). Blend nanofibers rich in fibroin, show branched and flatten morphology characterized by a bimodal diameter distribution. In spite of the lower viscosities of the solutions, nanofibers rich in keratin have more homogeneous diameter distribution, with a round cross section. This may be due to the higher conductivity that promotes nanofiber formation. Keratin rich nanofibers are also thinner than those rich in fibroin.

Particular behaviour was observed for the 50/50 blend solutions that showed the highest zero-shear viscosity and the longest relaxation time (i.e. lowest shear thinning onset). The resulting nanofibers were thin with respect to those produced with the other blend ratios, and the trend was repeated at different operating conditions (Zoccola et al., 2008). It had been supposed some interactions between keratin and fibroin promote the formation of entangled chain networks. It has also been pointed out that keratin/fibroin 50/50 blend nanofibers have excellent mechanical properties (Baek et al., 2007). These studies encourage further investigations into these types of natural matrices for use in the biomedical field, biotechnology and water purification.

## REFERENCES


1. Alemdar, Y. Iridag, M. Kazanci, 2005 Flow behavior of regenerated wool-keratin proteins in different mediums. International Journal of Biological Macromolecules., 35 151153 .

2. Aluigi, R. Innocenti, C. Tonin, C. Vineis, 2004 Studies on keratin/cellulose films from cuprammonium solution. Autex Research Journal, 4 175181 .

3. Aluigi, M. Zoccola, C. Vineis, C. Tonin, F. Ferrero, M. Canetti, 2007a Study on the structure and properties of wool keratin regenerated from formic acid. International Journal of Biological Macromolecules., 41 266273 .

4. Aluigi, A. Varesano, A. Montarsolo, C. Vineis, F. Ferrero, G. Mazzuchetti, C. Tonin, 2007b Electrospinning of keratin/PEO blend nanofibres. Journal of Applied Polymer Science, 104 863870 .

5. Aluigi, C. Vineis, A. Varesano, G. Mazzuchetti, F. Ferrero, C. Tonin, 2008 Structure and properties of keratin/PEO blend nanofibres. European Polymer Journal, 44 24652475 .

6. Aluigi, C. Vineis, C. Tonin, C. Tonetti, A. Varesano, G. Mazzuchetti, 2009 Wool keratin-based nanofibres for active filtration of air and water. Journal of Biobased Materials and Bioenergy, 3 311319 .

7. D. H. Baek, C. S. Ki, I. C. Um, Y. H. Park, 2007 Metal ion adsorbability of electrospun wool keratose/silk fibroin blend nanofiber mats. Fibers and Polymers, 8 271277 .

8. R. S. Barhate, S. Ramakrishna, 2007 Nanofibrous filtering media: filtration problems and solutions from tiny materials. Journal of Membrane Science, 296 18 .

9. J. R. Barone, W. F. Schmidt, C. F. Liebner, 2005 Thermally processed keratin films. Journal of Applied Polymer Science, 97 16441651 .

10. J. E. Bergsma, F. R. Rozema, R. R. M. Bos, W. C. de Bruijn, G. Boering, A. J. Pennings, 1995 In vivo degradation and biocompatibility study of in vitro pre-degraded as-polymerized polylactide particles. Biomaterials, 16 267274 .

11. M. Bryner, J. Armantrout, 2006 U.S. Patent 2006138710.

12. R. Cecil, 1963 In: The proteins, 1 Neurath H. (Ed.), Academic Press, New York, USA.

13. J. F. Cooley, 1902 U.S. Patent 692631.

14. F. H. C. Crick, 1953 The Fourier transform of a coiled-coil. Acta Crystographica, 6 685689 .

15. J. M. Deitzel, J. Kleinmeyer, J. K. Hirvonen, Tan. N. C. Beck, 2001 Controlled deposition of electrospun poly(ethylene oxide) fibers. Polymer, 42 81638170 .

16. N. P. Desai, J. A. Hubbel, 1991 Solution technique to incorporate polyethylene oxide and other water-soluble polymers into surfaces of polymeric biomaterials. Biomaterials, 12 144153 .

17. F. Dotti, A. Varesano, A. Montarsolo, A. Aluigi, C. Tonin, G. Mazzuchetti, 2007 Electrospun porous mats for high efficiency filtration. Journal of Industrial Textiles, 37 151162 .

18. L. M. Dowling, W. G. Crewther, D. A. D. Parry, 1986 Secondary structure of component 8c-1 of α-keratin. Biochemical Journal, 236 705712 .

19. J. Ellison, A. Phatak, D. W. Giles, C. W. Macosko, F. S. Bates, 2007 Melt blown nanofibers: fiber diameter distributions and onset of fiber break-up. Polymer, 48 33063316 .

20. Formhals, 1934 U.S. Patent 1975504.

21. R. D. B. Fraser, Rae. T. P. Mac, D. A. D. Parry, E. Suzuki, 1971 The structure of feather keratin. Polymer, 12 3556 .

22. R. D. B. Fraser, Rae. T. P. Mac, G. E. Rogers, 1972 In: Keratins: Their composition, structure and biosynthesis, Charles C. Thomas Publisher, Springfield, Illinois, USA.

23. M. E. Furth, A. Atala, M. E. Van Dyke, 2007 Smart biomaterials design for tissue engineering and regenerative medicine. Biomaterials, 28 50685073 .

24. R. H. Garret, C. M. Grisham, 2002 In: Biochemistry 2nd Edition, Higher Education Press, Beijing, China.

25. P. Gibson, H. Schreuder-Gibson, D. Rivin, 2001 Transport properties of porous membranes based on electrospun nanofibers. Colloid Surface A, 187-188, 469-481.

26. R. Gopal, S. Kaur, Z. , C. Chan, S. Ramakrishna, T. Matsuura, 2006 Electrospun nanofibrous filtration membrane. Journal of Membrane Science, 281 581586 .

27. S. Hamasaki, A. Tachinaba, D. Tada, K. Yamauchi, T. Tanabe, 2008 Fabrication of highly porous keratin sponges by freeze-drying in the presence of calcium alginate beads. Material Science and Engineering: C, 28 12501254 .

28. E. Homonoff, 2008 Nanofibrillated fibers: opening newmarkets to nanofibre usage. International Fibers Journal 23 2224 .

29. Z. M. Huang, Y. Z. Zhang, M. Kotaki, S. Ramakrishna, 2003 A review on polymer nanofibers by electrospinning and their applications in nanocomposites. Composites Science Technology, 63 22232253 .

30. P. D. Jeffrey, 1972 The molecular weights of the low-sulphur proteins from wool: A review. Journal of the Textile Institute, 63 91113 .

31. H. J. Jin, S. V. Fridrikh, G. C. Rutledge, D. L. Kaplan, 2002 Electrospinning Bombyx mori silk with poly(ethylene oxide). Biomacromolecules, 3 12331239 .

32. P. Kar, M. Misra, 2004 Use of keratin fiber for separation of heavy metals from water. Journal of Chemical Technology & Biotechnology, 79 13131319 .

33. K. Katoh, M. Shibayama, T. Tanabe, K. Yamauchi, 2004 Preparation and properties of keratin-poly(vinyl alcohol) blend fiber. Journal of Applied Polymer Science, 91 756762 .

34. S. Ki, E. H. Gang, I. C. Um, Y. H. Park, 2007 Nanofibous membrane of wool keratose/silk fibroin blend for heavy metal ion adsorption. Journal of Membrane Science, 302 2026 .

35. Kilic, F. Oruc, A. Demir, 2008 Effects of polarity on electrospinning process. Textile Research Journal, 78 532539 .

36. K. Lee, S. Kong, W. Park, W. Ha, I. Kwon, 1998 Effect of surface properties on the antithrombogenicity of silk fibroin/S-carboxymethyl kerateine blend films. Journal of Biomaterial Science Polymer Edition, 9 905914

37. J. Lee, M. J. Cuddihy, N. A. Kotov, 2008 Three-dimensional cell culture matrices: State of the art. Tissue Engineering B, 14 6186 .

38. Li, Y. Xia, 2004 Electrospinning of nanofibers: reinventing the wheel. Advanced Materials, 16 11511170 .

39. L. Li, M. W. Frey, T. B. Green, 2006 Modification of air filter media with nylon-6 nanofibers. Journal of Engineered Fibers and Fabrics, 1 124 .

40. J. A. Maclaren, B. Milligan, 1981 The chemical reactivity of the wool fibre, In: Wool science, 109127 , Science Press, Marrickville, Australia.

41. L. Martínez-Hernández, C. Velasco-Santos, M. De -Icaz, V. M. Castaño, 2007 Dynamical-mechanical and thermal analysis of polymeric composites reinforced with keratin biofibers from chicken feathers. Composites Part B: Engineering, 38 405410 .

42. Montefusco, 2005 The use of nonwovens in air filtration. Filtration and Separation, 42 3031 .

43. W. J. Morton, 1902 U.S. Patent 705691.

44. L. Pauling, R. B. Corey, 1953 Compound helical configurations of polypeptide chains: Structure of proteins of the α-keratin type. Nature, 171 5961 .

45. M. Perez, M. Swan, J. Louks, 2000 U.S. Patent 6110588.

46. Q. P. Pham, U. Sharma, A. G. Mikos, 2006 Electrospinning of polymeric nanofibers for tissue engineering applications: A review. Tissue Engineering, 12 11971211 .

47. Pourdeyhimi, N. Fedorova, S. Sharp, 2006 U.S. Patent 2006292355.

48. X. H. Qin, S. Y. Wang, 2006 Filtration properties of electrospinning nanofibers. Journal of Applied Polymer Science, 102 12851290 .

49. S. Ramakrishna, K. Fujihara, W. E. Teo, T. Yong, Z. , R. Ramaseshan, 2006 Electrospun nanofibers: Solving global issues. Materials Today, 9 4050 .

50. H. Reneker, A. L. Yarin, 2008 Electrospinning jets and polymer nanofibers. Polymer, 49 23872425 .

51. W. F. Schmidt, 1998 Innovative feather utilization strategies, Proceedings of the National Poultry Waste Management Symposium, 276282 , Springdale, AR, October 19-22, 1998.

52. P. M. M. Schrooyen, P. J. Dijkstra, R. C. Oberthür, A. Bantjes, J. Feijen, 2001 Stabilization of solutions of feather keratins by sodium dodecyl sulfate. Journal of Colloid and Interface Science, 240 3039 .

53. S. L. Shenoy, W. D. Bates, H. L. Frisch, G. E. Wnek, 2005 Role of chain entanglements on fiber formation during electrospinning of polymer solutions: good solvent, non-specific polymer-polymer interaction limit. Polymer, 46 33723384 .

54. Y. M. Shin, M. M. Hohman, M. P. Brenner, G. C. Rutledge, 2001 Experimental characterization of electrospinning: the electrically forced jet and instabilities. Polymer, 42 99559967 .

55. Shin, G. G. Chase, D. H. Reneker, 2005 The effect of nanofibers on liquid-liquid coalescence filter performance. AIChE Journal, 51 31093113 .

56. T. J. Sill, Recum. H. A. von, 2008 Electrospinning: Applications in drug delivery and tissue engineering. Biomaterials, 29 19892006 .

57. W. K. Son, J. H. Youk, T. S. Lee, W. H. Park, 2004 The effects of solution properties and polyelectrolyte on electrospinning of ultrafine poly(ethylene oxide) fibers. Polymer. 45 29592966 .

58. M. Spei, R. Holzem, 1990 Further thermoanalytical investigations of annealed keratins: The time and temperature dependence. Colloid and Polymer Science, 268 630635 .

59. Tachinaba, Y. Furuta, H. Takeshima, T. Tanabe, K. Yamauchi, 2002 Fabrication of wool keratin sponge scaffolds for long-term cell cultivation. Journal of Biotechnology, 93 165170 .

60. Tachinaba, S. Kaneko, T. Tanabe, K. Yamauchi, 2005 Rapid fabrication of keratin-hydroxyapatite hybrid sponges toward osteoblast cultivation and differentiation. Biomaterials, 26 297302 .

61. P. Taddei, P. Monti, G. Freddi, T. Arai, M. Tsukada, 2003 Binding of Co(II) and Cu(II) cations to chemically modified wool fibres: An IR investigation. Journal of Molecular Structure, 650 105113 .

62. T. Tanabe, N. Okitsu, K. Yamauchi, 2004 Fabrication and characterization of chemically crosslinked keratin films. Materials Science and Engineering: C, 24 441446 .

63. S. A. Theron, E. Zussman, A. L. Yarin, 2004 Experimental investigation of the governing parameters in the electrospinning of polymer solutions. Polymer, 45 20172030 .

64. O. P. Thompson, I. J. O'Donnel, 1959 Studies on oxidized wool: A comparison of the completeness of oxidation with peracetic and performic acids. Australian Journal of Biological Sciences, 12282 .

65. Tonin, A. Aluigi, Songia. M. Bianchetto, C. D'Arrigo, M. Mormino, C. Vineis, 2004 Thermoanalytical characterization of modified keratin fibres. Journal of Thermal Analalysi and.Calorimetry, 77 987996 .

66. Tonin, M. Zoccola, A. Aluigi, A. Varesano, A. Montarsolo, C. Vineis, F. Zimbardi, 2006 Study on the conversion of wool keratin by steam explosion. Biomacromolecules, 7 34993504 .

67. P. P. Tsai, H. Schreuder-Gibson, P. Gibson, 2002 Different electrostatic methods for making electret filters. Journal of Electrostatics, 54 333341 .

68. A. Tseng, A. Notargiacomo, T. P. Chen, 2005 Nanofabrication by scanning probe microscope lithography: A review. Journal of Vacuum Science and Technology B, 23 877894 .

69. Varesano, A. Aluigi, C. Vineis, C. Tonin, 2008 Study on the shear viscosity behaviour of keratin/PEO blends for nanofibre electrospinning. Journal of Polymer Science Part B: Polymer Physics, 46 11931201 .

70. Vasconcelos, G. Freddi, A. Cavaco-Paulo, 2008 Biodegradable materials based on silk fibroin and keratin. Biomacromolecules, 9 12991305 .

71. Bergen. W. von, 1963 In: Wool Handbook 3rd Edition, 1 John Wiley & Sons, Inc., New York, USA.

72. Wortmann, G. Zwiener, R. Sweredjiuk, F. Doppelmayer, F. J. Wortmann, 1999 Sorption of indoor air pollutants by sheep's wool: Formaldehyde as an example. Proceedings of the International Wool Textile Organization, Report CTF 3, Florence, Italy, June 1999.

73. Wortmann, S. Thome, R. Sweredjuk, G. Zwiener, 2005 Chemisorption of protein reactive indoor air pollutants by wool, Proceedings of the

International Wool Textile Organization Conference, Report CTF03, 17 , Hobart, Australia, April 2005.

74. Xie, S. Li, S. Zhang, 2001 Ionic liquids as novel solvents for the dissolution and blending of wool keratin fibres. Green Chemistry, 7 606608 .

75. X. N. Xie, H. J. Chung, C. H. Sow, A. T. S. Wee, 2006 Nanoscale materials patterning and engineering by atomic force microscopy nanolithography. Materials Science and Engineering R: Reports, 54 148 .

76. K. Yamauchi, A. Yamauchi, T. Kusunoki, A. Kohda, Y. Konishi, 1996 Preparation of stable aqueous solution of keratins and physiochemical and biodegradational properties of films. Journal of Biomedical Material Research, 31 439444 .

77. L. Yarin, S. Koombhongse, D. H. Reneker, 2001 Taylor cone and jetting from liquid droplets in electrospinning of nanofibers. Journal of Applied Physics, 90 48364846 .

78. Yin, S. Rastogi, A. E. Terry, C. Popescu, 2007 Self-organization of oligopeptides obtained on dissolution of feather keratins in superheated water. Biomacromolecules, 8 800806 .

79. S. Zhang, 2002 Emerging biological materials through molecular self-assembly. Biotechnology Advances, 20 321339 .

80. M. Zoccola, A. Aluigi, C. Vineis, F. Ferrero, M. Piacentino, C. Tonin, 2008 Study on cast membranes and electrospun nanofibers made from keratin/fibroin blends. Biomacromolecules, 9 28192825 .

# Chapter 6

# Fabrication of Pd Doped $WO_3$ Nanofiber as Hydrogen Sensor

**Alireza Nikfarjam [1], Somayeh Fardindoost [2] and Azam Iraji zad [2,3,*]**

[1]Faculty of New Sciences & Technologies, University of Tehran, P.O. Box 14395-1374, Tehran, Iran

[2]Department of Physics, Sharif University of Technology, Azadi Street, P.O. Box 11365-9161, Tehran, Iran

[3]Institute for Nanoscience and Nanotechnology, Sharif University of Technology, Azadi Street, P.O. Box 11155-8639, Tehran, Iran

## ABSTRACT

Pd doped $WO_3$ fibers were synthesized by electro-spinning. The sol gel method was employed to prepare peroxopolytungstic acid (P-PTA). Palladium chloride and Polyvinyl pyrrolidone (PVP) was dissolved in the sol Pd:$WO_3$ = 10% molar ratio. The prepared sol was loaded into a syringe connected to a high voltage of 18.3 kV and electrospun fibers were collected on the alumina substrates. Scanning electron microscope (SEM), X-ray powder diffraction (XRD) and X-ray photoelectron spectroscopy (XPS) techniques were used to analyze the crystal structure and chemical composition of the fibers after heat treatment at 500 °C. Resistance-sensing measurements exhibited a sensitivity of about 30 at 500 ppm hydrogen in air, and the response and recovery times were about 20 and 30 s, respectively, at 300 °C. Hydrogen gas sensing mechanism of the sensor was also studied.

## KEYWORDS

Pd doped $WO_3$ nanofiber; hydrogen sensor; electrospinning; sensing mechanism

## 1. INTRODUCTION

Hydrogen burns cleanly without releasing pollutants or greenhouse gases and therefore, in recent years, more attention has been paid to hydrogen as an excellent candidate for fossil fuel replacement in household and transportation applications. However, hydrogen is a highly flammable fuel with a wide combustion range of 4%–75%. Besides, it has a large diffusion coefficient of 0.61 $cm^2/s$ in air. Therefore, detection and leakage control of this gas is a challenging subject. Today, development of a hydrogen sensor with negligible power consumption and high stability, sensitivity and fast response is desired.

Reports of various technologies for hydrogen sensing include FETs [1,2], optical fibers [3,4], thermoelectric [5,6], Schottky diodes [7,8,9], surface acoustic wave devices [10,11] and metal oxides [12,13,14,15,16,17]. Metal oxide sensors are under intensive development and have been studied for decades [18,19]. In metal oxide sensors, the electrical resistance of the film changes when exposed to a target gas. Also, among metal oxides, $WO_3$ is widely used to detect many gases like $H_2$, $H_2S$, $NO_2$ and VOC [20,21,22,23]. Moreover, it has been shown that the addition of an appropriate amount of noble metal additives promotes chemical reactions by reducing activation energy between the surface of metal oxide and target gas. Also, this increases the response and selectivity as well as decreasing the maximum working temperature of the sensors. For hydrogen sensors, modification of metal oxides, especially $WO_3$, by metal additives such as Pt, Pd, or Au, using different techniques is under intensive investigation [24,25,26,27,28,29,30,31].

It is known that gas absorption depends on the surface area, which is affected by the preparation method. Increasing the surface to volume ratio to enhance gas sensing properties is currently under development by introducing different nanostructures such as nanowires, nanorods, nanobelts and nanofibers. Nanofibers have high density of surface sites that make them excellent candidates for gas sensing applications [32].

Among different fabrication methods, electrospinning is a simple and cost effective method for preparing one dimensional materials. Recently, several researches have reported fabrication of metal oxide nanofibers [33,34,35,36,37,38,39] in which metal oxide nanoparticles can be added to a polymer matrix to make nanocomposite fibers. Heat treatment removes polymer and leaves metal oxide material with fiber shape.

Metal oxide fibers could be synthesized by combining sol-gel and electrospinning methods. Adding catalysts to the solution (sol) produces nanofibers with improved catalytic properties for selective gas detection at lower temperatures. Polymer matrix can be decomposed through heat treatment and leaves metal oxide fiber with lengths from several hundred nanometers to few micrometers.

Preparing one dimensional $WO_3$ fibers by sol-gel method was first reported by Lu *et al.* [40]. Others used pure $WO_3$ fiber for detecting ammonia, $NO_2$ and CO gasses [41,42]. In our previous work [29], the sensitivity of both pure and Pd-modified $WO_3$ thin films was measured towards hydrogen as a function of operating temperatures in the range of 30 °C to 350 °C. A systematic increase in sensitivity was observed for the modified films in the whole temperature range. The sensitivity increased due to a reduction of the activation energy between the $WO_3$ surface and the hydrogen gas in the presence of Pd (PdO). It was seen that the sensitivity of Pd:W = 10% films have considerable amounts at room temperature and the working temperature (the temperature which has maximum sensitivity) decreased by increasing the Pd concentrations.

Since high response and recovery times were observed at low working temperatures. Given the results of our previous experiments on Pd modified $WO_3$ films, we decided to study structure effect on modification of sensing behavior. Therefore, a combination of electrospinning and the sol gel method was studied to compare the result of the sensing behavior of Pd:$WO_3$ thin film with the electrospun Pd doped $WO_3$ nanofiber samples with the same ratio (Pd:W = 10%). In this work, we prepared $WO_3$ nanoparticles by the sol-gel method and then Pd and PVP were added to the sol. To remove polymer and make $WO_3$ nanoparticles with fiber shape, samples were annealed in air at 500 °C for 1 h. The Pd doped $WO_3$ nanofibers showed fast response time and high sensitivity toward hydrogen gas at low temperature.

## 2. EXPERIMENTAL SECTION

In our sol-gel route, Peroxopolytungstic acid (P-PTA) sol was prepared according to the Kudo route [43]. Five grams of tungsten wire and 20 mL of $H_2O_2$ (30%) was stirred at room temperature for 48 h until the whole tungsten wire was dissolved. Unreacted $H_2O_2$was removed by using platinum net. Then 20 mL ethanol was added and the solution was heated at 80 °C which resulted in a sol color change from clear to orange.

For the activation process we used Pd catalyst. Since in our previous report, we obtained best sensitivity for Pd:W = 10% molar ratio [29], we chose this ratio for adding $PdCl_2$ salt (Merk) directly to the P-PTA sol.

To fabricate fibers, 1.0 g polyvinyl pyrrolidone (PVP) (Aldrich) was added into the P-PTA sol and was stirred for one day. The resulting solution was loaded into a plastic syringe and its needle was connected to a high-voltage DC supply up to 30 kV. A voltage of 18.3 kV was applied between the needle and grounded target. $Al_2O_3$ substrates were placed on a piece of flat aluminum with 20 cm distance from the tip of the needle to collect fibers. The prepared samples were annealed in air at 500 °C for 1 h.

X-ray powder diffraction (XRD) analysis was recorded by a Philips X'pert instrument operating with CuKα radiation (λ = 1.54178 Å) at 40 kV/40 mA. A Philips XL30 model Scanning electron microscope (SEM) was used to obtain the SEM images. For the X-ray photoelectron spectroscopy (XPS) experiment, an Al anode X-ray source was employed with a concentric hemispherical analyzer (Specs model EA10 plus) to analyze the energy of the emitted photoelectrons. Gas sensing properties of the as-fabricated sensor were measured using a static test system. A data logger was used to collect the data and the whole system was automatically controlled by computer.

For measuring the electrical resistance, Au/Ti comb-like interdigitated electrodes with 150 nm thickness were evaporated on the sample surface through a mask. The samples were placed in a small stainless steel chamber with several electrical feed-through, gas inlet and gas outlet. For measuring sensitivity, a constant dc voltage of 5 V was applied to the circuit; including the sensor in series with a constant resistor. Voltage variation across the resistor was measured. During the measurements, the fibers were warmed by a heater located on the back of the substrates. We used dry air as the reference gas and hydrogen as the target gas. Sensitivity was defined as $S = (Vrg - Vra)/Vra$, where *Vrg* and *Vra* are the sensor voltage in the presence and absence of hydrogen, respectively. The measurement setup consists of mass flow controllers and a data acquisition system for recording resistance changes. We define the response time as the time taken for the sensor's resistance to undergo a 90% variation with respect to its equilibrium value and recovery time as the time taken to achieve 60% variation to reach its initial value in air after the removal of $H_2$. The gas sensing performance was tested at hydrogen concentration of 500 ppm in the range of 30 °C to 300 °C.

# 3. RESULTS AND DISCUSSION

## 3.1. Structure Analysis

SEM images of as-spun PVP composite fibers and the sintered fibers at 500 °C for 1 h are displayed in Figure 1a,b, respectively.Figure 1a shows the used Aluminum foil surface and some PVP composite fibers with an average diameter of about 500 nm. The heating process compresses the fibers and reduces their average thickness to about 150 nm while keeping their morphology as is shown in Figure 1b. Heat treatment at 500 °C removes the polymer and leaves metal oxide material with fiber shapes. SEM images show continuous fibers with lengths of several tens micrometers. By electrospinning, a huge number of fibers electrospun over the substrate and form as a film. To observe better by SEM, we made a sample with low density fibers (Figure 1a,b). Figure 1c shows another sample with high density of fibers (sample for gas sensing experiments). Both high fibrous surface area as well as high porosity increase effective sensitive surface area.

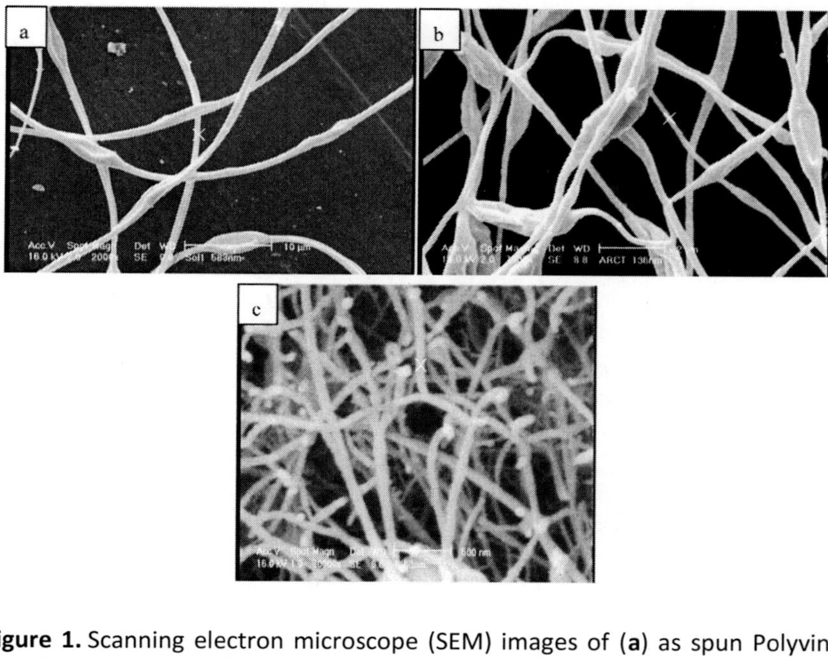


**Figure 1.** Scanning electron microscope (SEM) images of (**a**) as spun Polyvinyl pyrrolidone (PVP)/Pd/ peroxopolytungstic acid (P-PTA) fiber; (**b**) $WO_3$ fiber, after sintering at 500 °C for 1 h; (**c**) sintered sample with high density of fibers.

As presented in Figure 2, heat treatment at 500 °C resulted in the formation of crystals of both monoclinic $WO_3$ and oxidized Pd. The evidence of PdO after annealing can be observed by XPS analysis. Data in Figure 3a indicates that the surface of the annealed Pd doped $WO_3$ nanofiber is composed of tungsten, oxygen, carbon contamination, and small amounts of palladium. Figure 3b, shows the peaks related to palladium oxide formation after the annealing process.

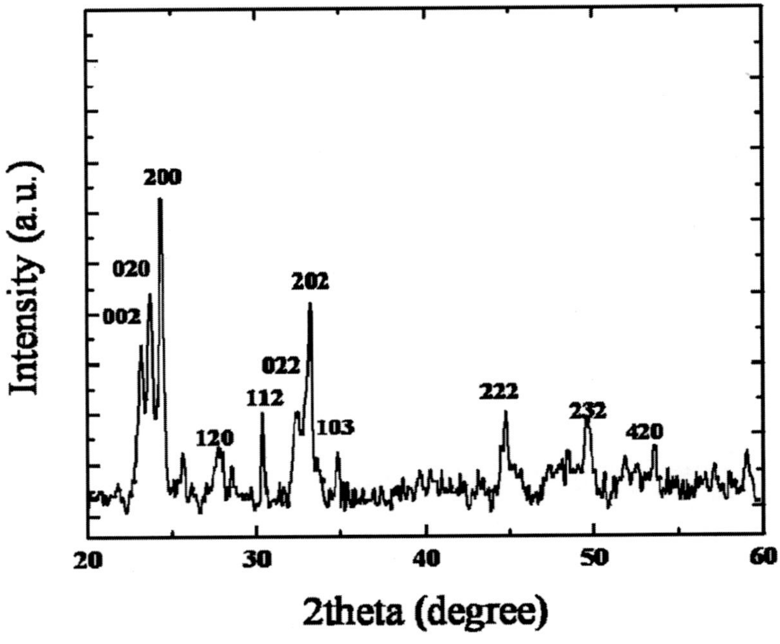


**Figure 2.** The X-ray powder diffraction (XRD) pattern of the Pd doped $WO_3$ nanofiber sintered at 500°C.

## 3.2. Hydrogen Sensing Measurements

Hydrogen sensing properties of the Pd-modified $WO_3$ samples were investigated while they were exposed to 500 ppm $H_2$ at different operating temperatures. Figure 4 shows the changes in the voltage of a series resistor *vs.* time (transient response). The used electrical circuit is shown in the inset of Figure 4. It presents fast response and recovery times of about 20 and 30 s, respectively.

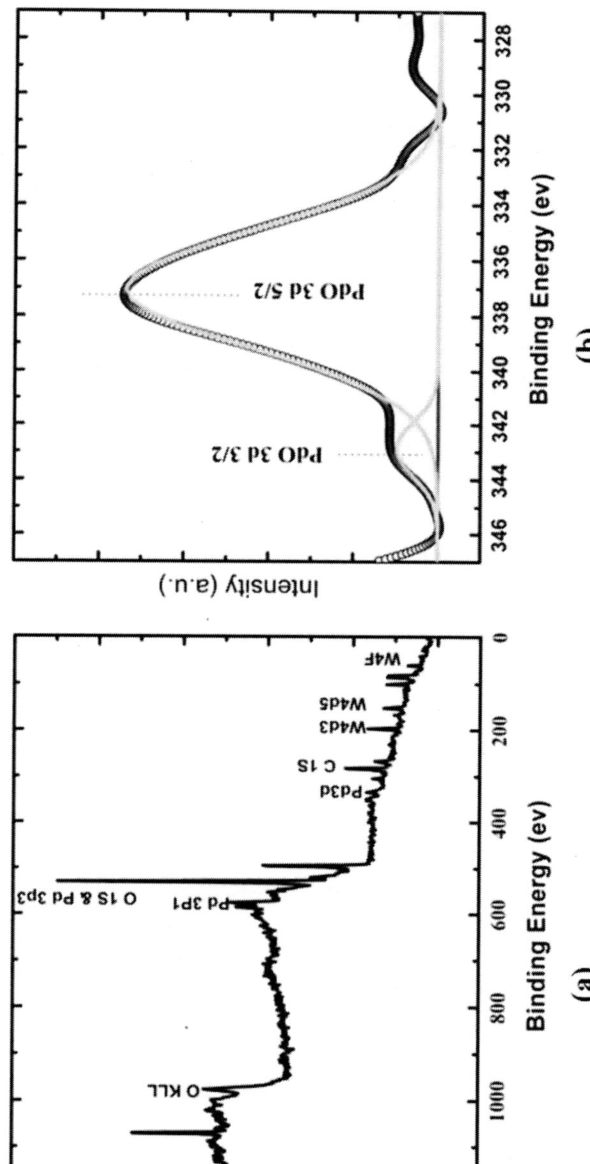


**Figure 3.** The X-ray photoelectron spectroscopy (XPS) patterns of the Pd doped $WO_3$nanofiber that was heat treated at 500 °C, (**a**) Peak survey and (**b**) Deconvolution in the Pd 3d region.

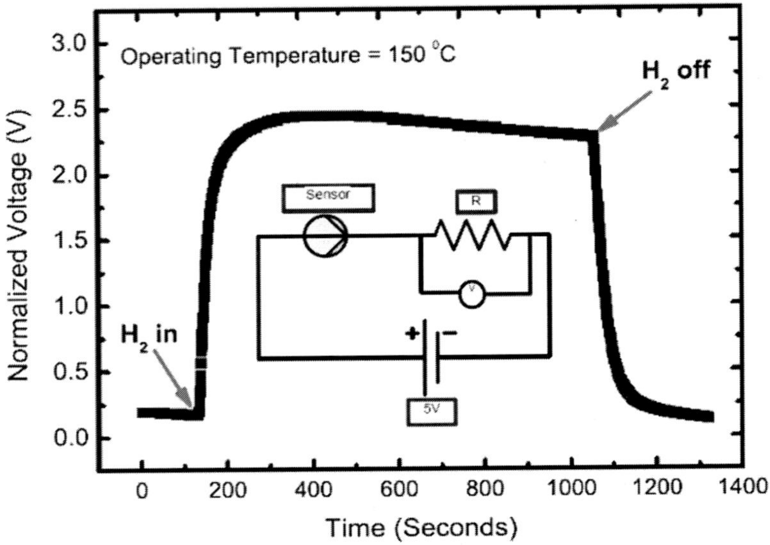


**Figure 4.** Normalized voltage variation across the series resistor as output signal

We measured the sensitivity of the samples as a function of operating temperature in the range of 75 °C to 300 °C. As shown inFigure 5, the sensitivity is noticeable at a temperature of about 75 °C and it is enhanced by increasing the operating temperature to 300 °C. This high sensitivity may be attributed to the reduction of the activation energy between the $WO_3$ surface and hydrogen gas in the presence of Pd (PdO).

Response and recovery times of the Pd-modified $WO_3$ fiber for 500 ppm hydrogen are presented in Figure 6. It presents a decaying behavior with considerable fast response and recovery times at high temperatures (300 °C). At low working temperature the response time is higher due to slow desorption of the formed water molecules on the surface especially at temperatures below 100 °C.

Our results demonstrated faster response and recovery times in the order of 160–190 s for low temperature sensing which is so considerable compare to 60 min recovery time for Pd:$WO_3$ = 10% in the form of thin film [29]. It can be attributed to the facilitate occupancy and desorption of $H_2$ molecules at grain boundaries on the large surface area of nanofibers in a porous network. However, the obtained results of the response magnitudes and the working temperature are not as good as our previous work on thin film sensors.

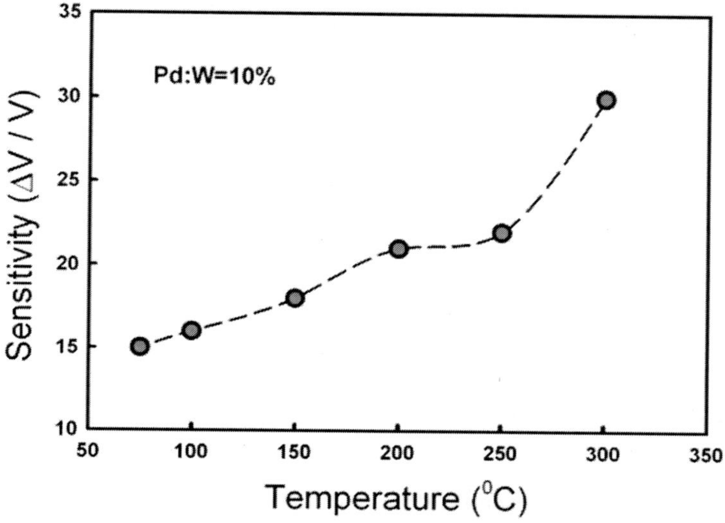


**Figure 5.** The sensitivity of the Pd-modified $WO_3$ nanofiber towards 500 ppm hydrogen as a function of operating temperature.

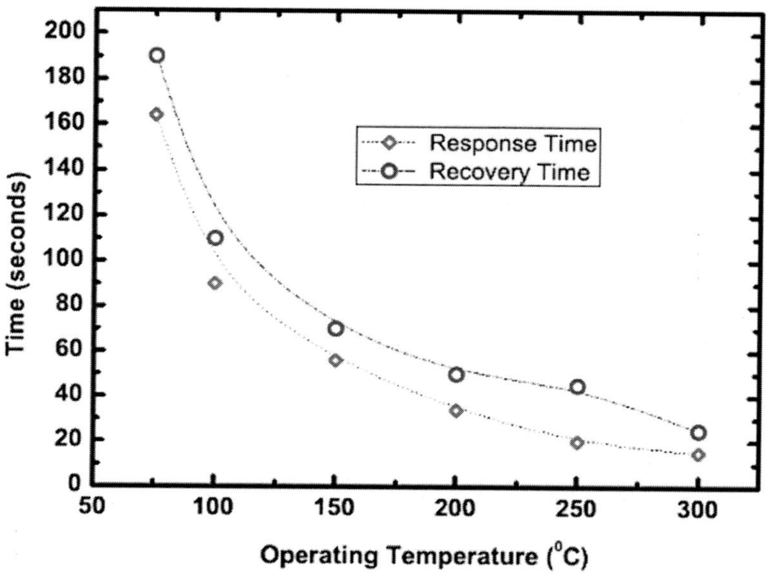


**Figure 6.** The response and recovery times of the Pd-modified $WO_3$ nanofiber towards 500 ppm hydrogen as a function of operating temperature.

## 3.3. Hydrogen Sensing Mechanism

The $WO_3$ fibers have been made of small grains and grain boundary barriers are existed between neighboring grains located in a fiber structure. This structure has an increased surface area as well as providing electronic transmission channels for electron transferring between $WO_3$ particulates. Since $WO_3$ nanofibers in this study are continuous with $WO_3$ grains, we expect sensing mechanism of the single $WO_3$ fiber is principally similar to that of metal oxide films. So, grain size has important effect on sensing performances [44,45,46]. However, despite fibers being continuous media, there are a great number of fiber to fiber connections, where electron current passing through these connection potential barriers has an important role in total current passing through the fibers' film.

In addition, regarding our previous work [29], presence of Pd makes a systematic increase in the sensitivity in the whole temperature ranges. This can be attributed to the effect of Pd catalyst in performing electronic sensitization mechanism and reducing the activation energy between the $WO_3$ surface and the hydrogen gas [47]. As shown here in Figure 7a and b, the $O_2$ molecules chemisorb and dissociate on the surface of the $WO_3$ grains. Then they trap and extract electrons from the conduction band of n-type semiconductor ($WO_3$) due to their high electron affinity. A depletion or space-charge region develops on the surface of the grain. This increases the potential barrier height between neighboring grains and also widens it, which is called energy band bending as shown in Figure 7a. This makes electron transferring between grains difficult which decreases the conductance of the fiber. By exposing fibers to hydrogen gas, reaction with the adsorbed oxygen molecule takes place. This re-injects captured electrons to the depletion region and the reverse effect will take place (Figure 7b).

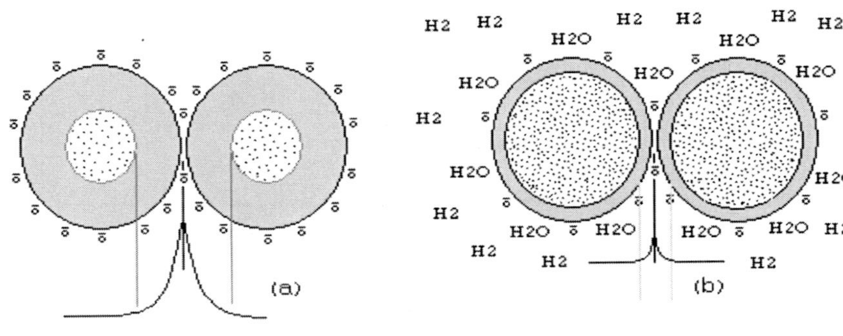


**Figure 7.** Hydrogen sensing mechanism of $WO_3$ neighboring grains (**a**) before and (**b**) after hydrogen exposure.

# 4. CONCLUSIONS

Pd doped $WO_3$ fibers were synthesized by combining electrospinning and sol-gel methods and were characterized by SEM, XRD and XPS methods. The sensor fabricated from these fibers exhibited high sensitivity and rapid response/recovery to hydrogen at 300 °C. The highest sensitivity was about 30 when the sensor was exposed to 500 ppm, and the response and recovery times were about 20 and 30 s, respectively. The $WO_3$ fiber sensors showed fast response times toward hydrogen even at low operating temperatures. These results demonstrate that Pd doped $WO_3$ fibers can be used as the sensing material for fabricating high performance hydrogen sensors. Besides, hydrogen gas sensing mechanism of the sensor was studied and a model proposed.

# REFERENCES


1. Yamaguchi, T.; Kiwa, T.; Tsukada, K.; Yokosawa, K. Oxygen interference mechanism of platinum–FET hydrogen gas sensor.*Sens. Actuators A Phys.* **2007**, *136*, 244–248.

2. Higuchi, T.; Nakagomi, S.; Kokubun, Y. Field effect hydrogen sensor device with simple structure based on GaN. *Sens. Actuators B Chem.* **2009**, *140*, 79–85.

3. Villatoro, J.; Moreno, D.L.; Monzón-Hernández, D. Optical fiber hydrogen sensor for concentrations below the lower explosive limit. *Sens. Actuators B Chem.* **2005**, *110*, 23–27.

4. Slaman, M.; Dam, B.; Schreuders, H.; Griessen, R. Optimization of Mg-based fiber optic hydrogen detectors by alloying the catalyst. *Int. J. Hydrog. Energy* **2008**, *33*, 1084–1089.

5. Nishibori, M.; Shin, W.; Izu, N.; Itoh, T.; Matsubara, I.; Yasuda, S.; Ohtani, S. Robust hydrogen detection system with a thermoelectric hydrogen sensor for hydrogen station application. *Int. J. Hydrog. Energy* **2009**, *34*, 2834–2841.

6. Huang, H.; Luan, W.; Zhang, J.S.; Qi, Y.S.; Tu, S.T. Thermoelectric hydrogen sensor working at room temperature prepared by bismuth–telluride P–N couples and Pt/g-$Al_2O_3$. *Sens. Actuators B Chem.* **2008**, *128*, 581–585.

7. Ito, K.; Kojima, K. Hydrogen detection by Schottky diodes. *Int. J. Hydrog. Energy* **1982**, *7*, 495–497.

8. Tang, W.M.; Lai, P.T.; Xu, J.P.; Chan, C.L. Enhanced hydrogen sensing characteristics of MI SiC Schottky-diode hydrogen sensor by trichloroethylene oxidation. *Sens. Actuators A Phys.* **2005**, *119*, 63–67.

9. Tsai, T.H.; Chen, H.I.; Lin, K.W.; Hung, C.W.; Hsu, C.H.; Chen, L.Y.; Chu, K.-Y.; Liu, W.-C. Comprehensive study on hydrogen sensing properties of a Pd–AlGaN-based Schottky diode. *Int. J. Hydrog. Energy* **2008**, *33*, 2986–2992.

10. Ippolito, S.J.; Kandasamy, S.; Kalantar-Zadeh, K.; Wlodarski, W. Layered SAW hydrogen sensor with modified tungsten trioxide selective layer. *Sens. Actuators B Chem.* **2005**, *108*, 553–557.

11. Jakubik, W.P. Investigations of thin film structures of $WO_3$ and $WO_3$ with Pd for hydrogen detection in a surface acoustic wave sensor system. *Thin Solid Films* **2007**, *515*, 8345–8350.

12. Comini, E. Metal oxide nano-crystals for gas sensing. *Anal. Chim. Acta* **2006**, *568*, 28–40.

13. Aroutiounian, V. Metal oxide hydrogen, oxygen, and carbon monoxide sensors for hydrogen setups and cells. *Int. J. Hydrog. Energy* **2007**, *32*, 1145–1158.

14. Adamyana, A.Z.; Adamyana, Z.N.; Aroutiouniana, V.M.; Arakelyana, A.H.; Touryanb, K.J.; Turner, J.A. Sol-gel derived thin film semiconductor hydrogen gas sensor. *Int. J. Hydrog. Energy* **2007**, *32*, 4101–4108.

15. Korotcenkov, G. Metal oxides for solid-state gas sensors: What determines our choice? *Mater. Sci. Eng. B* **2007**, *139*, 1–23.

16. Adamyan, A.Z.; Adamyan, Z.N.; Aroutiounian, V.M. Study of sensitivity and response kinetics changes for SnO2 thin-film hydrogen sensors. *Int. J. Hydrog. Energy* **2009**, *34*, 8438–8443.

17. Boon-Bretta, L.; Bousek, J.; Morettoa, P. Reliability of commercially available hydrogen sensors for detection of hydrogen at critical concentrations: Part II—Selected sensor test results. *Int. J. Hydrog. Energy* **2009**, *34*, 562–571.

18. Sakai, G.; Matsunaga, N.; Shimanoe, K. Theory of gas-diffusion controlled sensitivity for thin film semiconductor gas sensor.*Sens. Actuators B Chem.* **2001**, *80*, 125–131.

19. Eranna, G.; Joshi, B.C.; Runthala, D.P.; Gupta, R.P. Oxide materials for development of integrated gas sensors—A comprehensive review. *Crit. Rev. Solid State Mater. Sci.* **2004**, *29*, 111–188.

20. Ippolito, S.J.; Kandasamy, S.; Kalantar-Zadeh, K.; Wlodarski, W. Hydrogen sensing characteristics of $WO_3$ thin film conductometric activated by Pt and Au catalysts. *Sens. Actuators B Chem.* **2005**, *108*, 154–158.

21. Ionescu, R.; Hoel, A.; Granqvist, C.G.; Llobet, E.; Heszler, P. Ethanol and $H_2S$ gas detection in air and in reducing and oxidizing ambience: Application of pattern recognition to analyze the output from temperature-modulated nanoparticulate $WO_3$gas sensors. *Sens. Actuators B Chem.* **2005**, *104*, 124–131.

22. Penza, M.; Martucci, C.; Cassano, G. $NO_x$ gas sensing characteristics of $WO_3$ thin films activated by noble metals (Pd, Pt, Au) layers. *Sens. Actuators B Chem.* **1998**, *50*, 52–59.

23. Luo, Sh.; Fu, G.; Chen, H. Gas-sensing properties and complex impedance analysis of Ce-added $WO_3$ nanoparticles to VOC gases. *Solid-State Electron.* **2007**, *51*, 913–919.

24. Veith, G.M.; Lupini, A.R. Magnetron sputtering of gold nanoparticles onto $WO_3$ and activated carbon. *Catal. Today* **2007**, *122*, 248–253.

25. Cabot, A.; Arbiol, J.; Morante, J.R. Analysis of the noble metal catalytic additives introduced by impregnation of as obtained $SnO_2$ sol-gel nanocrystals for gas sensors. *Sens. Actuators B Chem.* **2000**, *70*, 87–100.

26. Opara, U.; Ovec, K.; Orel, B.; Georg, A.; Wittwer, V. The gasochromic properties of sol-gel $WO_3$ films with sputtered Pt catalyst. *Pergamon* **2000**, *68*, 541–551.

27. Ruiz, A.; Arbiol, J.; Cirera, A.; Cornet, A.; Morante, J.R. Surface activation by Pt-nanoclusters on titania for gas sensing applications. *Mater. Sci. Eng. C* **2002**, *19*, 105–109.

28. Moreno, D.L.; Monzón-Hernández, D. Effect of the Pd–Au thin film thickness uniformity on the performance of an optical fiber hydrogen sensor. *Appl. Surf. Sci.* **2007**, *253*, 8615–8619.

29. Fardindoost, S.; Iraji zad, A.; Rahimi, F.; Ghasempour, R. Pd doped $WO_3$ films prepared by sol-gel process for hydrogen sensing. *Int. J. Hydrog. Energy* **2010**, *35*, 854–860.

30. Epifani, M.; Abriol, J.; Pellicer, E.; Comini, E.; Siciliano, P.; Faglia, G.; Morante, J.R. Synthesis and gas sensing properties of Pd-Doped $SnO_2$ nanocrystals. A case study of general morphology for doping metal oxides nanocrystals. *Cryst. Growth Des.***2008**, *8*, 1774–1778.

31. Malyshev, V.V.; Pislyakov, A.V. Investigation of gas-sensitivity of sensor structures to hydrogen in a wide range of temperature, concentration and humidity of gas medium. *Sens. Actuators B Chem.* **2008**, *134*, 913–921.

32. Huang, Z.M.; Zhang, Y.Z.; Kotaki, M.; Ramakrishna, S. A review on polymer nanofibers by electrospinning and their applications in nanocomposites. *Compos. Sci. Technol.* **2003**, *63*, 2223–2253.

33. Wang, Z.; Liu, L. Synthesis and ethanol sensing properties of Fe-doped $SnO_2$ nanofibers. *Mater. Lett.* **2009**, *63*, 917–919.

34. Song, X.; Liu, L. Characterization of electrospun ZnO–$SnO_2$ nanofibers for ethanol sensor. *Sens. Actuators A* **2009**, *154*, 175–179.

35. Song, X.; Zhang, D.; Fan, M. A novel toluene sensor based on ZnO-$SnO_2$ nanofiber. *Appl. Surf. Sci.* **2009**, *255*, 7343–7347.

36. Park, J.A.; Moon, J.; Lee, S.J.; Lim, S.C.; Zyung, T. Fabrication and characterization of ZnO nanofibers by electrospinning.*Curr. Appl. Phys.* **2009**, *9*, S210–S212.

37. Liu, L.; Zhang, T.; Li, S.; Wang, L.; Tian, Y. Preparation, characterization, and gas-sensing properties of Pd-doped $In_2O_3$nanofibers. *Mater. Lett.* **2009**, *63*, 1975–1977.

38. Zhang, H.; Li, Z.; Liu, L.; Wang, C.; Wei, Y.; MacDiarmid, G.A. $Mg^{2+}$/$Na^{+}$-doped rutile TiO2 nanofiber mats for high-speed and anti-fogged humidity sensors. *Talanta.* **2009**, *79*, 953–958.

39. Wang, Y.; Ramos, I.; Santiago-Avilés, J.J. Detection of moisture and methanol gas using a single electrospun tin oxide nanofiber. *IEEE Sensors J.* **2007**, *7*, 1347–1348.

40. Lu, X.; Liu, X.; Zhang, W.; Wang, C.; Wei, Y. Large-scale synthesis of tungsten oxide nanofibers by electrospinning. *J. Colloid Interface Sci.* **2006**, *298*, 996–999.

41. Wang, G.; Ji, Y.; Huang, X.; Yang, X.; Gouma, P.I.; Dudley, M. Fabrication and characterization of polycrystalline $WO_3$nanofibers and their application for ammonia sensing. *J. Phys. Chem. B* **2006**, *110*, 23777–23782.

42. Piperno, S.; Passacantando, M.; Santucci, S.; Lozzi, L.; La Rosa, S. $WO_3$ nanofibers for gas sensing applications. *J. Appl. Phys.* **2007**, *101*, 124504–124507.

43. Kudo, T.; Okamoto, H.; Matsumoto, K.; Sasaki, Y. Peroxopolytungstic acids synthesized by direct reaction of tungsten or tungsten carbide with hydrogen peroxide. *Inorg. Chim. Acta* **1986**, *111*, L27–L28.

44. Choi, S.-W.; Park, J.Y.; Kim, S.S. Dependence of gas sensing properties in ZnO nanofibers on size and crystallinity of nanograins. *J. Mater. Res.* **2011**, *26*, 1662–1665.

45. Choi, S.-W.; Park, J.Y.; Kim, S.S. Growth behavior and sensing properties of nanograins in CuO nanofibers. *Chem. Eng. J.* **2011**, *172*, 550–556.

46. Park, J.Y.; Asokan, K.M.; Choi, S.-W.; Kim, S.S. Growth kinetics of nanograins in $SnO_2$ fibers and size dependent sensing properties. *Sens. Actuators B Chem.* **2011**, *152*, 254–260.

47. Arbiol, J. Metal Additives Distribution in $TiO_2$ and $SnO_2$ Semiconductor Gas Sensor Nanostructured Materials. Ph.D. Thesis, Barcelona University, Barcelona, Spain, July 2001..

# Chapter 7

# Nanofiber Reinforced Composite Polymer Electrolyte Membranes

**A. Kumar[1] and M. Deka[1]**

[1] Department of Physics, Tezpur University,, India

## 1. INTRODUCTION

The path breaking studies of Wright and Armand on ionically conducting polymers, called "polymer electrolytes" in the 1970s have opened an innovative area of materials research with potential applications in the power sources industry (Fenton et al., 1973). The main applications of the polymer electrolytes are in rechargeable lithium batteries as an alternative to liquid electrolytes (Chen et al., 2002, Lobitz et al., 1992). The advantages such as no leakage of electrolyte, higher energy density, flexible geometry and improved safety hazards have drawn the attention of many researchers on the development of lithium polymer batteries and other electrochemical devices such as supercapacitors, electrochromic windows, and sensors (Gray, 1991). In batteries being a separator membrane polymer electrolyte must meet the following requirements.

1 high ionic conductivity

2 high cationic transference number

3 good dimensional stability

4 high electrochemical stability and chemical compatibility with both Li anode and cathode material and

5 good mechanical stability.

The need for high ionic conductivity arises from the fact that at what rate or how fast energy from a Li-battery can be drained, which largely depends on the extent of ionic mobility in the electrolyte and hence on ionic conductivity. For battery applications, along with high ionic conductivity the electrolyte material must be dimensionally stable since the polymer electrolyte will also function as separator in the battery, which will provide electrical insulation between the cathode and the anode. This implies that it must be possible to process polymer electrolyte into freestanding film with adequate mechanical strength. Requirement of high cationic transport number rather than anionic is also important in view of the battery performance because concentration gradients caused by the mobility of both cations and anions in the electrolyte arise during discharging, which may result in premature battery failure.

Recent advances in nanotechnology have made materials and devices easier to be fabricated at the nanoscale. Nanofibres and nanowires with their huge surface area to volume ratio, about a thousand times higher than that of a human hair, have the potential to significantly improve current technology and find applications in new areas. Nanofibers in particular, have been used for a wide range of applications such as tissue engineering (Bhattarai et al., 2004), filter media (Suthar & Chase, 2001), reinforcement in composites (Chatterjee & Deopura, 2002) and micro/nano-electro-mechanical systems (MEMS/NEMS) (Sundararajan et al., 2002). Such fibers can be made from various materials such as polymers, carbon and semiconductors into the form of continuous nanofibers, nanofibrous networks or short nanowires and nanotubes. On the other hand polymer nanocomposite formation through the reinforcement of particles having dimensions less than 100 nm into polymer matrix finds important applications for the development of advanced materials. This reinforcement occurs for particles of spherical shape as well as for plate-like and fibrous nanofillers, the latter with their high aspect ratio exhibits a stronger effect. As a result, nanofibers because of their excellent mechanical properties coupled with very high aspect ratio theoretically are the ideal candidates for reinforcing a polymer matrix. The main challenge in this case lies in the fact of increased specific surface area due to the small size of the reinforcing particles, forces per unit mass resulting from interactions between this surface and surrounding area become more pronounced. As a result nanofibers cannot be easily dispersed in substances of different surface energy. However a strong interface between the reinforcing phase and the host matrix is always desirable in order to achieve desirable properties.

## 2. TYPES OF POLYMER ELECTROLYTES

The development of polymer electrolytes has passed mainly three stages namely: (a) solid polymer electrolytes, (b) gel polymer electrolytes and (c) composite polymer electrolytes. In dry solid polymer electrolytes the polymer host itself is used as a solid solvent along with lithium salt and does not contain any organic liquids. The most commonly studied polymer electrolyte membranes are complexes of Li salts with a high molecular weight polyethylene oxide (PEO) (Ahn et al., 2003). PEO excels as a polymer host because of its high solvating power for lithium ions and its compatibility with the lithium electrode (Algamir & Abraham, 1994). However, it is also known that the high conductivity ($10^{-3}$-$10^{-4}$ S/cm) of most PEObased polymer electrolytes requires operation in the temperature range of 80-100 °C (Ahn et al., 2003, Algamir & Abraham, 1994, Abraham, 1993, Kovac et al., 1998), below which these electrolytes suffer from low conductivity values in the range of $10^{-7}$-$10^{-8}$ S/cm because of the high crystallinity of PEO (Kovac et al., 1998). Gel electrolytes are formed by incorporating an electrolyte solution into polymer matrix (Algamir & Abraham, 1994, Tarascon et al., 1996). Since the electrolyte molecules such as ethylene carbonate (EC), propylene carbonate (PC), diethyl carbonate (DEC) can solvate ions, coordinating polymers like PEO are no longer necessary. The ion conduction in these electrolytes takes place through the liquid electrolytes where the host polymer mostly provides the structural support. The gel polymer electrolyte systems based on poly(methyl methacrylate) (PMMA) (Rajendran & Uma, 2000, Rajendran et al., 2001, Stephan et al., 2000, Feuillade & Perche, 1975) have been proposed for lithium battery application particularly because of their beneficial effects on stabilization of the lithium–electrolyte interface (Appetecch et al., 1995). However, reasonable conductivity achieved of such plasticized film is offset by poor mechanical properties at high plasticizer content. Rajendran et al. (Rajendran & Uma, 2000; Rajendran et al., 2001) were able to improve the mechanical property of PMMA by blending with poly(vinyl alcohol) (PVC). However, a decrease of ionic conductivity was observed due to higher viscosity and lower dissociability of lithium salt. Recently poly(vinylidenefluoride) (PVdF) as a host has drawn the attention of many researchers due to its high anodic stability and high dielectric constant ($\varepsilon = 8.4$) which helps in greater ionization of lithium salts (Choe et al., 1995). Unfortunately, PVdF-based polymer electrolytes suffer due to syneresis; a phenomenon by which the liquid component separates out from the host matrix in due course or upon application of pressure leading to battery leak and related safety problems. Very recently, poly(vinylidenefluoride-co-hexafluoropropylene) {P(VdF-HFP)} based systems have drawn the attention of many researchers because of its various appealing properties like high dielectric constant, low crystallinity and glass transition temperature (Song et al., 2000; Wu et al., 2006, Stephan et al., 2006; Stephan & Nahm, 2006; Stolarska et al., 2007; Michael & Prabaharan,

2004). P(VdF-HFP) has excellent chemical stability due to VdF unit and plasticity due to HFP unit (Aravindan & Vickraman, 2007). However, gelled or plasticized P(VdF-HFP) based electrolytes exhibit drawbacks, such as increased reactivity with lithium metal electrode, solvent volatility and poor mechanical properties at high degree of plasticization (Jacob et al., 2003). In order to retain the mechanical properties of polymer gel electrolytes, the gel films have to be hardened either by chemical or physical curing (high energy radiation), which results in high processing costs. Alternatively, the addition of inert oxides to the polymer electrolytes has recently become an ever increasing attractive approach, due to the improved mechanical stability, enhanced ionic conductivity and electrode-electrolyte interface stability (Croce et al., 1998, Quartarone et al., 1998). This type of electrolyte is known as composite polymer electrolyte. The increase in ionic conductivity in composite polymer electrolytes depends on the concentration and particle size of the inert solid phases. Generally, smaller the particle sizes of the oxides, the larger the conductivity enhancement. However, due to the absence of exact structure–property correlations in the polymer electrolyte systems, a complete understanding of the ion conduction phenomenon is still lacking. Nonetheless, to explain the mechanistic aspects of ion transport in micro/nanocomposite polymer electrolyte systems, a working hypothesis has been suggested accordingly to which, the dispersion of submicron or nano-size filler particles having large surface area, into the polymer host lowers the degree of crystallinity, which may also be thought to be due to Lewis acid–base interaction between ceramic surface states and polymer segments (Croce et al., 1998, Golodnitsky et al., 2002). Hence, in addition to the usual space charge effects of the dispersoid particles, the increased amorphosity would also support the conductivity enhancement in terms of increased ionic mobility through the amorphous phase. Essentially because of this idea, nanocomposite polymer electrolytes wherein nanosized inert solid particles are added to the polymer electrolytes are presently the focus of many studies, both practical as well as theoretical.

It has been reported that addition of nanoscale inorganic fillers, such as alumina ($Al_2O_3$), silica ($SiO_2$), titania ($TiO_2$) etc. to the polymer electrolytes resulted in the improvements of transport properties as well as mechanical and electrochemical properties (Kim et al., 2001; Kim et al., 2002; Scrosati et al., 2000). A novel composite micro-porous polymer electrolyte membrane based on optimized composition of P(VdF-HFP)-$ZrO_2$ was prepared by a preferential polymer dissolution process. The incorporation of $ZrO_2$ nanoparticles in the P(VdF-HFP) matrix, improved the ionic conductivity due to the availability of a large amount of oxygen vacancies on $ZrO_2$ surface which may act as the active Lewis acidic sites that interact with ions (Kalyana et al., 2007). A gel polymer electrolyte based on the blend of poly(methyl methacrylate-co-acrylonitrile-co-lithium methacrylate) (PMAML) and poly(vinylidene fluoride-co-hexafluoropropylene) P(VdF-HFP)) was prepared and characterized. The highest ionic conductivity achieved in the system was $2.6\times10^{-3}$ S $cm^{-1}$ at ambient temperature (Wang & Tang, 2004). Highly conducting porous polymer electrolytes comprised of P(VdF-HFP), metal oxide ($TiO_2$, MgO, ZnO)/or mesoporous zeolite (MCM-41, SBA-15), ethylene carbonate (EC), propylene carbonate (PC) and $LiClO_4$ were fabricated with a simple direct evaporation method. These polymer composite electrolytes were stable up to 5.5 V (versus $Li/Li^+$) and the lithium ion cells assembled with these polymer electrolytes show a good performance at a discharge rate below *C*/2 (Wu et al., 2006). Composite polymer electrolyte (CPE) membranes, comprising P(VdF-HFP), aluminum oxyhydroxide ($AlO[OH]_n$) of two different sizes 7 μm/14 nm and $LiN(C_2F_5SO_2)_2$ as the lithium salt were prepared using a solution casting technique. The incorporation of nanofillers greatly enhanced the ionic conductivity and the compatibility of the composite polymer electrolyte (Stephan et al., 2006). Nano $SiO_2$–P(VDF-HFP) composite porous membranes were prepared as the matrix of porous polymer electrolytes through *in situ* composite method based on hydrolysis of tetraethoxysilane and phase inversion. It is found that the *in situ* prepared nano silica was homogeneously dispersed in the polymeric matrix, enhanced conductivity and electrochemical stability of porous polymer electrolytes, and improved the stability of the electrolytes against lithium metal electrodes (He et al., 2005). Composite polymer membranes comprising of P(VdF-HFP) /$Al_2O_3$ were prepared by phase inversion technique with poly(ethylene glycol) (PEG) as an additive. The polymer membrane prepared with a weight ratio of PVdF-HFP (40):PEG (40):$Al_2O_3$ (20) showed maximum protonic conductivity due to the combined effect of inert filler and its porous nature (Kumar et al., 2007). The protonic conductivity of silica polymerized *in situ* within a P(VdF-HFP) matrix has been studied. The conductivity linearly increases with the silica content (Carrière et al., 2001). Gel polymer electrolyte membranes composed of P(VdF-HFP) and surface modified aluminum or titanium oxide were prepared according to the so-called Bellcore process. The ionic conductivity of polymer membrane increased by more than one order of magnitude upon the addition of filler into polymer host (Stolarska et al., 2007). Nanocomposite polymer electrolyte (NCPE) membranes of P(VdF-HFP) matrix with ethylene carbonate and diethyl carbonate mixtures as plasticizing agents, $SiO_2$ nanoparticles as filler and complexed with $LiPF_3(CF_3CF_2)_3$ were prepared by solvent casting technique. NCPE membranes containing 2.5 wt% of $SiO_2$ exhibited enhanced conductivity of 1.13 mS $cm^{-1}$ at ambient temperature (Arvindan & Vickraman, 2007). Various amounts of nanoscale rutile $TiO_2$ particle were used as fillers in the preparation of P(VdF-HFP)-based porous polymer electrolytes. Physical, electrochemical and transport properties of the electrolyte films were investigated in terms of surface morphology, thermal and crystalline properties, swelling behavior after absorbing electrolyte solution, chemical and electrochemical stabilities, ionic conductivity, and compatibility with lithium electrode (Kim et al., 2003).

In the present chapter we report novel composite polymer electrolytes by incorporating dedoped (insulating) polyaniline nanofibers instead of nanoparticles into P(VdF-HFP)-(PC+DEC)-$LiClO_4$ gel polymer electrolyte system and PEO-P(VdF-HFP)-$LiClO_4$ blend electrolyte system. The concentration of dedoped polyaniline nanofibers has been varied and its effects on ionic transport in both the systems have been investigated.

# 3. SYNTHESIS TECHNIQUES OF NANOFIBERS

## 3.1. Electrospinning

Electrospinning has been recognized as an efficient technique for the fabrication of polymer nanofibers. Various polymers have been successfully electrospun into ultrafine fibers in recent years mostly in solvent solution and some in melt form. In terms of the flexibility of the process, electrospinning is able to fabricate continuous nanofibres from a huge range of materials. Of the major classes of materials, electrospinning is able to produce nanofibres of polymers, composites, semiconductors and ceramics (Huang et al., 2003, Chronakis, 2005). The formation of nanofibers through electrospinning is based on the uniaxial stretching of a viscoelastic solution. There are basically three components in a eletrospun setup: a high voltage power supply, a capillary tube with a needle and a metal collector. A typical electrospinning set up is shown in Fig. 1. When a sufficiently high voltage is applied to a liquid droplet, the body of the liquid becomes charged, and electrostatic repulsion counteracts the surface tension and droplet is stretched, at a critical point a stream of liquid erupts from the surface. This point of eruption is known as the Taylor cone (Taylor, 1969). If the molecular cohesion of the liquid is sufficiently high, stream breakup does not occur (if it does, droplets are electrosprayed) and a charged liquid jet is formed. As the jet dries out in flight, the mode of current flow changes from ohmic to convective as the charge migrates to the surface of the fibre. The jet is then elongated by a whipping process caused by electrostatic repulsion initiated at small bends in the fibre, until it is finally deposited on the grounded collector. The elongation and thinning of the fibre resulting from this bending instability leads to the formation of uniform fibres with nanometer-scale diameters (Li & Xia, 2004). The main advantage of the electrospinning nanomanufacturing process is that it is cost effective compared to that of most bottom-up methods. The nanofibers prepared from Electrospinning process are often uniform and continuous and do not require expensive purification unlike submicrometer diameter whiskers, inorganic nanorods and carbon nanotubes (Dzenis, 2004). Polymer nanofibers mats are being considered for use in composite materials reinforcement, sensors, filtration, catalysis, protective clothing, biomedical applications, space applications such as solar cells, and micro- and nano optoelectronic device such as LEDs and photocells. Carbon nanofibers made from polymeric precursors further expand the list of possible uses of electrospun nanofibers (Li & Xia, 2004, Subbiah et al., 2005).

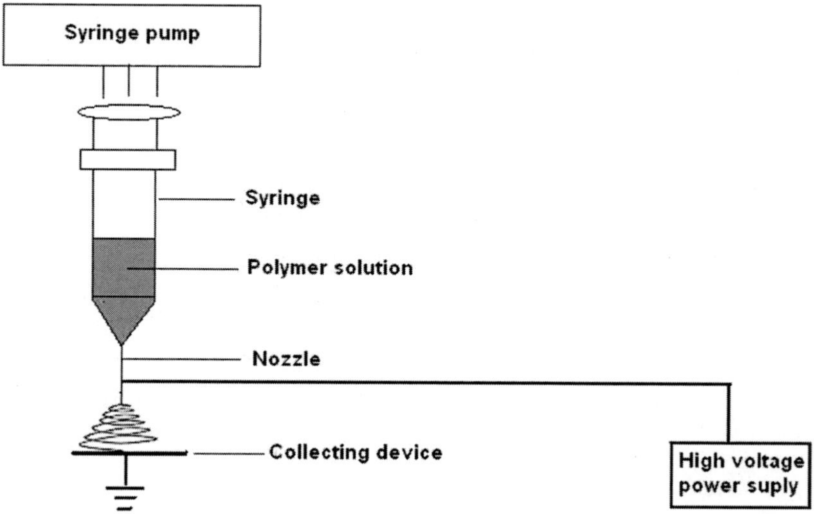


**Figure 1.** Shematic diagram of electrospinning setup

## 3.2. Template Synthesis

An effective way to produce nanometer fibers (or nano-tubes) is based on the use of membrane-template techniques (Martin, 1961, Delvaux et al., 2000, Steinhart et al., 2002). Membranes, with nanochannels generated by fission-fragment tracks or by electrochemical etching of aluminum metal, are used as templates for either chemical or electrochemical deposition of conductive polymers (Pathasarathy & Martin, 1994), metal (Van de Zande et al., 1997), semiconductor (Klein et al., 1993), and other materials for the generation of nanofibers or nanotubes. Since the nanochannels on membranes are very uniform in size, the diameter and the aspect ratio of the nanofibers synthesized by the membranetemplate technique can be precisely controlled. This greatly facilitates the interpretation of optical data and the processing of these fibers (or tubes) into 2-D nanostructured materials (de Heer et al., 1995). Single-crystal semiconductor nanofibers can also be grown catalytically by metallorganic vapor phase epitaxy and laser ablation vapor-liquid-solid techniques ( Morales & Lieber, 1998). The synthesis of these one-dimensional structures with diameters in the range of 3 to 15 nm holds considerable technological promise for optoelectronic device applications.

## 3.3. Phase Separation

This method is normally used to synthesisze the polymer nanofibers. In phase separation, a polymer is first mixed with a solvent before undergoing gelation. The main mechanism in this process is the separation of phases due to physical incompatibility. One of the phase – which is that of the solvent – is then extracted, leaving behind the other other phase. In nutshell phase separation technique involves five basic steps (Ma & Zhang, 1999)

- Dissolution of polymer.
- Liquid–liquid phase separation process.
- Polymer gelation (controls the porosity of nanoscale scaffolds at low temperature).
- Extraction of solvent from the gel with water.
- Freezing and freeze-drying under vacuum.

Gelation is the most critical step that controls the porous morphology of the nanofibrous foams. The duration of gelation vary with polymer concentration and gelation temperature. At low gelation temperature nanoscale fiber networks are formed, whereas high gelation temperature results in the formation of a platelet-like structure due to the nucleation of crystals and their growth. This formation of platelet-like structure is overcome by increasing the cooling rates, which produce uniform nanofibers. However, gelation condition or polymer concentration does not affect the average diameter of fibers significantly. An increase in polymer concentration results in the decrease of porosity and increase of mechanical properties (Young's modulus and tensile strength). Other process parameters such as types of polymer and solvent, and thermal treatment also influence the morphology of the nanofibrous scaffolds (Zhang et al, 2005). The advantage of the phase separation process is that it is a relatively simple procedure and the requirements are very minimal in terms of equipment compared with the previously discussed techniques of electrospinning and template synthesis.

## 3.4. Interfacial Polymerization

Another effective and versatile way to synthesize nanofibers of conductive polymers is the interfacial polymerization technique. Generally conductive polymers like polyaniline, polypyrrole, PEDOT etc. are synthesized by this technique. This method was originally developed by Huang (Huang, 2003), where they synthesized nanofibers of polyaniline (PAni). Interfacial polymerization does not depend on any specific template or dopant. Highquality polyaniline nanofibers are obtained even when common mineral acids such as hydrochloric, sulfuric and nitric acids are used as dopants. They found it is the nature for PANI to form nanofibrillar morphology. In order to obtain pure PANI nanofibers, the secondary growth of the initially formed nanofibers must be suppressed. The interfacial polymerization of PAni involves the polymerization of aniline at the interface between two immiscible liquids, where the newly formed PAni nanofibers diffuse away from the interface to the aqueous solution because of their hydrophilic nature. This makes more reaction sites available at the interface and avoids further growth of the PAni nanofibers. S. Goel et. al. (Goel et al., 2007) reported the synthesis of polypyrrole nanofibers in the presence of different dopants including hydrochloric acid (HCl), ferric chloride (FeCl3), ptoluene sulfonic acid (p-TSA), camphor sulfonic acid (CSA), and polystyrene sulfonic acid (PSSA) using a simple interfacial oxidative polymerization method. They observed that that the electrical conductivity of PPy nanostructures depends upon the nature of dopant (PPy-p-TSA > CSA > HCl > FeCl3 > PSSA), PPy-p-TSA nanofibers showing the highest electrical conductivity of $6 \times 10^{-2}$ S/cm.

# 4. P(VDF-HFP)-(PC+DEC)-LICLO$_4$-DEDOPED POLYANILINE NANOFIBERS COMPOSITE GEL POLYMER ELECTROLYTE SYSTEM:

## 4.1. Preparation

The host copolymer poly(vinylidenefluoride-co-hexafluoropropylene) {P(VdF-HFP)} ($M_w \approx 400000$) and salt lithium perchlorate ($LiClO_4$) were received from Aldrich, USA.

These two raw materials were heated at 50º C and 100º C respectively before use to remove the moisture. Organic solvents propylene carbonate (PC) and diethyl carbonate (DEC) were used without further treatment as obtained from EMerck. Dedoped polyaniline (PAni) nanofibers were synthesized by the interfacial polymerization technique (Huang, 2006). The interfacial polymerization reaction was carried out in 30ml glass vials. 1M amount of aniline was dissolved in 10ml of organic solvent carbon tetrachloride ($CCl_4$). Ammonium peroxydisulfate {$(NH_4)_2S_2O_8$} (0.25M) was dissolved in 10 ml of double distilled water and dopant acid (HCl). The polyaniline nanofibers were dedoped with 1M NaOH. The electronic conductivity of PAni nanofibers was measured with Keithley 2400 LV souecemeter. The electronic conductivity of doped nanofibers is of the order of $10^{-4}$ $Scm^{-1}$, whereas after dedoping with NaOH solution the electronic conductivity was found to be of the order of $10^{-11}Scm^{-1}$. This confirms the insulating nature of dedoped polyaniline nanofibers.

P(VdF-HFP)-(PC+DEC)-$LiClO_4$ - x wt. % dedoped PAni nanofiber (x= 0, 2, 4, 6, 8,10) membranes were prepared by conventional solution casting technique. Predetermined amounts of P(VdF-HFP), $LiClO_4$ and (PC+DEC) were dissolved in acetone in the ratio of 6:1.5:1.5:1 (by weight). Subsequently the solution was stirred at 50ºC for 12 hours. PC has high dielectric constant ($\varepsilon$ = 64.6) but has high viscosity ($\eta$ = 2.53), whereas DEC has low dielectric constant ($\varepsilon$ = 2.82) but has low viscosity ($\eta$ = 0.748). Combination of PC and DEC (1:1 by volume) solvent was used as optimization for high dielectric constant ($\varepsilon$ = 33.71) and low viscosity ($\eta$ = 1.639) to achieve high ionic conductivity. After 12 hours of mixing at 50ºC the dedoped polyaniline nanofiber was added in the gel polymer solution and allowed to stir for another 12 hours. The viscous solution thus obtained was cast onto Petri dish and allowed to dry at room temperature. Different membranes were synthesized by varying the concentration of dedoped polyaniline nanofibers.

The ionic conductivity of the nanocomposite polymer electrolyte films was determined by ac impedance measurements using a Hioki 3532-50 LCR Hitester in the frequency range from 42 Hz to 5MHz. The temperature dependence of ionic conductivity was also measured by heating the samples from room temperature (25 °C) to 60 °C. The nature of conductivity of nanofibers dispersed gel polymer electrolytes was determined by transference number measurements using Wagner polarization technique with polymer electrolyte membrane between graphite blocking electrodes. The transference number was found to be = 0.98 indicating that conductivity was essentially ionic in nature. The interfacial stability of nanocomposites polymer electrolytes was studied by fabricating stainless steel/polymer electrolyte/stainless steel cells at room temperature and monitored for 20 days. X-ray diffraction patterns of the prepared membranes were obtained by Rigaku miniflex diffractometer at room temperature. Surface morphology of the composite electrolytes was studied by using Scanning Electron Microscope (SEM) (JEOL 6390 LV). The size of PAni nanofibers was determined by TEM (JEOL-TEM-100 CXII).

## 4.2. Results and Discussion

### 4.2.1. Tem Studies

Fig. 2 shows the TEM micrograph of PAni nanofibers. From the figure it is observed that nanofiber is composed of randomly packed polymer chains. As the PAni nanofibers are synthesized by interfacial polymerization, overgrowth of polyaniline on the nanofiber scaffolds does not take place and nanofibrillar morphological units are formed. The diameter and length of the fibers are found to be 20 nm to 30 nm and more than 1000 nm respectively. These high aspect ratio (> 50) nanofibers were used in polymer electrolytes as a filler to increase the ionic conductivity and electrochemical properties.

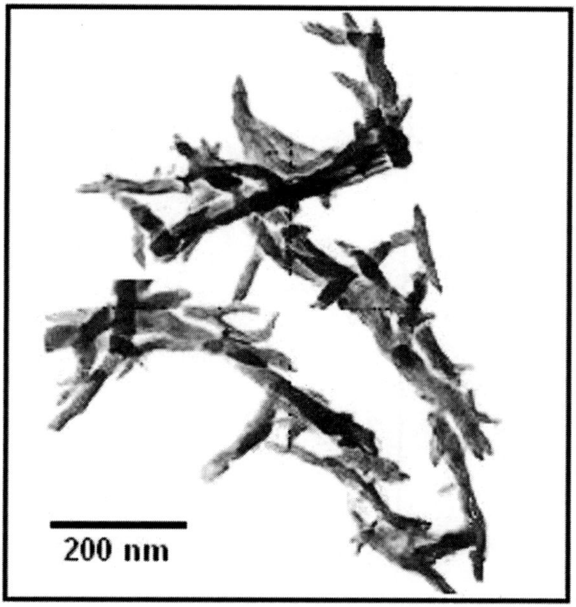


**Figure 2.** TEM micrograph of dedoped Pani nanofibers

### 4.2.2. Ionic Conductivity Studies

Fig. 3a shows the impedance plot of P(VdF-HFP)-(PC+DEC)-$LiClO_4$ polymer electrolyte and Fig. 3bpresents the complex impedance spectra of P(VdF-HFP)-

(PC+DEC)-$LiClO_4$-x% dedoped polyaniline nanofiber (x = 2, 4, 6, 8 and 10) composite polymer electrolytes. The impedance spectra comprise a distorted semicircular arc in the high frequency region followed by a spike in the lower frequency region (Aravindan & Vickraman, 2008). The high frequency semicircle is due to the bulk properties and the low frequency spike is due to the electrolyte and electrode interfacial properties. The impedance spectra can be modeled as an equivalent circuit having a parallel combination of a capacitor and a resistor in series or parallel with a constant phase element (CPE) (Marzantowicz et al., 2005). The impedance of CPE is given by

$$Z_{CPE} = k\,(j\omega)^{-p} \quad \text{where } 0 < p < 1 \tag{1}$$

When p=0, Z is frequency independent and k is just the resistance and when p=1, Z = k/jω = - jk/ ω, the constant $k_1$ now corresponds to the capacitance. When p is between o and 1, the CPE acts in a way intermediate between a resistor and a capacitor. The use of series CPE terms tilts the spike and parallel CPE terms depress the semicircle. The bulk electrical resistance value ($R_b$) is calculated from the intercept at high frequency side on the Z' axis. The ionic conductivity is calculated from the relation $\sigma = l/R_b r^2 \pi$; where l and r are thickness of polymer electrolyte membrane and radius of the sample membrane discs and $R_b$ is the bulk resistance obtained from complex impedance measurements. The value of $\sigma_{ionic}$ of plasticized nanocomposite polymer electrolytes was evaluated from complex impedance spectra and expressed as a function of nanofiber concentration at room temperature (25 °C) as shown in Fig. 4. It is observed that the $\sigma_{ionic}$ increases with increase in nanofiber concentration. Maximum conductivity was found to be $6.31 \times 10^{-3}$ $Scm^{-1}$ at room temperature for x = 6 wt. % dedoped polyaniline nanofiber filler, which is by over one order of magnitude higher as compared to that ($2.5\times10^{-4}$ $Scm^{-1}$) for polymer electrolyte without nanofibers. However as the filler (dedoped nanofiber) concentration increases beyond 6 wt. %, the ionic conductivity decreases. It is known that plasticized polymer electrolytes show highly porous structure created by plasticized rich phase (Aravindan & Vickraman, 2007). When dedoped polyaniline nanofiber is incorporated in the porous membrane, the movement of $Li^+$ ion through the pores is facilitated by the filler due to formation of conduction path resulting in higher conductivity. As the nanofiber content increases from 2 wt. % to 6 wt. % the porous structure is remarkably widened leading to entrapment of large volume of liquid electrolyte in the pores, which results in increase in ionic conductivity. Moreover the reorganization of P(VdFHFP) chains is prevented due to high aspect ratio (> 50) nanofibers leading to increase in amorphicity with increasing concentration of nanofibers which is consistent with XRD results. However, XRD results show that beyond 6 wt. % the nanofibers get phase separated out from the polymer matrix and form insulating aggregation (Wieczorek et al.,

1996), which impede the $Li^+$ ion motion resulting in decrease in ionic conductivity. Fig. 5 shows the conductivity versus temperature inverse plots of polymer electrolyte membranes in the temperature range from 25 °C to 60 C. Highest ionic conductivity of $2.2 \times 10^{-2} Scm^{-1}$ has been found at 60 °C with 6 wt. % of dedoped polyaniline nanofibers. The figure shows that the ionic conduction in nanocomposites polymer electrolytes obey the Vogel-Tamman-Fulcher (VTF) relation (Quartarone et al., 1998)

$$\sigma = \sigma_0 \exp(-B/k(T-T_0)) \quad (2)$$

where B is a constant with dimensions that of the energy, $T_0$ is the idealized glass transition temperature at which the probability of configurational transition tends to become zero and is generally regarded as having a value between 20 to 50 K below glass transition temperature ($T_g$) of the polymer. Since $T_g$ of P(VdF-HFP) is -100 °C, $T_0$ will be far below the temperature regions of measurements from room temperature (25 °C) to 60 °C. Therefore, VTF behavior (2) can be modeled as Arhenius behavior as shown lnσ vs. 1000/T plots in Fig. 5. As expected the increase in temperature leads to increase in ionic conductivity because as the temperature increases the polymer expands to produce more free volume, which leads to enhanced ionic mobility and polymer segmental mobility. The enhancement of ionic conductivity by the dedoped polyaniline nanofiber can be explained by the fact that the nanofiber inhibits the recrystallization kinetics, helping to retain the amorphous phase down to relatively low temperatures (Rhoo et al., 1997).

### 4.2.3. Interfacial Stability

Compatibility of nanocomposites polymer electrolyte with electrode materials is an important factor for polymer battery applications. In order to improve the interfacial stability of polymer electrolytes before and after incorporating dedoped polyaniline nanofiber, the ionic conductivity was measured by fabricating stainless steel/polymer electrolyte membrane/stainless steel cells at room temperature and monitored for 20 days. Polymer electrolytes without nanofiber and containing 6 wt. % of dedoped polyaniline nanofiber have been selected to observe the effect of nanofiber on interfacial stability and the results are shown in Fig. 6.

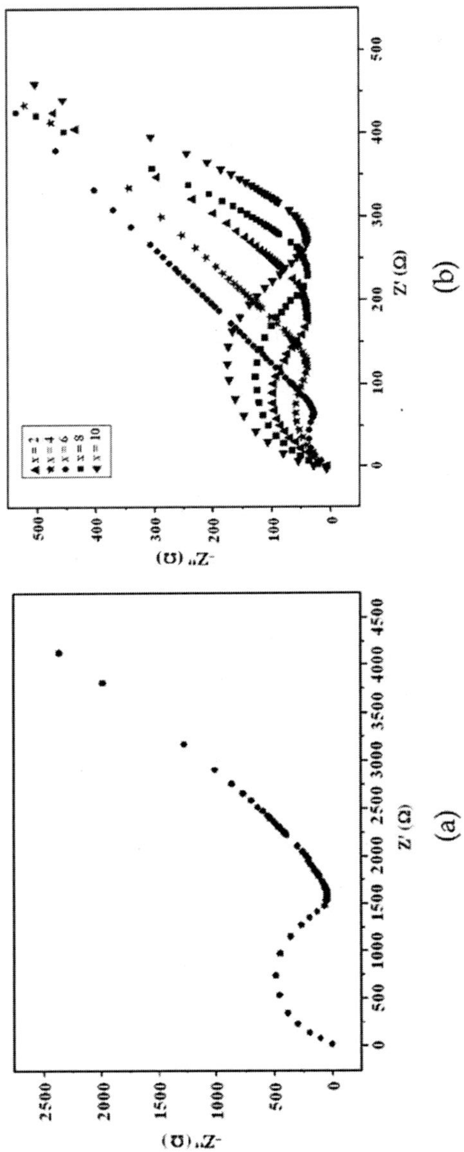


**Figure 3.** a) Complex impedance spectrum of P(VdF-HFP)-(PC+DEC)-$LiClO_4$ gel polymer electrolyte without incorporating dedoped nanofibers.(b) Complex impedance spectra of P(VdF-HFP)-(PC+DEC)-$LiClO_4$-x% dedoped polyaniline nanofibers (x = 2, 4, 6, 8 and 10)

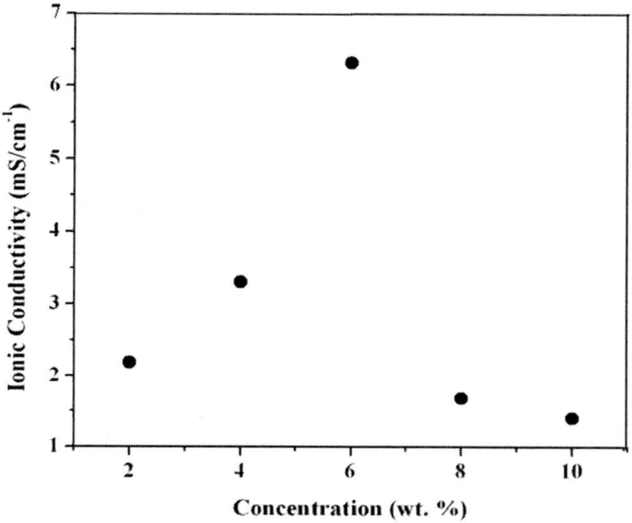


**Figure 4.** Variation of ionic conductivity of P(VdF-HFP)-(PC+DEC)-$LiClO_4$-x% dedoped polyaniline nanofibers with nanofiber concentration.

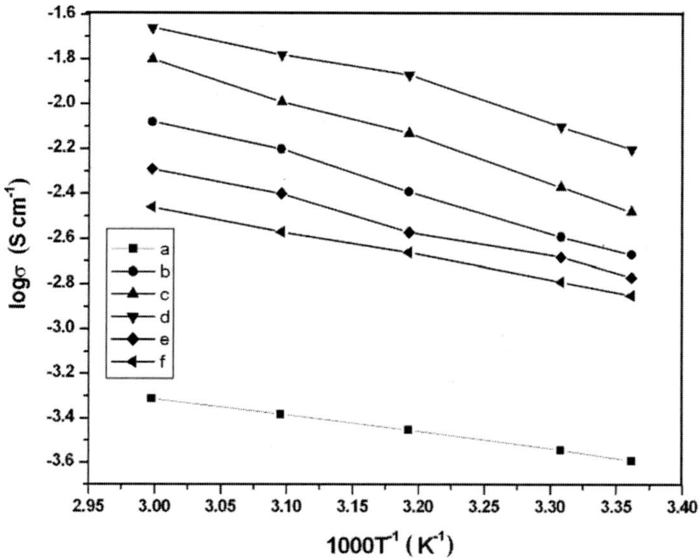


**Figure 5.** Logσ vs. temperature inverse curves of P(VdF-HFP)-(PC+DEC)-$LiClO_4$-x% dedoped polyaniline nanofibers membranes (a) x=0, (b) x=2, (c) x=4, (d) x=6, (e) x=8, (f) x=10.

It reveals that ionic conductivity of both the electrolytes decreased with time but decrease of ionic conductivity in the polymer electrolyte without nanofiber is much larger as compared to that of polymer electrolytes containing nanofibers. This result confirms that the interfacial stability of the polymer electrolyte containing nanofibers is better than that of without nanofibers. This can be attributed to the fact that when nanofiber is added passivation of polymer electrolyte due to reaction with electrode material decreases. High aspect ratio (> 50) nanofibers get accumulated on the surface of the electrode and effectively impede the electrode electrolyte reaction (Zhang et al., 2004). Fig. 7schematically depicts that when dedoped (insulating) nanofibers are incorporated in the polymer electrolyte the electrolyte does not make direct contact with the electrode, which increases the interfacial stability.

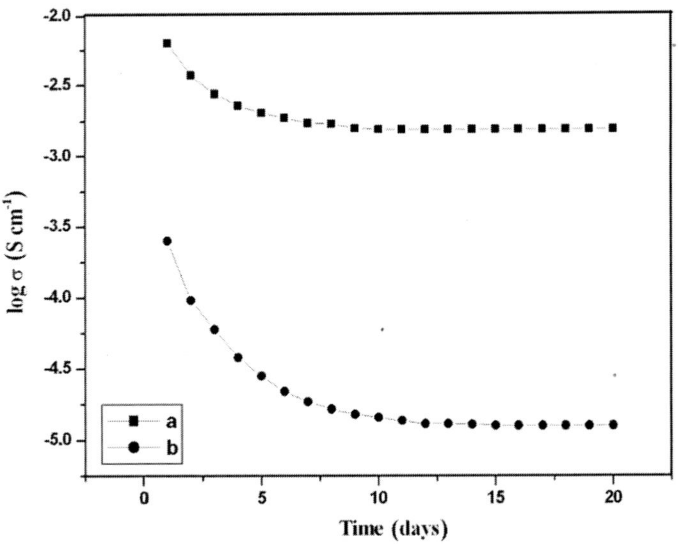


**Figure 6.** Interfacial stability of (a) P(VdF-HFP)-PC+DEC-$LiClO_4$-6% dedoped polyaniline nanofibers polymer electrolyte and (b) P(VdF-HFP)-PC+DEC-$LiClO_4$ gel polymer electrolyte

## 4.2.4. Xrd Analysis

X-ray diffraction patterns of pure P(VdF-HFP) and dedoped polyaniline nanofibers are presented inFig. 8a& 8b). In Fig. 8a the peaks at 2θ=20º and 38 correspond to (020) and (202) crystalline peaks of P(VdF) (Abbrent et al, 2001). This is a confirmation of partial crystallization of the PVdF units in the copolymer to give an overall semi-crystalline morphology for P(VdF-HFP). High intensity peaks at 2θ=20º and 2θ=23º are observed in the XRD pattern of dedoped polyaniline nanofiber (Fig. 8b). Fig. 9(a-f)shows the XRD patterns of P(VdF-HFP)-(PC+DEC)-$LiClO_4$-x% dedoped polyaniline nanofiber composite polymer electrolytes. It is observed that the addition of dedoped polyaniline nanofibers in polymer electrolytes increases the broadening and decreases the intensity of XRD peaks. This is due to fact that addition of nanofibers prevents polymer chain reorganization causing significant disorder in the polymer chains which promotes the interaction between them (Kumar et al., 2007).

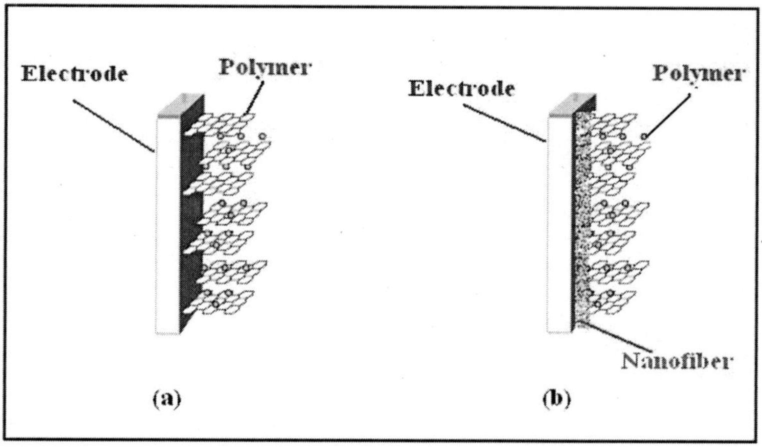


**Figure 7.** Schematic representation of electrode/polymer electrolyte interfacial passivation (a) without and (b) with dedoped polyaniline nanofibers. without nanofibers

## 4.2.5. Morphological Study

Scanning electron micrographs of P(VdF-HFP)-(PC+DEC)-$LiClO_4$-x% dedoped polyaniline nanofiber gel composite polymer electrolytes are shown in Fig. 10(a-f). It is observed that the polymer electrolytes show highly porous structure with uniform pore distribution. Addition of fillers (dedoped polyaniline nanofibers) resulted in improved morphology, since the fillers occupied the pores along with the plasticizers. It is observed that pore distribution on the surface of the polymer electrolytes becomes denser with the increase of filler (dedoped polyaniline nanofibers) content at first, and reaches the maximum when the weight ratio of filler is about 6 wt. % (Fig. 10d), subsequently decreases when filler content increases further {Fig. 10(e-f)}. In composite gel polymer electrolytes the porous structure gives conducting pathways for $Li^+$ movement (Arvindan & Vickraman, 2007). The high aspect ratio of nanofibers remarkably increases the pore density and widens the porous structure of the polymer electrolytes (Xi et al., 2006). The above phenomenon is possibly due to the fact that the dedoped nanofibers try to occupy the pores in the gel polymer electrolyte and in the process pore distribution becomes denser. Highly porous structure leads to better connectivity of the liquid electrolyte through the pores accounting for the increase in ionic conductivity. Highly porous surface morphology of the polymer electrolytes is effectively formed on account of the interaction of dispersed dedoped (insulating) nanofibers with polymer component as well as the affinity with solvent molecules (Kim et al., 2003). However beyond 6 wt. % of filler content the nanofibers get phase separated from the P(VdF-HFP) matrix and form insulating clusters. These phase separated insulating percolation clusters impede ion movement and hence ionic conductivity decreases. The phenomenon of phase separation is strongly supported by XRD results as discussed in section 3.4.

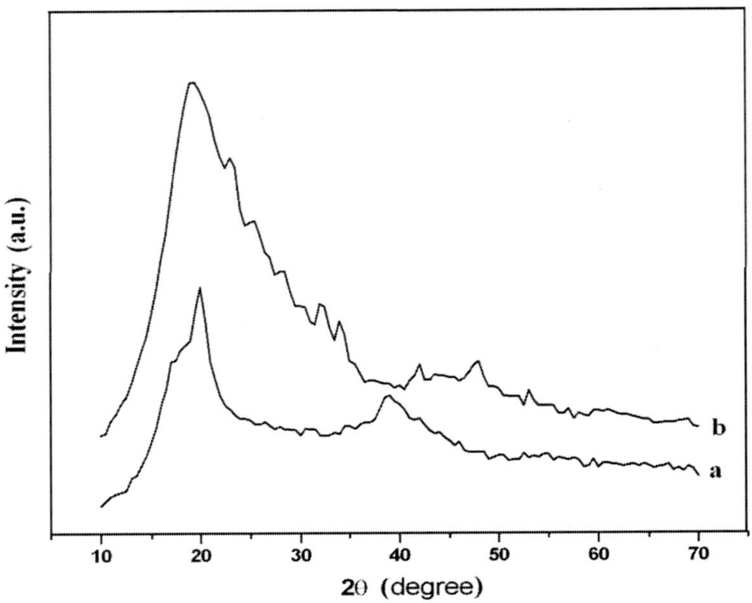


**Figure 8.** XRD patterns of pure (a) P(VdF-HFP) and (b) Dedoped polyaniline nanofibers.

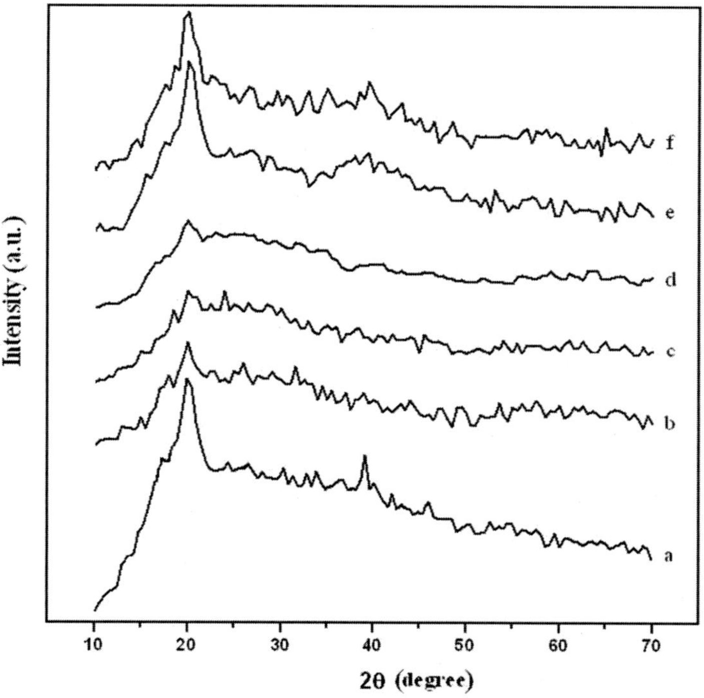


**Figure 9.** XRD patterns of P(VdF-HFP)-(PC+DEC)-$LiClO_4$-x% dedoped polyaniline nanofibers membranes (a) x=0, (b) x=2, (c) x=4, (d) x=6, (e) x=8, (f) x=10.

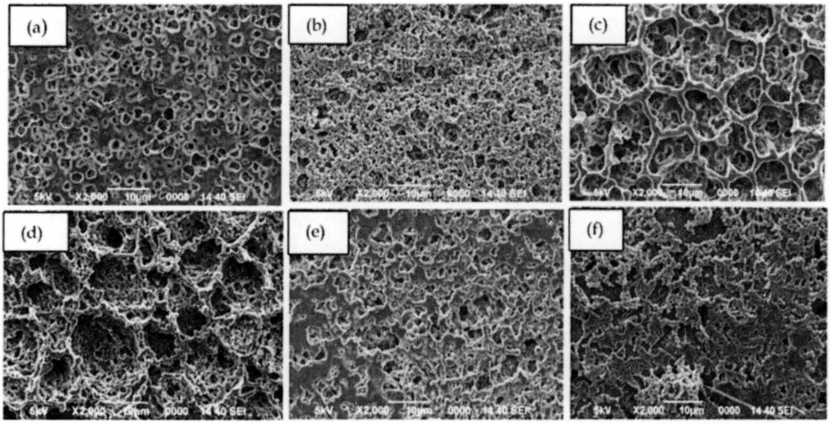


**Figure 10.** SEM micrographs of P(VdF-HFP)-(PC+DEC)-$LiClO_4$-x% dedoped polyaniline nanofibers membranes (a) x=0, (b) x=2, (c) x=4, (d) x=6, (e) x=8, (f) x=10.

### 4.2.6. Ftir Studies

FTIR is a powerful tool to characterize the chain structure of polymers and has led the way in interpreting the reactions of multifunctional monomers including rearrangements and isomerizations (Pavia et al., 2001). FTIR spectra of P(VdF-HFP), $LiClO_4$, dedoped polyaniline nanofibers and polymer complexes are shown in Fig. 11. The symmetric and asymmetric C-H stretching vibration of pure P(VdF-HFP) is observed at 3000cm$^{-1}$. Frequency 1633 cm$^{-1}$ is assigned to C=O, C=C bonding. Frequencies 1486 cm$^{-1}$ and 1404 cm$^{-1}$ are assigned to –$CH_3$ asymmetric bending and C–O stretching vibration of plasticizer propylene carbonate and diethyl carbonate. Frequencies 12861066 cm$^{-1}$ are assigned to –C–F– and –$CF_2$– stretching vibration. Frequency 881 cm$^{-1}$ is assigned to vinylidene group of polymer. The vibrational peaks of PVdF and $LiClO_4$ are shifted to (1786, 1401, 882, 837 cm$^{-1}$) and (1633, 1154, 626 cm$^{-1}$) in the polymer electrolyte respectively. The C–N stretching vibration of secondary amine in polyaniline nanofiber arises at 1289 cm$^{-1}$. The ammonium ion displays broad absorption in the frequency region 33503050 cm$^{-1}$ because of N-H stretching vibration. The N-H bending vibration of secondary aromatic amine of polyaniline nanofiber occurs at 1507 cm$^{-1}$. The frequency 1637 cm$^{-1}$ of polyaniline nanofiber is assigned to C=C of aromatic ring. As mentioned earlier the C-H symmetric and asymmetric bending frequencies are observed at 3000 cm$^{-1}$ in pure P(VdF-HFP). However, after incorporating dedoped polyaniline nanofibers the corresponding bands in the electrolyte systems show a large shift (3021 cm$^{-1}$) in higher frequency region which is a characteristic of highly disordered conformation (Porter et al., 1987). This is in good agreement with the XRD results.

# 5. PEO/P(VDF-HFP)-LICLO$_4$-DEDOPED POLYANILINE NANOFIBERS COMPOSITE SOLID POLYMER ELECTROLYTE SYSTEM:

## 5.1. Preparation

The host polymer PEO ($M_W$ = 6, 00,000), the copolymer P(VdF-HFP) ($M_W$ = 4,00,000) and salt lithium perchlorate ($LiClO_4$) were received from Aldrich, USA. All the raw materials were heated at 50 C under vacuum. Organic solvents acetonitrile and acetone were used as received from Emerk to prepare thin polymer electrolyte membranes by solution casting technique. Appropriate amount of PEO and salt $LiClO_4$ (O/Li = 8) were dissolved in acetonitrile and then mixed together, stirred and heated at 50 C. P(VdF-HFP) was fixed at 40 wt. % of

PEO for all samples, was stirred in presence of acetone at 50 C. Subsequently both the polymer solutions were mixed, stirred and heated at 50 C for 12-14 hours. Dedoped polyaniline nanofibers were then added in the blend polymer solutions and allowed to stir for another 7-8 hours. The viscous solution thus obtained was cast onto Petri dish and allowed to dry at room temperature. This procedure provided mechanically stable, free standing and flexible membranes. The blend based composite polymer electrolyte membranes used in this study were denoted as PEO-$LiClO_4$-P( VdF-HFP)-x% dedoped polyaniline nanofibers (x = 0, 5, 10, 15, 20, 25).

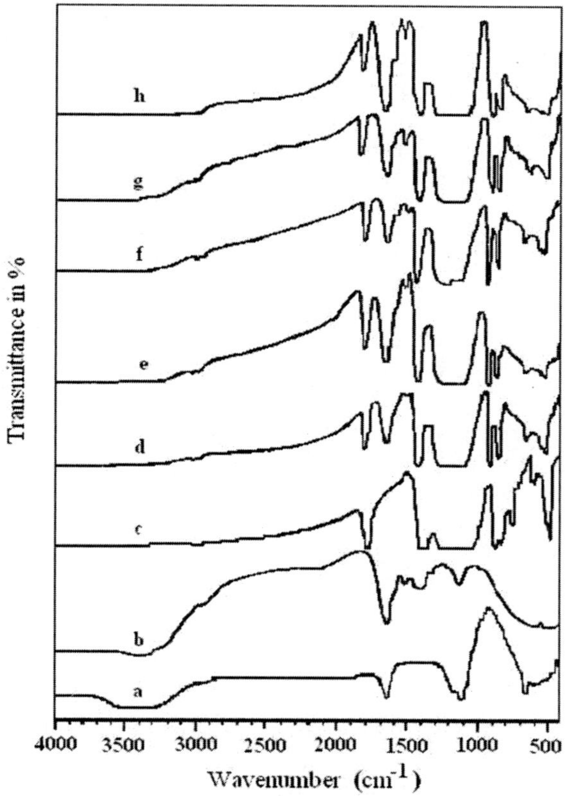


**Figure 11.** FTIR spectra of (a) $LiClO_4$, (b) dedoped polyaniline nanofibers, (c) P(VdF-HFP), (d) P(VdF-HFP)-(PC+DEC)-$LiClO_4$-2% dedoped polyaniline nanofibers, (e) P(VdF-HFP)-(PC+DEC)-$LiClO_4$-4 % dedoped polyaniline nanofibers, (f) P(VdF-HFP)-(PC+DEC)-$LiClO_4$-6 % dedoped polyaniline nanofibers, (g) P(VdF-HFP)-(PC+DEC)-$LiClO_4$-8 % dedoped polyaniline nanofibers, (h) P(VdF-HFP)-(PC+DEC)-$LiClO_4$-10 % dedoped polyaniline nanofibers composite polymer electrolyte systems.

## 5.2. Results and Discussion

### 5.2.1. X-Ray Diffraction Studies

X-ray diffraction patterns of pure PEO, P(VdF-HFP) and dedoped polyaniline nanofibers are presented in Fig. 12(a-c). High intensity peaks at 2θ=20º and 2θ=23º are observed in the XRD pattern of dedoped polyaniline nanofibers. In Fig. 12b the peaks at 2θ=20º and 38 correspond to (020) and (202) crystalline peaks of P(VdF-HFP). PEO shows a characteristic peak at 2θ=20º. Fig. 13 shows the XRD patterns of PEO-P(VdF-HFP)-$LiClO_4$-x% dedoped polyaniline nanofibers composite polymer electrolytes. It is observed that when P(VdF-HFP) is blended with PEO, no additional peak appears, only the intensity of crystalline peaks decreases suggesting that the amorphicity increases (Leo et al., 2002). When dedoped polyaniline nanofibers are incorporated in the PEO-P(VdF-HFP)-$LiClO_4$ the intensity further decreases as shown in Fig. 13(b-f). The degree of crystallinity is determined by a method described elsewhere (Saikia et al., 2006). It is observed that the degree of crystallinity decreases with increasing nanofibers concentration and reaches a minimum at 15 wt. % nanofibers concentration. This reduction in crystallinity upon addition of nanofibers is attributed to the suppression of the reorganization of polymer chains by the nanofibers (Scrosati et al., 2001). However, at higher concentration of nanofibers (>15 wt.%), the degree of crystallinity increases with increasing nanofibers concentration indicating that crystalline phase starts increasing above 15 wt.% of nanofibers concentration due to reorganization of polymer chains in PEO-P(VdF-HFP)-$LiClO_4$ electrolyte system. At 20 wt.% and 25 wt.% of nanofibers concentration an additional peak appears at 2θ=23º, which can be assigned to dedoped polyaniline nanofibers suggesting that above 15 wt.% polyaniline nanofibers get phase separated from the PEO-P(VdF-HFP)-$LiClO_4$ polymer electrolyte phase.

### 5.2.2. Ionic Conductivity Studies

The complex impedance plots for PEO-P(VdF-HFP)-$LiClO_4$ polymer electrolyte membranes with different concentration of polyaniline nanofibers are presented in Fig. 14(a-f). The variation of ionic conductivity with increasing concentration of nanofibers is shown in Fig. 15. It is observed that the $\sigma_{ionic}$ increases with increase in concentration of nanofibers. Maximum conductivity was found to be $3.1 \times 10^{-4}$ $Scm^{-1}$ at room temperature for 15 wt. % dedoped polyaniline nanofiber fillers, which is over seven times higher as compared to that ($4.5 \times 10^{-5}$ $Scm^{-1}$) for polymer electrolyte without nanofibers.

However, as the filler (dedoped nanofibers) concentration increases beyond 15 wt. %, the ionic conductivity decreases. The enhancement up to 15 wt. % of nanofibers concentration seems to be correlated with the fact that the dispersion of dedoped polyaniline nanofibers to PEO-P(VdF-HFP) prevents polymer chain reorganization due to the high aspect ratio (>50) of nanofibers, resulting in reduction in polymer crystallinity, which gives rise to an increase in ionic conductivity. The increase in ionic conductivity may also result from Lewis acid-base interaction (Rajendran & Uma, 2000, Croce et al., 2001, Chung et al., 2001, Stephan & Nahm, 2006). In the present composite polymer electrolytes, the oxygen atom in PEO has two lone pair of electrons and nitrogen atom in PAni nanofibers has one lone pair of electrons, which act as strong Lewis base centers and $Li^+$ cations as strong Lewis acid giving rise to numerous acid-base complexes in the composite polymer electrolyte. This allows mobile ions to move more freely either on the surface of the nanofibers or through a low density polymer phase at the interface, which results in enhanced ionic conductivity. The reduction in crystallinity upon addition of polyaniline nanofibers up to 15 wt. % is consistent with XRD results. Enhancement in ionic conductivity can also be attributed to the creation of polymer-filler interface. The filler-polymer interface is a site of high defect concentration providing channels for faster ionic transport (Kumar & Scanlon, 1994) and the structure and chemistry of filler-polymer interface may have even more important role than the formation of amorphous phase in the electrolyte.

On the other hand, the decrease in ionic conductivity for concentration of nanofibers higher than 15 wt. % can be attributed to the blocking effect on the transport of charge carriers resulting from the phase separation of nanofibers (Kim & Park, 2007). Besides, above 15 wt. % of nanofibers concentration a depressed semicircle is seen in the impedance spectra, which is characteristic of a system where more than one conduction processes are present simultaneously (Kurian et al., 2005). SEM micrographs show that, at higher concentration of nanofibers (15 wt. %), a two phase microstructure is observed. This could be attributed to the fact that at higher concentration of nanofibers, uniform dispersion of nanofibers in PEO-P(VdF-HFP) matrix is difficult to achieve due to formation of phase-separated morphologies. This is expected to affect the conductivity of the system, since a large concentration of $Li^+$ cations are trapped in the phase separated nanofibers. Thus the decrease of ionic conductivity above 15 wt. % nanofibers content can be attributed to the effect of phase separation, which is consistent with the XRD and SEM results. Fig. 16 shows the conductivity versus temperature inverse plots of polymer electrolyte films in the temperature range from 25 °C to 80 °C. All the samples show a break point at around 60 °C, near the melting temperature of PEO, reflecting the well-known transition from PEO crystalline to amorphous phase.

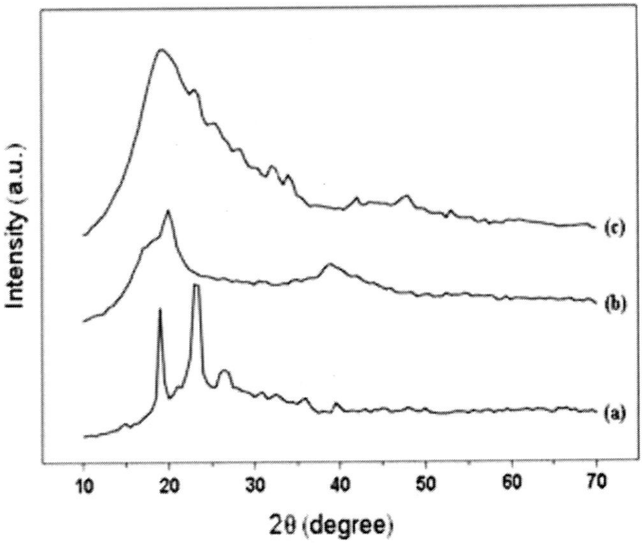


**Figure 12.** XRD patterns of (a) PEO, (b) P(VdF-HFP), (c) dedoped polyaniline nanofibers.

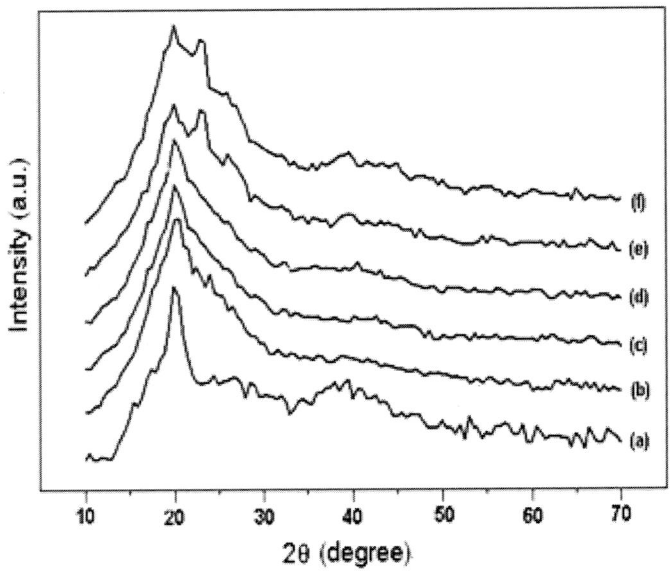


**Figure 13.** XRD patterns of PEO-P(VdF-HFP)-$LiClO_4$-x% dedoped polyaniline nanofibers polymer electrolyte membranes (a) x = 0, (b) x = 5, (c) x = 10, (d) x = 15, (e) x = 20 and (f) x = 25.

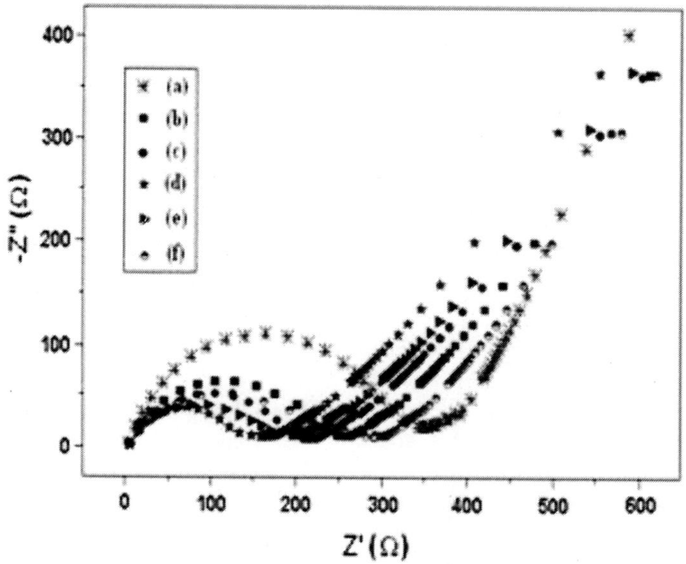


**Figure 14.** Complex impedance spectra of PEO-P(VdF-HFP)-$LiClO_4$-x% dedoped polyaniline nanofibers polymer electrolyte membranes (a) x = 0 (b) x = 5, (c) x = 10, (d) x = 15, (e) x = 20 and (f) x = 25.

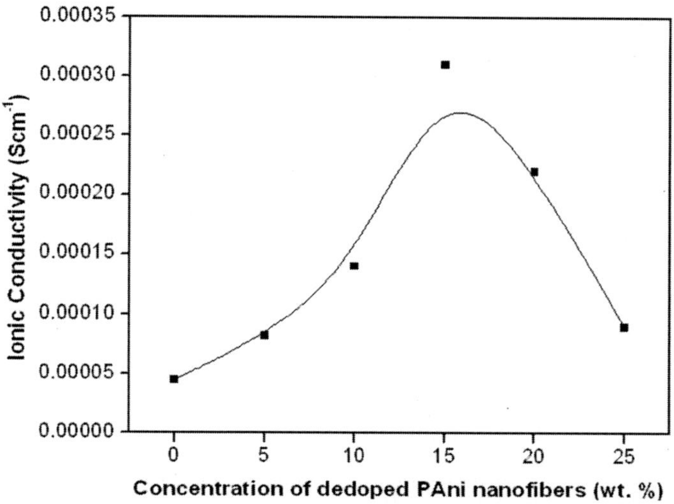


**Figure 15.** Variation of ionic conductivity with different concentration of dedoped polyaniline nanofibers.

As expected the increase in temperature leads to increase in ionic conductivity because as the temperature increases the polymer chains flex at increased rate to produce larger free volume, which leads to enhanced polymer segmental and ionic mobilities. The enhancement of ionic conductivity by the dedoped polyaniline nanofibers can be explained by the fact that the nanofibers inhibit the recrystallization kinetics, helping to retain the amorphous phase down to relatively low temperatures (Rhoo et al, 1997).

## 5.2.3. Morphological Studies

The SEM micrographs for PEO-P(VdF-HFP)-$LiClO_4$-x% dedoped polyaniline nanofibers membranes are presented in Fig. 17(a-f). In general three-four phases are known to coexist in the PEO based polymer electrolytes viz. crystalline PEO phase, crystalline PEO-Li salt complex phase and amorphous PEO phase. It is observed that below 15 wt. % nanofibers concentration (Fig. 17 a-c), the surface morphology is granular and smooth, which could be attributed to the reduction of PEO crystallinity due to complexation with lithium salt and polyaniline nanofibers. At 15 wt% concentration of nanofibers, rough morphology and sharp interfaces are observed (Fig. 17d) which may facilitate lithium ion conduction along the interface (Saikia & Kumar, 2005).

Fig. 17e shows that at 20 wt% of nanofiber concentration a two phase microstructure is observed due to phase segregation of nanofibers. Phase separation becomes more prominent at 25 wt% of nanofibers as shown in Fig. 17f. The nanofibers get phase separated from the PEO-P(VdF-HFP) polymer matrix and form domain like regions, which may act as physical barriers to the effective motion of the ions leading to decrease in ionic conductivity.

# 6. CONCLUSIONS

Stable dedoped PAni nanofiber-P(VdF-HFP) composites can be readily prepared by interfacial polymerization followed by solution casting method. The TEM result shows that polyaniline nanofibers of diameter 20-30 nm in size are formed by interfacial polymerization. The ionic conductivity of dedoped PAni nanofiber-P(VdF-HFP) based composite electrolyte was influenced by increase in PAni content. It obeys the Arrhenius Law of conductivity. The dedoped PAni nanofiber-P(VdF-HFP) based composite electrolyte has the highest ionic conductivity of $6.3\times 10^{-3}$ $Scm^{-1}$ at room temperature. The crystallinity is found to decrease in case of composites. Increase in ionic conductivity can be attributed to the fact that dispersed nanofibers improve the porous structure of the gel polymer electrolytes forming better connectivity for ion motion through the liquid electrolyte. This is confirmed by SEM results. However at higher filler content conductivity decreases because the dedoped nanofibers start forming insulating clusters that encumber ion movement. The interfacial stability of the nanofibers dispersed polymer gel electrolyte membranes is observed to be better than that of gel polymer electrolytes without nanofibers. XRD analysis reveals that amorphicity increases upon addition of nanofibers which could be attributed to the reduction in chain reorganization of polymer by nanofibers.

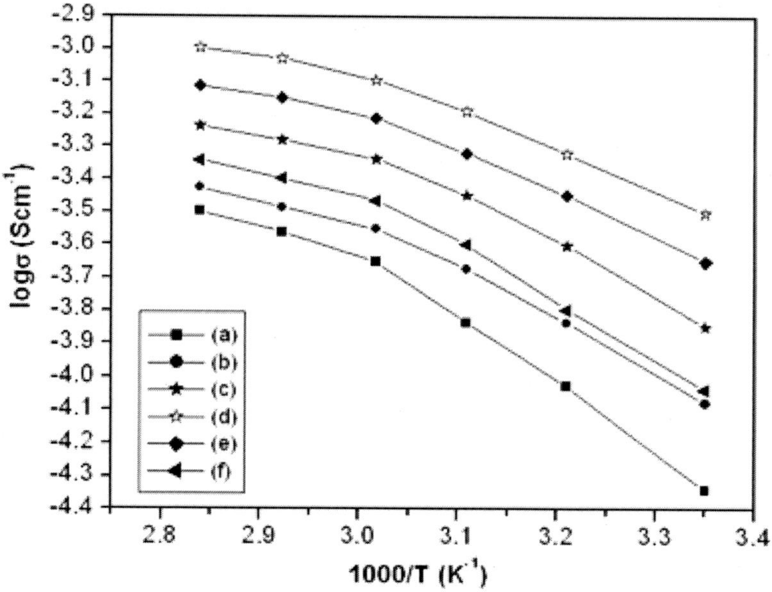


**Figure 16.** logσ vs. temperature inverse curve PEO-P(VdF-HFP)-$LiClO_4$-x% dedoped polyaniline nanofibers polymer electrolyte membranes (a) x = 0, (b) x = 5, (c) x = 10, (d) x = 15, (e) x = 20 and (f) x = 25

In the PEO-P(VdF-HFP)-$LiClO_4$ polymer electrolyte the XRD, SEM and conductivity results show that the conductivity of increases when dedoped polyaniline nanofibers are added as a filler upto a concentration of 15 wt%. At higher concentration (> 15 wt%) the polyaniline nanofibers get phase separated from the polymer matrix. The three moieties PEO, P(VdF-HFP) and dedoped polyaniline nanofiber no longer remain a miscible uniform phase but nanofibers get phase separated.

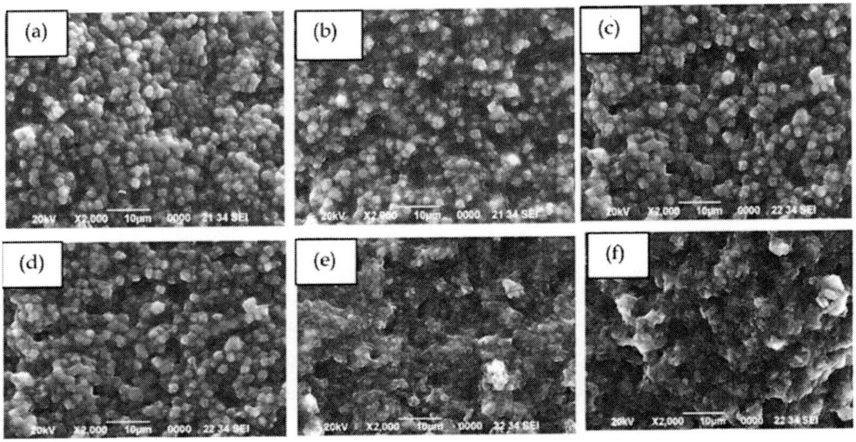


**Figure 17.** SEM micrographs of PEO-P(VdF-HFP)-$LiClO_4$-x% dedoped polyaniline nanofibers polymer electrolytes (a) x = 0, (b) x = 5, (c) x = 10, (d) x = 15, (e) x = 20 and (f) x = 25.

Both the electrolyte systems show that by using dedoped (insulating) polyaniline nanofibers as fillers the ionic conductivity can be enhanced. It is also observed that there is a certain critical concentration above which the dedoped polyaniline nanofiber phase gets separated out from the electrolyte which thereby reduces the ionic conductivity. A probable explaination for this effect can be stated as when the concentration of the dedoped nanofibers is increased agglomeration occurs, the nanofiber phase gets separated out from the electrolyte and some domain like structures are formed which is evident from the SEM images. These domain like structures caused due to the increase in concentration of the dedoped nanofibers create barriers in the conducting path and hinders the ionic movement thereby reducing the ionic conductivity.

## REFERENCES

1. S. Abbrent, J. Plestil, D.;. Hlavata, J. Lindgren, J. Tegenfeldt, A. Wendsjo, 2001 Crystallinity and morphology of PVdF-HFP-based gel electrolytes. Polymer, 42 1407 1416 .

2. K. M. Abraham, 1993 Highly conductive polymer electrolytes, In: Applications of electroactive polymers, Scrosati, B. (Ed.), 75 112 , Chapman and Hall, London.

3. J. H. ;. Ahn, G. Wang, Liu. H. K. X.;, S. X. Dou, 2003 Nanoparticle-dispersed PEO polymer electrolytes for Li batteries. J. Power Sources, 119121 422 426 .

4. M. Algamir, K. M. Abraham, 1994 Lithium batteries, new materials, developement and prospectives, In: Industrial chemistry library, 5 Pistoia, G. (Ed.), 93-136, Elsevier, Amsterdam.

5. G. Appetecchi, Crose. F. B.;, B. Scrosati, 1995 Kinetics and stability of the lithium electrode in poly(methylmethacrylate)-based gel electrolytes. Electrochim. Acta, 40 991 997 .

6. V. Aravindan, P. Vickraman, 2007 Polyvinylidenefluoride-hexafluoropropylene based nanocomposite polymer electrolytes(NCPE) complexed with LiPF3(CF3CF2)3. Eur. Polym. J. 43 5121 5127 .

7. V. Aravindan, P. Vickraman, 2008 Characterzation of SiO2 and Al2O3 incorporated PVdF-HFP based composite polymer electrolytes with LiPF3(CF3CF2)3. J.Appl. Polym. Sci., 108 1314 1322 .

8. S. Bhattarai, Bhattarai. R.;, Yi. H. N.;, Hwang. P. H. K.;, D. I. Cha, H. Y. Kim, 2004 Novel biodegradable electrospun membrane: scaffold for tissue engineering. Biomaterials, 25 2595 2602 .

9. D.;. Carrière, P.;. Barboux, F.;. Chaput, O. Spalla, J. P. Boilot, 2001 Enhanced connectivity in hybrid polymers. Solid State Ionics, 145 141 147 .

10. Chatterjee, B. L. Deopura, 2002 Carbon nanotubes and nanofibre: an overview. Fiber Polym., 3(4), 134 139 .

11. H. Chen, Chiu. C. W.;, Wu. H. Y.;, Shen. I. W. . D.;, F. C. Chang, 2002 Solid state electrolyte nanocomposites based on poly (ethylene oxide), poly (oxypropelene) diamine, mineral clay and lithium perchlorate. Polymer, 43 5011 5016 .

12. H. Choe, Giaccami. S.;, Alamgir. M. J.;, K. M. Abraham, 1995 Preparation and characterization of poly(vinyl sulfone)- and poly(vinylidene fluoride)-based electrolytes. Electrochim. Acta, 40 2289 2293 .

13. S. Chronakis, 2005 Novel nanocomposites and nanoceramics based on polymer nanofibers using electrospinning process-a review. J. Mater. Process. Technol. 167 283 293 .

14. S. Chung, Wang. H.;, Persi. Y.;, Croce. L.;, Greenbaum. S. F.;, Scrosati. B. G.;, E. Plichta, 2001 Enhancement of ion transport in polymer electrolytes by addition of nano-scale inorganic oxides. J. Power Sources 97-98, 644-648.

15. F.;. Croce, G. Appetecchi, Persi. L. B.;, B. Scrosati, 1998 Nanocomposite polymer electrolytes for lithium batteries. Nature, 394 456 458 .

16. F.;. Croce, L.;. Persi, B.;. Scrosati, F.;. Serraino-Fiory, E. Plichta, M. A. Hendrickson, 2001 Role of the ceramic fillers in enhancing the transport properties of composite polymer electrolytes. Electrochim. Acta, 46 2457 2461 .

17. W. De Heer, Bacsa. W. A.;, Chatelain. S.;, Gerfin. A.;, Humphrey. T.;-Baker, Forro. L. R.;, D. Ugarte, 1995 Aligned nanotube films: production and optical and electronic properties. Science, 268 845 847 .

18. M.;. Delvaux, J.;. Duchet, P. Stavaux, Legras. R. Y.;, S. Demoustier-Champagne, 2000 Chemical and electrochemical synthesis of polyaniline micro- and nano-tubules. Synth. Met., 113, 275280.

19. Y. Dzenis, 2004 Spinning continuous fibers for nanotechnology. Science, 304 1917 1919 .

20. B. Fenton, Parker. J. M. E.;, P. V. Wright, 1973 Complexes of alkali metal ions with poly(ethylene oxide). Polymer, 14, 589.

21. G. Feuillade, Ph. Perche, 1975 Ion-conductive macromolecular gels and membranes for solid lithium cells. J. Appl. Electrochem., 5, 63 69 .

22. S.;. Goel, N. A. Mazumdar, A. Gupta, 2009 Synthesis and characterization of polypyrrole nanofibers with different dopants. Polymers for Advanced Technologies, DOI: 10.1002/pat.1516.

23. D.;. Golodnitsky, G. Ardel, E. Peled, 2002 Ion-transport phenomena in concentrated PEO-based composite polymer electrolytes. Solid State Ion. 147 141 155 .

24. F. M. Gray, 1991 Solid polymer electrolytese fundamental and technological applications, VCH, London, NewYork.

25. X.;. He, Q.;. Shi, X.;. Zhou, C. Wan, C. Jiang, 2005 In situ composite of nano SiO2-P(VDF-HFP) porous polymer electrolytes for Li-ion batteries; Electrochim. Acta, 51 1069 1075 .

26. J. Huang, 2006 Syntheses and applications of conducting polymer polyaniline nanofibers. Pure Appl. Chem., 78 15 27 .

27. Z. Huang, Zhang. Y. M.;, Kotaki. M. Z.;, S. Ramakrishna, 2003 A review on polymer nanofibers by electrospinning and their applications in nanocomposites. Compos. Sci. Technol. 63 2223 2253 .

28. M. M. Jacob, Hackett. E. E.;, E. P. Giannelis, 2003 From nanocomposite to nanogel polymer electrolytes. J. Mater. Chem. 13 1 5 .

29. Sundaram. N. Kalyana, Vasudevan. T. T.;, A. Subramania, 2007 Synthesis of ZrO2 nanoparticles in microwave hydrolysis of Zr (IV) salt solutions-Ionic conductivity of PVdF-co-HFP-based polymer electrolyte by the inclusion of ZrO2 nanoparticles. Journal of Physics and Chemistry of Solids, 68 264 271 .

30. K. Kim, Ko. J. M.;, Park. N. M.;, Ryu. K. S. G.;, S. H. Chang, 2003 Characterization of poly(vinylidenefluoride-co-hexafluoropropylene)-based polymer electrolyte filled with rutile TiO2 nanoparticles. Solid State Ionics, 161(1-2), 121-131.

31. K.M. Kim, N.G Park, K. S. Ryu, S. H. Chang, 2002 Characterization of poly(vinylidenefluoride-co-hexafluoropropylene)-based polymer electrolyte filled with TiO2 nanoparticles. Polymer, 43 3951 3957 .

32. K. Kim, Ryu. K. M.;, Kang. S. S.;, Chang. S. H. G.;, I. J. Chung, 2001 The effect of silica addition on the properties of poly(vinylidene fluoride)-co-hexafluoropropylene)-based polymereElectrolytes. Macromol. Chem. Phys., 202(6), 866-872.

33. S. Kim, S. J. Park, 2007 Preparation and electrochemical behaviors of polymeric composite electrolytes containing mesoporous silicate fillers. Electrochim. Acta 52 3477 3484 .

34. J. Klein, Herrick. I. I. R. D.;, Palmer. D.;, Sailor. M. D.;, Brumlik. C. J. J.;, C. R. Martin, 1993 Electrochemical fabrication of cadmium chalcogenide microdiode arrays. Chem. Mater., 5 902 904 .

35. M.;. Kovac, M. Gaberscek, J. Grdadolnik, 1998 The effect of plasticizer on the microstructural and electrochemical properties of a (PEO)nLiAl(SO3 Cl)4 system. Electrochim. Acta, 44 863 870 .

36. B. Kumar, L. G. Scanlon, 1994 Polymerceramic composite electrolytes. J. Power Sources, 52 261 268 .

37. G. Kumar, Kim. G.;, Nahm. K. S. P.;, R. N. Elizabeth, 2007 Structural characterization of PVdF-HFP/ PEG/Al2O3 proton conducting membranes for fuel cells. J. Membr. Sci., 303 126 131 .

38. R.;. Kumar, A.;. Subramania, Sundaram. N. Kalyana, Vijaya. T.;, G. Kumar, I. Bhaskaran, 2007 Effect of MgO nanoparticles on ionic conductivity and electrochemical properties of nanocomposite polymer electrolyte. J. Membr. Sci., 300 104 110 .

39. M.;. Kurian, M. Galvin, Trapa. P. E.;, Sadoway. D. R. E.;, A. M. Mayes, 2005 Singleion conducting polymer-silicate nanocomposite electrolytes for lithium battery applications. Electrochim. Acta, 50 2125 2134 .

40. C. Leo, Rao. G. V. S. J.;, B. V. R. Chowwdari, 2002 Studies on plasticized PEO-lithium triflate-ceramic filler composite electrolyte system. Solid State Ionics, 148 159 171 .

41. P.;. Lobitz, H.;. Fullbier, A. Reich, K. Ambrahsat, 1992 Polymer solid electrolytes-PEO and alkali halides: A modified preparative technique. Solid State Ionics, 58 49 54 .

42. P. X. Ma, R. Zhang, 1999 Synthetic nano-scale fibrous extracellular matrix. Synthetic nano-scale fibrous extracellular matrix. J. Biomed. Mater. Res., 46 60 72 .

43. C. R. Martin, 1994 Nanomaterials: A membrane-based synthetic approach. Science 266 1961 1966 .

44. M.;. Marzantowicz, J. Dygas, Jenninger. W. R.;, I. Alig, 2005 Equivalent circuit analysis of impedance spectra of semicrystalline polymer. Solid State Ionics, 176 2115 2121 .

45. M. S. Michael, S. R. S. Prabaharan, 2004 Rechargeable lithium battery employing a new ambient temperature hybrid polymer electrolyte based on PVK+PVdF-HFP (copolymer). J. Power Sources, 136 408 415 .

46. M. Morales, C. M. Lieber, 1998 A laser ablation method for the synthesis of crystalline semiconductor nanowires. Science, 279 208 211 .

47. R. V. Pathasarathy, C. R. Martin, 1994 Synthesis of polymeric microcapsule arrays and their use for enzyme immobilization. Nature, 369 298 301 .

48. D. Pavia, Lampman. G. M. L.;, G. S. Kriz, 2001 Introduction to spectroscopy, third ed., Harcourt College Publ., USA.

49. M. Porter, Bright. T. D.;, Allara. D. L. B.;, C. E. D. Chidsey, 1987 Spontaneously organized molecular assemblies. 4. Structural characterization of n-alkyl thiol monolayers on gold by optical ellipsometry,

infrared spectroscopy, and electrochemistry. J.Am. Chem.Soc., 109 3559 3568 .

50. E.;. Quartarone, P. Mustarelli, A. Magistris, 1998 PEO-based composite polymer electrolytes. Solid State Ionics, 110 1 14 .

51. S.;. Rajendran, R. Kannan, O. Mohendran, 2001 An electrochemical investigation on PMMA/PVdF blend-based polymer electrolytes. Mater. Lett., 49 172 179 .

52. S. Rajendran, T. Uma, 2000 Conductivity studies on PVC/PMMA polymer blend electrolyte. Mater. Lett., 44 242 247 .

53. H. Rhoo, Kim. H. J.;, Park. J. M. T.;, T. S. Hwang, 1997 Ionic conduction in plasticized PVC/PMMA blend polymer electrolytes. Electrochim. Acta, 42 1571 1579 .

54. D.;. Saikia, A. M. Hussain, Kumar. P.;, Singh. F. A.;, D. K. Avasthi, 2006 Ionic conduction studies in Li 3+ ion irradiated (PVDF-HFP)-(PC+DEC)-LiCF3SO3 gel polymer electrolytes. NIMB, 244 230 234 .

55. D. Saikia, A. Kumar, 2005 Ionic transport in P(VDF-HFP)-PMMA-LiCF3SO3-(PC+DEC)-SiO2 composite gel polymer electrolyte. Eur. Polym. J., 41 563 568 .

56. B.;. Scrosati, F. Croce, S. Panero, 2001 Progress in lithium polymer battery R&D. J. Power Sources, 100 93 100 .

57. B.;. Scrosati, F. Croce, L. Persi, 2000 Impedance spectroscopy study of PEObased nanocomposite polymer electrolytes. J. Electrochim Soc, 147 1718 1721 .

58. J. Song, Wang. Y. Y. Y.;, C. C. Wan, 2000 Conductivity study of porous plasticized polymer electrolytes based on poly(vinylidene fluoride). A comparison with polypropylene separators. J. Electrochem. Soc., 147 3219 3225 .

59. M.;. Steinhart, J. Wendolff, Greiner. A. H.;, R. Wehrspohn, Nielsch. B.;, Choi. J. K.;, U. Gosele, 2002 Polymer Nanotubes by Wetting of Ordered Porous Templates. Science, 296, 1997.

60. M. Stephan, K. S. Nahm, 2006 Review on composite polymer electrolytes for lithium batteries. Polymer, 47 5952 5964 .

61. Stephan, Nahm. K. M.;, Kumar. T. S.;, Kulandainathan. M. P.;, Ravi. G. A.;, J. Wilson, 2006 Nanofiller incorporated poly(vinylidene fluoride-

hexafluoropropylene) (PVdF-HFP) composite electrolytes for lithium batteries. J. Power Sources, 159 1316 1321 .

62. Stephan, Nahm. K. M.;, Prem. S.;, T.;. Kumar, M. Kulandainathan, Ravi. G. A.;, J. Wilson, 2006 Nanofiller incorporated poly(vinylidene fluoride-hexafluoropropylene) (PVdF-HFP) composite electrolytes for lithium batteries; J. Power Sources, 159 1316 1321 .

63. Stephan, Renganathan. N. M.;, Kumar. T. G.;, Pitchumani. P.;, Shrisudersan. J. R.;, N. Muniyandi, 2000 Ionic conductivity studies on plasticized PVC/PMMA blend polymer electrolyte containing LiBF4 and LiCF3SO3. Solid State Ionics, 130 123 132 .

64. M.;. Stolarska, L.;. Niedzicki, R.;. Borkowska, A. Zalewska, W. Wieczorek, 2007 Structure, transport properties and interfacial stability of PVdF/HFP electrolytes containing modified inorganic filler. Electrochim. Acta, 53 1512 1517 .

65. T.;. Subbiah, G. Bhat, Tock. R. W. S.;, S. Parameswaran, S. S. Ramkumar, 2005 Electrospinning of nanofibers. J. Appl. Polym. Sci. 96 557 569 .

66. S.;. Sundararajan, B. Bhushan, T. Namazu, Y. Isono, 2002 Mechanical property measurements of nanoscale structures using an atomic forcemicroscope. Ultramicroscopy, 91(1-4):111-118.

67. Suthar, G. Chase, 2001 Nanofibres in filter media. Chem. Eng., 726 26 28 .

68. J. Tarascon, Gozdz. A. M.;, Schmutz. S.;, Shokoohi. F. C.;, P. C. Warren, 1994 Performance of Bellcore's plastic rechargeable Li-ion batteries. Solid State Ionics, 86 49 54 .

69. G. I. Taylor, 1969 Electrically driven jets. Proc R. Soc. London, Ser A, 313 453 475 .

70. D. Li, Y. Xia, 2004 Electrospinning of nanofibers: Reinventing the wheel?. Adv. Matter., 16 1151 1170 .

71. M. Van de Zande, Bohmer. M. I.;, Fokkink. L. G. J. R.;, C. Shonenberger, 1997 Aqueous gold sols of rod-shaped particles. J. Phys. Chem.B, 101 852 854 .

72. Z. Wang, Z. Tang, 2004 A novel polymer electrolyte based on PMAML/PVDF-HFP blend. Electrochim. Acta, 49 1063 1068 .

73. W.;. Wieczorek, J. R. Stevens, Z. Florjanczyk, 1996 Composite polymer based solid electrolytes. The Lewis acidbased approach. Solid State Ionics, 85 67 72 .

74. Wu, Lu. M. G.;, Tsai. C. C. I.;, H. J. Chuang, 2006 PVdF-HFP/ metal oxide nanocomposites: The matrices for highconducting, low-leakage porous polymer electrolytes. J. Power Sources, 159 295 300 .

75. J.;. Xi, X.;. Qiu, J.;. Li, X.;. Tang, W. Zhu, L. Chen, 2006 PVDFPEO blends based microporous polymer electrolyte: effect of PEO on pore configurations and ionic conductivity. J. Power Sources, 157 501 506 .

76. S. Zhang, Ervin. M. S.;, Xu. K. H.;, T. R. Jow, 2004 Microporous poly (acrylonitrile-methyl methacrylate) membrane as a separator of rechargeable lithium battery. Electrochim. Acta, 49 3339 3345 .

77. Y.;. Zhang, H.;. Ouyang, C. Lim, Ramakrishnan. S. T.;, Z. M. Huang, 2005 Electrospinning of gelatin fibers and gelatin/PCL composite fibrous scaffolds Electrospinning of gelatin fibers and gelatin/PCL composite fibrous scaffolds. J. Biomed. Mater. Res. Part B: Appl. Biomater., 72B, 156 165 .

# Chapter 8

# Preparation of Ag/HBP/PAN Nanofiber Web and Its Antimicrobial and Filtration Property

**Li-Rong Yao,[1,2] Xiao-Mo Song,[3] Guang-Yu Zhang,[2] Shan-Qing Xu,[2] Ye-Qun Jiang,[2] De-Hong Cheng,[1] and Yan-Hua Lu[1]**

[1]Liaoning Provincial Key Laboratory of Functional Textile Materials, Eastern Liaoning University, Dandong 118001, China

[2]School of Textile and Clothing, Nantong University, Nantong 226019, China

[3]Department of Chemical Environment and Life, Dalian University of Technology, Dalian 116024, China

## ABSTRACT

To widen the application of nanofibers web in the field of medical health materials, a new Ag/amino-terminated hyperbranched polymer (HBP)/ polyacrylonitrile (PAN) nanofiber web with excellent antimicrobial activity and filtration property was produced with Ag/HBP dispersion solution and PAN nanofiber. Ag/HBP dispersion solution was prepared with HBP as reducer and stabilizer, and Ag/HBP/PAN nanofiber was prepared by modifying electrospun PAN nanofiber with Ag/HBP aqueous solution. The characterization results showed that spherical Ag nanoparticles were prepared and they had a narrow distribution in HBP aqueous solution. The results of Ag/HBP/PAN nanofiber characterized with SEM and EDS showed that the content of silver nanoparticles on the surface of PAN nanofiber was on the increase when the treating temperature rose. The bacterial reduction rates of HBP-treated PAN nanofiber against S. aureus and E. coli were about 89%, while those of the Ag/HBP/PAN nanofiber against S. aureus and E. coli were 99.9% and 99.96%, respectively, due to the cooperative effects from the amino groups in HBP and Ag nanoparticles. Moreover, the small pores and high porosity in Ag/HBP/PAN nanofiber web resulted in high filtration efficiency (99.9%) for removing smaller particles (0.1 μm~0.7 μm), which was much higher than that of the gauze mask.

# 1. INTRODUCTION

Nanofibers, which have small pores, a great surface area, and high porosity, are widely used in filtration, Li-ion batteries, wound dressing, and other fields [1–5]. Electrospinning is a useful one-step and straightforward process for fabrication of nanofibers: one is the treatment of multiple working fluids synchronously on one spinneret to create nanofibers with complicated structures such as core-shell or multiple layers [6, 7], and the other is after treatment or transition of the electrospun nanofibers for functional applications [8].

Polyacrylonitrile (PAN) is a kind of polymer material with good stability, mechanical properties, and abundant functional cyanogroups (–CN) on its macromolecular chain. The preparation of PAN nanofibers by electrospinning has attracted wide attention, including preparation conditions of PAN nanofibers [9], PAN nanocomposite, and so forth [10–12]. The nanofiber or nanocomposite web was the useful selective layer for removing smaller particles. Compared to the microfiltration (MF), ultrafiltration (UF), and reverse osmosis (RO) membranes have porosities from 5 to 35%; the nanofiber membranes could easily be made into MF and UF membranes with porosities in range of 80 to 90% [13]. In addition, the interconnectivity of the pores eliminates the inefficiency of the dead-end pores that are commonly formed in conventional membranes. These characteristics would significantly increase flux and reduce transmembrane pressure drop.

Recently, polymer nanofibers loaded metal nanoparticles have attracted wide interest. Compared with other metals, silver has a unique catalytic performance, high conductivity, and antibacterial, so, more attention has been received. In recent years, electrospun PAN nanofibers loaded silver nanoparticles have been prepared using different methods. For example, by the silver mirror reaction (SMR) method, the silver nanoparticles were uniformly dispersed on the surface of PAN nanofibers [14]. And some researchers prepared the silver nanoparticles by electrospinning mixed solution of $AgNO_3$ and PAN, using chemical reduction method to reduce $Ag^+$ into

$Ag^0$ [15–18]. Using the ion exchange method to get silver nanoparticles from $Ag_2CO_3$, we finally got PAN composite nanofibers [19]. Atmospheric plasma treatment and high temperature were some other methods to prepare electrospun PAN nanofibers loaded silver nanoparticles [20, 21]. These PAN nanofibers loaded silver nanoparticles have excellent antibacterial and conductive property [22–25]. However, the silver nanoparticles prepared in the above research had lower stability because there were no protectors outside them, and the weak combination between silver nanoparticles and PAN nanofiber was another shortcoming.

Hyperbranched polymers (HBP) are a subfamily of dendritic polymers and can be used as effective polymers due to their unique sphere structure network, low viscosity, and functional terminal groups, such as hydroxyl, amine, epoxide, anhydride, carboxylic, and isocyanate groups, especially for amino-terminal HBP which has shown excellent adsorption to several organic polymers and metal-binding behavior [26]. HBP has an excellent dispersion and stabilization in aqueous, which can be used as a reducer and stabilizer for preparing nanoparticles [27].

In our work, HBP was synthesized and used to prepare silver nanoparticles aqueous solution, and a new Ag/HBP/PAN nanofiber web with excellent antibacterial and filtration property was successfully prepared. At first, the silver nanoparticles with a mean size of about 10 nm were prepared by mixing $AgNO_3$ and HBP solution under stirring; the nanosilver solution was used to treat PAN nanofiber web; the concentration of nanosilver solution and the treating temperature were mainly investigated, which were the key factors affecting the content of Ag nanoparticles on PAN nanofibers. Then, the antibacterial activity and filtration properties were also investigated.

# 2. EXPERIMENTAL SECTION

## 2.1. Materials

Polyacrylonitrile (PAN, white powder; Mw = 85,000), silver nitrate ($AgNO_3$), N,N-dimethylformamide (DMF), ethylenediamine, methyl acrylate, methanol, and cotton gauze were purchased from Sinopharm Chemical Reagent Co., Ltd. (China) and were of analytical grade. In all the experiments, the chemicals were of analytical grade and all the reagents were used as received without any further purification.

## 2.2. Preparation of Ag/HBP Solution

Hyperbranched polymer (HBP) was synthesized and Ag/HBP solution was prepared according to our previous work in patent [28]. Firstly, ethylenediamine (30.5 g, 0.5 mol) was added to a 500 mL three-necked round-bottomed glass flask equipped with nitrogen gas protection and magnetic stirring, and methyl acrylate (43.05 g, 0.5 mol) in methanol (100 mL) was added dropwise to the flask with a magnetic agitator and cooled with an ice bath. Then, the mixture was removed from the ice bath and left for stirring for further 6 h at room temperature. The mixture was transferred into an eggplant-shaped flask for automatic rotary vacuum evaporation, and the temperature was raised to 150°C in oil bath for 4 h under low pressure; after removing the methanol the yellow viscid HBP was obtained.

Then, 0.01 mol/L $AgNO_3$ (20 mL) solution was added dropwise to 2 g/L HBP (100 mL) solution with constant stirring at 50°C for 60 min, and light yellow semitransparent Ag/HBP hybrid was prepared.

## 2.3. Characterization of Ag/HBP

Transmission electron microscopy (TEM) images were obtained using an FEI Company (USA) Tecnai G220 transmission electron microscope operating at 300 kV accelerating voltage. Samples were prepared by evaporating the solution of nanoparticles onto a 200-mesh copper grid, which was coated with a carbon support film. The Zetasizer Nano ZS system (90Plus Zeta) from Brookhaven Company (USA) was used to test the size distribution of Ag/HBP. UV-vis spectra from 200 to 900 nm were obtained using a UV/Visible Absorption Spectrophotometry.

## 2.4. Preparation of Ag/HBP/PAN Antibacterial Nanofiber Web

The concentration of 12 wt.% PAN solution was prepared in solvent DMF, and the electrospinning was carried out by applying high voltage between the polymer solution contained in the syringe and the metallic collector; the syringe in this experiment can move from right to left. For all of the electrospinning experiments in this study, aluminum foil acceptor was placed on the collector for the deposition of fibers. The voltage used was 15 kV, the feed rate was 0.4 mL/h, and the needle collector distance was 15 cm. The thickness of nanofiber web was controlled by changing the spinning time.

The PAN nanofiber web prepared as mentioned above was immersed in the Ag/HBP solution at 30°C, 60°C, and 90°C in a water bath for 120 min, respectively; the liquor-to-fabric ratio was 50 : 1. Another sample was immersed in HBP solution at the same liquor-to-fabric ratio as a control sample. The treated samples were air-dried at room temperature for subsequent characterization.

## 2.5. Characterization of Ag/HBP/PAN Nanofiber Web

A Hitachi scanning electron microscope (model S-4800, LaB6 gun, Kevex X-ray EDS) was used to observe the morphology of the Ag nanoparticles on the surface of the samples. The diameter of PAN nanofiber was measured by image processing software (Image-Pro Plus 5.0); the Brunauer-Emmett-Teller (BET) surface area was measured at 77 K using a nitrogen adsorption-desorption Micromeritics ASAP-2020 analyzer (Micromeritics Co., USA). The porosity of PAN nanofiber web was calculated by the following equation:

$$\varepsilon = \left(1 - \frac{m}{Z \times A \times \rho}\right) \times 100\%, \tag{1}$$

where is the porosity of the relevant fibrous membrane; , , and are the mass, mean thickness, and area of per unit measured membrane, respectively. is the density of polymer raw material and the density of PAN is 1.184 g/cm$^3$. The nanofiber web was immersed in liquid nitrogen for several minutes and then cut off under liquid nitrogen. The thickness of nanofiber web was measured under optical microscope (Olympus CX31, Japan); the mean thickness was calculated after measurements for 5 times at different cross section.

## 2.6. Antimicrobial Activity of Ag/HBP/PAN Nanofiber Web

The antimicrobial activity of Ag/HBP/PAN nanofiber web was tested against E. coli and S. aureus by using a shake-flask method according to the literature [29]. The antimicrobial activity of Ag/HBP/PAN nanofiber web was tested against E. coli and S. aureus by using a shake-flask method according to FZ/T 73023-2006 (China). This method is especially designed for specimens treated with nonreleasing antibacterial agents under dynamic contact conditions. A sample of fabric (0.5 g), cut into small pieces with size of around 0.5 × 0.5 cm$^2$, was dipped into a flask containing 70 mL 0.3 mM phosphate-buffered saline

(PBS, monopotassium phosphate, pH 7.2) culture solution with cell concentration of $1 \times 10^5$–$4 \times 10^5$ colony-forming units (CFU)/mL. The flask was then shaken at 150 rpm on a rotary shaker at 24°C for 18 h. From each incubated sample, 1 mL solution was taken, diluted, and distributed onto an agar plate. All plates were incubated at 37°C for 24 h, and the colonies formed were counted. The percentage reduction was determined as follows:

$$\text{Reduction in CFU } (\%) = \frac{C - C'}{C} \times 100, \tag{2}$$

where and are the bacterial colonies of the PAN nanofiber web and the Ag-treated PAN nanofiber web, respectively.

## 2.7. Filtration Property of Ag/HBP/PAN Nanofiber Web

The Ag/HBP/PAN nanofiber web was placed between two layers of cotton gauze to form a sandwich structure; then the filtration of them and cotton gauze with 14 layers were tested. The aerosol filtration efficiency and airflow resistance were measured with a Filter Media Test System (AFC-131, TOPAS Corp., Germany). NaCl aerosol particles were prepared by an aerosol particle generator (SAG-410, TOPAS Corp., Germany). The neutralized NaCl aerosol particles were fed into a filter holder and down through the filter with an effective area of 100 $cm^2$. The diameter of sample is 11.5 cm; the test was conducted at room temperature with the continuous air flow rate at the surface of sample fixed at 5.34 cm/s.

# 3. RESULTS AND DISCUSSION

## 3.1. Preparation of Ag/HBP

The schematic illustration for fabrication of Ag/HBP aqueous solution was given in Figure 1. Hyperbranched polymer (HBP) was synthesized from ethylenediamine and methyl acrylate by melt polycondensation under vacuum. The Ag/HBP aqueous solution was synthesized by mixing $AgNO_3$ aqueous solution and HBP aqueous solution under vigorous stirring at 50°C. Abundant amino groups (primary amine, secondary amine, and quaternary amine) inside or at the terminal of HBP macromolecule have reduction performance to reduce $Ag^+$to silver nanoparticles, and $NH_2$ groups outside the silver nanoparticles play roles of both reductant and stabilizer, which can protect nanosilver particles from oxidation and provide higher security during use [30].

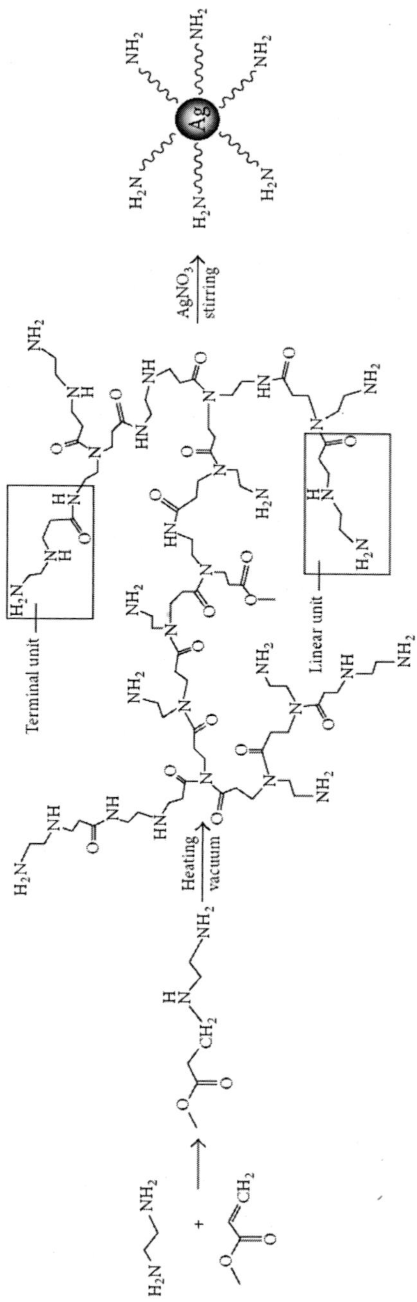


**Figure 1.** Scheme process of preparing Ag/HBP nanoparticle.

## 3.2. Characterization of Ag/HBP

The micromorphology and size distribution of the Ag nanoparticles are showed in Figure 2. The image of TEM shows that the diameter of Ag nanoparticles is about 10 nm and the nanosilvers show a standardized spherical particle and are well dispersed (Figure 2(a)), and the dates from Figure 2(b) further confirm that the Ag nanoparticles prepared have a narrow particle size distribution and the size is about 7 nm~13 nm. Results of UV-vis adsorption spectra in Figure 2(c) prove the characteristic adsorption peak of Ag/HBP at 407.09 nm, confirming the nanocrystalline character of Ag nanoparticles.

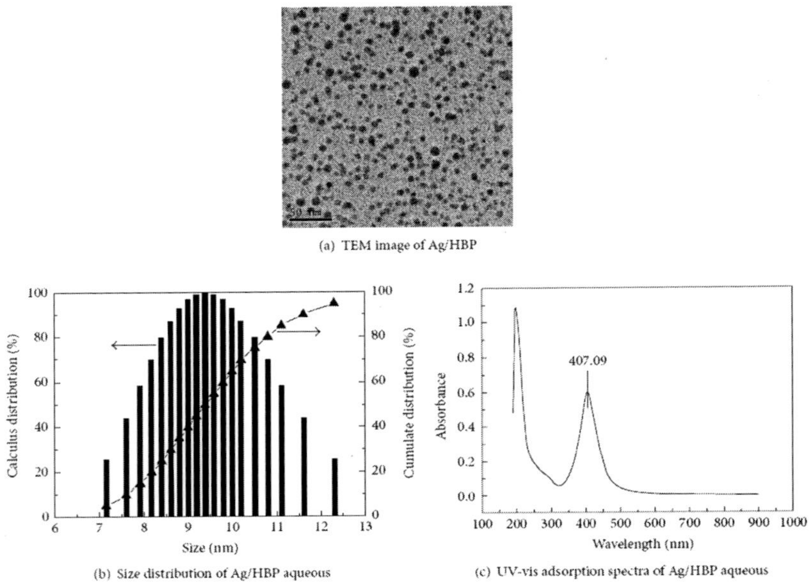


(a) TEM image of Ag/HBP

(b) Size distribution of Ag/HBP aqueous

(c) UV-vis adsorption spectra of Ag/HBP aqueous

**Figure 2:** Morphology and size distribution of Ag/HBP.

## 3.3. Preparation of Ag/HBP/PAN Nanofiber

The process of preparing Ag/HBP/PAN nanofiber web is present in Figure 3. As shown in Figure 3, three steps are needed for preparing Ag/HBP/PAN nanofiber. Firstly, the concentration of 12 wt.% PAN solution was used for electrospinning and uniform PAN nanofiber with mean diameter of 200 nm–300 nm was prepared. The second step is to synthesize Ag/HBP solution by mixing HBP and $AgNO_3$ under 50°C water bath. Then PAN nanofiber web prepared as mentioned above is immersed in the Ag/HBP solution; the concentration of Ag/HBP solution and treating temperature are investigated as the main factors affecting the fabrication of Ag/HBP/PAN nanofiber web.

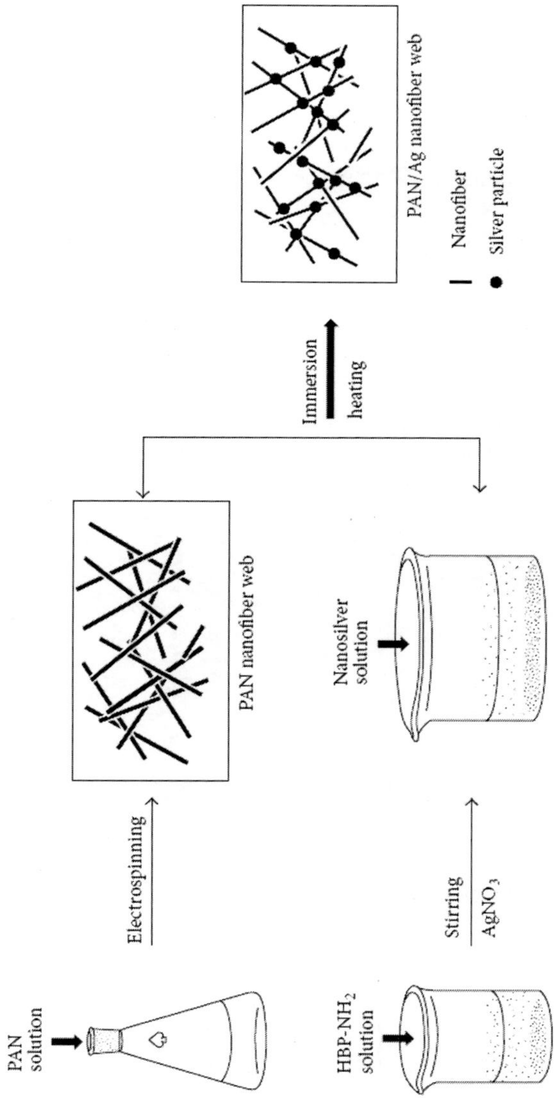


**Figure 3:** Scheme process of preparing Ag/HBP/PAN nanofiber.

## 3.4. Characterization of Ag/HBP/PAN Nanofiber

Figure 4 shows the morphology of PAN and Ag-treated PAN nanofibers. As shown in Figure 4, the low treating temperature (30°C) is used, and no significant increase of silver nanoparticles on the surface of PAN nanofibers is observed in SEM images [Figures 4(b), 4(c), and 4(d)] when the concentration of Ag/HBP solution increased from 0.02 wt.% to 0.08 wt.%, which indicated that treating temperature plays a much more important role to fix the silver nanoparticles onto the surface of PAN nanofiber.

(a) PAN nanofiber

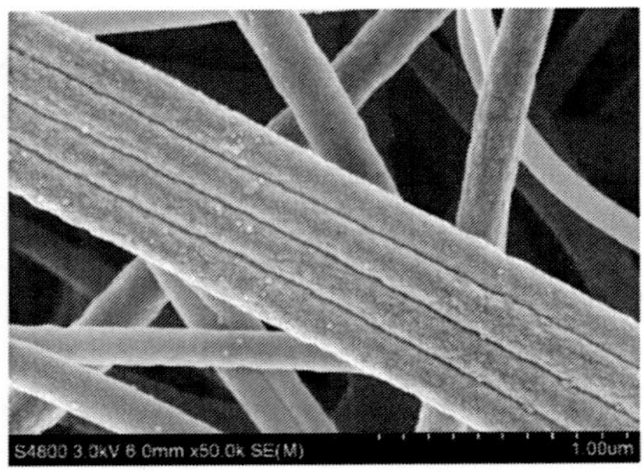


(b) Ag/HBP/PAN nanofiber, 0.02 wt.%

(c) Ag/HBP/PAN nanofiber, 0.05 wt.%

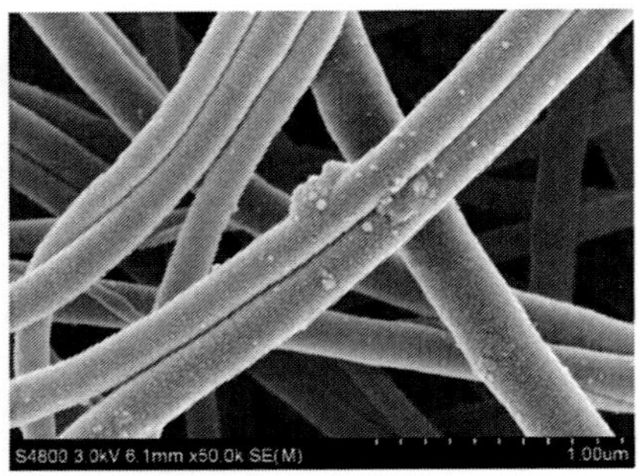


(d) Ag/HBP/PAN nanofiber, 0.08 wt.%

**Figure 4:** The morphology of Ag/HBP/PAN nanofiber (temperature of water bath is 30°C).

The SEM pictures of PAN nanofibers loaded silver nanoparticles are shown in Figure 5. From images of Figure5, no obvious changes are observed on morphology, fiber diameter, and pore size of PAN nanofiber web before and after treatment with Ag/HBP solution. However, the silver nanoparticles on the surface of PAN nanofiber have a significant increase when the processing temperature rose from 30°C to 90°C. This indicates that the assembly procedure was temperature dependent. It is believed that the increased electrostatic interaction was responsible for the improved assembly ability. Specifically, with increasing the temperature, the amino groups ($-NH_2$) capped on surface of the silver nanoparticles are highly cationized, endowing the nanoparticle surface with much higher positive charges. Thus the electrostatic interaction between negatively charged PAN nanofiber and positively charged Ag nanoparticles was strengthened. Besides, by the increase of temperature, the functional cyanogroup ($-CN$) on PAN nanofiber and the amino group ($-NH_2$) outside the silver nanoparticles have higher activity to combine with each other, and the silver nanoparticles will permeate through the pores inside nanofiber web which have more opportunity to contact with the cyanogroups on PAN nanofiber.

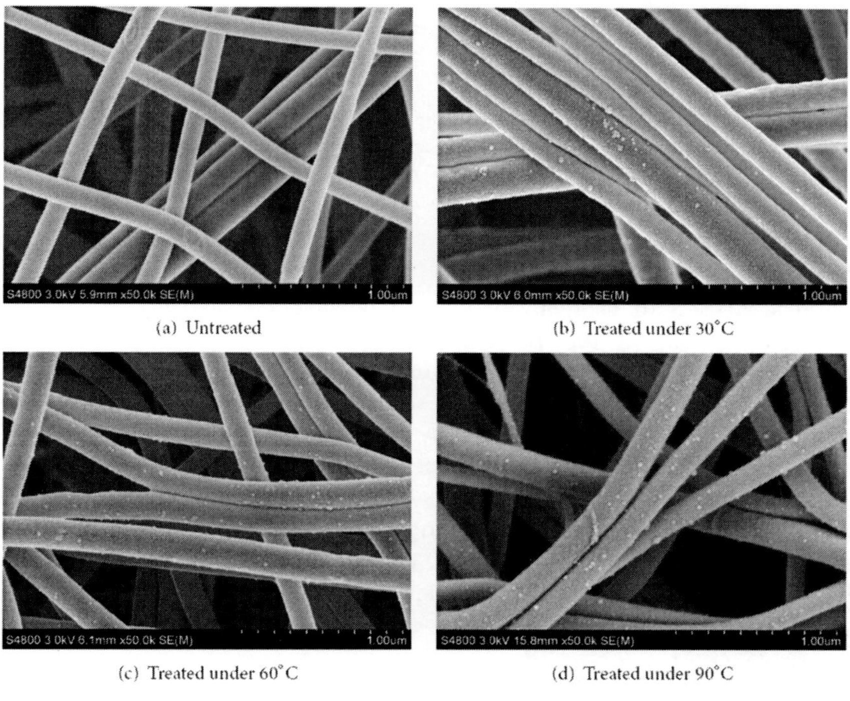


(a) Untreated (b) Treated under 30°C (c) Treated under 60°C (d) Treated under 90°C

**Figure 5:** The morphology of Ag/HBP/PAN nanofiber (concentration of silver solution is 0.05 wt.%).

The schematic diagram of hydrogen bonds formation between cyanogroup (–CN) and amino group (–$NH_2$) is shown in Scheme 1. Compared to the silver nanoparticles in situ reduced on the surface of fiber, this kind of hydrogen bonding provides stronger combining power between silver nanoparticles and PAN nanofibers. In general, there are two ways to prepare Ag-containing antibacterial materials; this means that the Ag particles are placed inside of the fiber or fixed on the surface of fiber. For the disposable hygiene products or breathing mask, the Ag particles on surface of fiber show rapid and efficient antimicrobial activity.

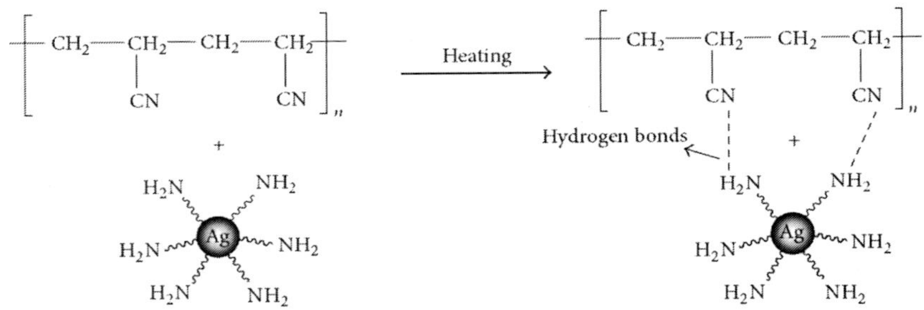


**Scheme 1:** Schematic diagram of bonding formation.

## 3.5. EDS Analysis

EDS test is used to characterize element change on the surface of PAN nanofiber and prove the reaction between Ag/HBP solution and PAN nanofibers indirectly, as shown in Figure 6. The results show that carbon, nitrogen, oxygen, and silver are the principle elements in Ag/HBP/PAN nanofiber web. The EDS analysis provides direct evidence that the silver nanoparticles are fixed on PAN nanofiber surface, which is also identified by SEM images in Figure 5. As shown in Figure 6, the silver content on PAN nanofiber was 2.45 wt.%, 8.41 wt.%, and 10.49 wt.%, respectively, when the treating temperature increased from 30°C to 90°C. The increase of Ag content on PAN nanofiber plays an important role in improving its antimicrobial activity.

## 3.6. Antimicrobial Activity

The antimicrobial activity of the Ag/HBP/PAN nanofiber web is evaluated using S. aureus and E. coli and with cotton gauze as control sample. The antimicrobial activity of PAN nanofiber, HBP-treated nanofiber, and Ag/HBP-treated nanofiber (0.05 wt.%, 90°C) is presented in Table 1.

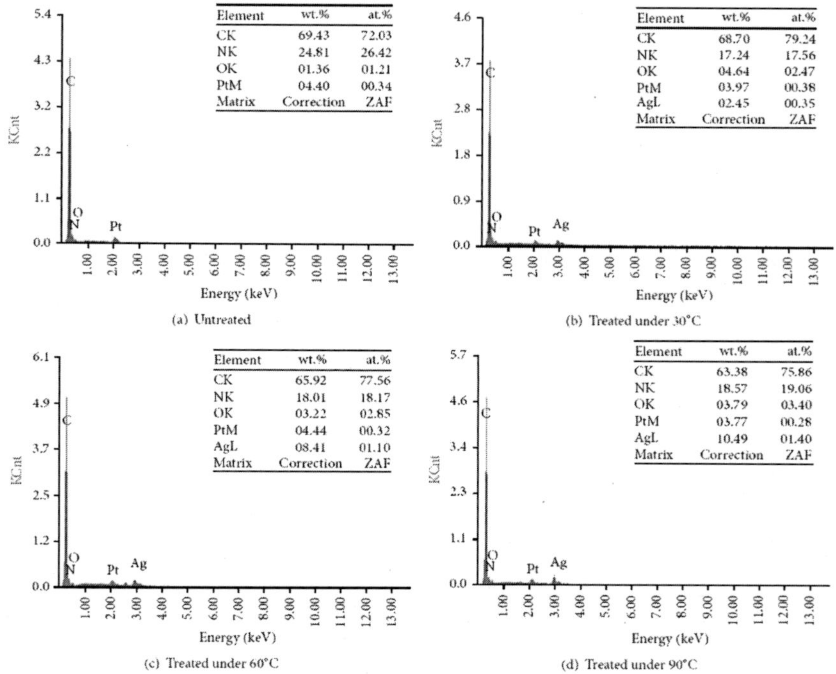


**Figure 6:** EDS of Ag/HBP/PAN nanofibers (concentration of silver solution is 0.05 wt.%).

Almost the same surviving cells shown in Table 1 indicate that PAN nanofiber shows few actions against S. aureus or E. coli. The PAN nanofiber web treated with HBP aqueous solution shows antibacterial activity and the bacterial reduction rates against S. aureus and E. coli are about 89%. The reason may be ascribed to the abundant amino groups on the PAN nanofiber, which tend to deteriorate the cell membranes of bacteria and induce leakage of intracellular components from bacterial cells, thereby inhibiting their effective growth [29]. After treatment with silver nanoparticles, the Ag/HBP/PAN nanofiber web exhibits excellent antibacterial activity and the bacterial reduction rates against S. aureus and E. coli both reached 99.9%, which may be due to the cooperative antimicrobial activity from the amino groups in HBP and Ag nanoparticles on PAN nanofibers.

**Table 1:** Antimicrobial activity of Ag/HBP/PAN nanofiber web.

| Samples | Antimicrobial activity | | | |
|---|---|---|---|---|
| | *S. aureus* | | *E. coli* | |
| | Surviving cells (CFU/mL) | % reduction | Surviving cells (CFU/mL) | % reduction |
| Cotton gauze | $1.61 \times 10^6$ | | $6.56 \times 10^6$ | |
| PAN nanofiber | $1.54 \times 10^6$ | 0 | $6.32 \times 10^6$ | 0 |
| HBP-treated nanofiber | $1.76 \times 10^5$ | 88.57 | $6.71 \times 10^5$ | 89.38 |
| Ag/HBP-treated nanofiber | $1.48 \times 10^3$ | 99.90 | $2.33 \times 10^3$ | 99.96 |

## 3.7. Filtration Property

Figure 7 shows the filtration efficiency of different filters. A lower face velocity was chosen in this paper because the filter we prepared may be used as a mask (5.34 cm/s). The dates in Figure 7 show that the filtration efficiency of cotton gauze differs enormously from nanofiber web for removing smaller particles (particle size lower than 0.7 μm), a clear U glyph filtration efficiency curve of cotton gauze is observed, the filtration efficiency of cotton gauze is lower than 80% when the particle size is smaller than 0.7 μm, and the particles about 0.4 μm are the most difficult to filter out for common filter such as gauze mask (filtration efficiency: 36.1%); however, the filtration efficiency of Ag/HBP/PAN nanofiber web easily reached 99.9%. The excellent filtration efficiency of nanofiber web is mainly due to the ultrafine fiber (about 180 nm), small pore diameter (less than 1 μm), high porosity (86.4%), and BET (25.3 $m^2/g$) (shown in Table 2). The thickness of nanofiber web was controlled by spinning time in our experiment, which almost has no effects on its filtration efficiency.

The pressure drop of the filters is shown in Figure 8. Clearly, higher pressure drop is observed for Ag/HBP/PAN nanofiber web. We know that small pore diameter, high porosity, and thickness will improve the filtration efficiency of filters. However, the factors mentioned above are also the key reasons for increasing the pressure drop. Therefore, smaller thickness of nanofiber web is necessary for decreasing the pressure drop when its mechanical property is enough during use. The pressure drop of Ag/HBP/PAN nanofiber web is a little higher than the PAN nanofiber web, which may be attributed to the pore change in nanofiber web during the process of Ag-containing treatment.

**Table 2:** Characterization of nanofiber web.

| Samples | Spinning time (min) | Thickness (μm) | Fiber diameter (nm) | BET ($m^2/g$) | Porosity (%) |
|---|---|---|---|---|---|
| PAN | 20 | 88.5 | 176.4 ± 42.5 | 26.7 | 85.9 |
| Ag/HBP/PAN-1 | 20 | 81.2 | 188.1 ± 31.6 | 25.3 | 86.4 |
| Ag/HBP/PAN-2 | 40 | 153.9 | 181.7 ± 26.9 | 24.4 | 84.8 |
| Ag/HBP/PAN-3 | 60 | 230.6 | 184.4 ± 30.3 | 25.7 | 85.5 |

# 4. CONCLUSIONS

The amino-terminated hyperbranched polymer (HBP) was synthesized in this paper at first; then the HBP was used as reductant and stabilizer to prepare Ag/HBP aqueous solution. The silver nanoparticles were prepared with a mean diameter of about 10 nm and they have an outstanding disperse ability and stability in aqueous solution. The uniform PAN nanofiber web was prepared by electrospinning and treated with Ag/HBP solution under water bath. The SEM and EDS showed that it was difficult for silver nanoparticles to be fixed on the surface of PAN nanofiber under low temperature (30°C) and the content of silver nanoparticles on PAN nanofiber increased from 2.45 wt.% to 10.84 wt.% when temperature rose to 90°C. The antibacterial activity and the bacterial reduction rates of S. aureus and E. coli of HBP-treated PAN nanofiber are about 89%; however, the Ag/HBP/PAN nanofiber showed better antibacterial property and it had 99.9% bacterial reduction ofStaphylococcus aureus and 99.96% bacterial reduction of Escherichia coli, respectively, which may be due to the cooperative effects from the amino groups in HBP and Ag nanoparticles. The filtration efficiency of cotton gauze differs enormously from nanofiber web for removing smaller particles (particle size lower than 0.7 µm), a clear U glyph filtration efficiency curve of cotton gauze was observed, and the filtration efficiency of gauze mask (14 layers) was lower than 80% and as low as 36.1% for the particles of 0.4 µm; however, the filtration efficiency of PAN nanofiber web easily reached 99.9%. Compared to lower pressure drop of gauze mask (about 20 Pa), a further cut in pressure drop of Ag-containing nanofiber web is the primary mission in our future work.

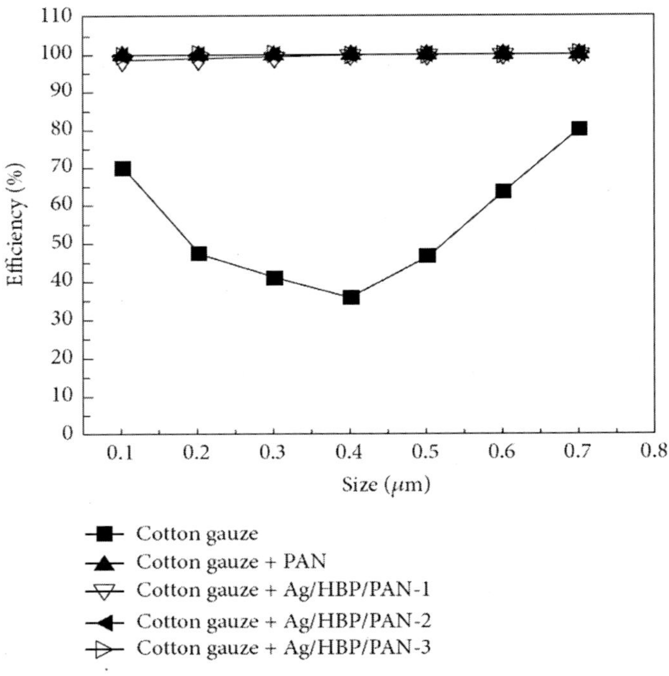


**Figure 7:** Filtration efficiency of filters.

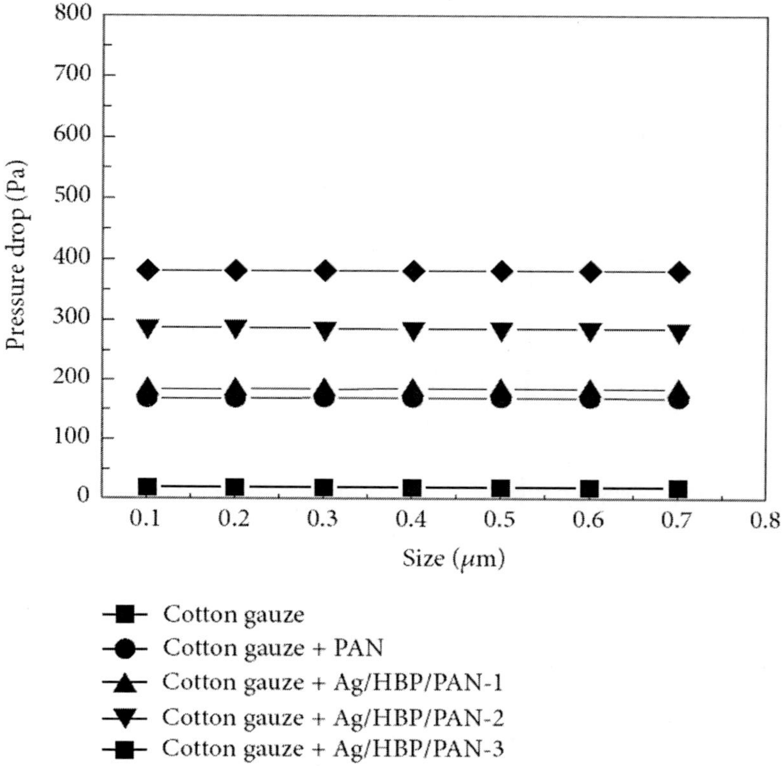


**Figure 8.** Pressure drop of filters.

## COMPETING INTERESTS

The authors declare that there are no competing interests.

## ACKNOWLEDGMENTS

This work was supported by Liaoning Provincial Key Laboratory of Functional Textile Materials, National Natural Science Fund (51503105 and 51343002), College Students Innovation and Entrepreneurship Training Program in Jiangsu Province (201410304028Z), and Liaoning BaiQianWan Talents Program.

# REFERENCES

1. S. M. S. Shahabadi, A. Kheradmand, V. Montazeri, and H. Ziaee, "Effects of process and ambient parameters on diameter and morphology of electrospun polyacrylonitrile nanofibers," Polymer Science Series A, vol. 57, no. 2, pp. 155–167, 2015.

2. M. Yanilmaz and X.-W. Zhang, "Polymethylmethacrylate/polyacrylonitrile membranes via centrifugal spinning as separator in Li-ion batteries," Polymers, vol. 7, no. 4, pp. 629–643, 2015.

3. M. A. Al-Omair, "Synthesis of antibacterial silver-poly(ε-caprolactone)-methacrylic acid graft copolymer nanofibers and their evaluation as potential wound dressing," Polymers, vol. 7, no. 8, pp. 1464–1475, 2015.

4. J.-J. Xu, B.-H. Xu, D. Shou, X.-J. Xia, and Y. Hu, "Preparation and evaluation of vancomycin-loaded N-trimethyl chitosan nanoparticles," Polymers, vol. 7, no. 9, pp. 1850–1870, 2015.

5. V. Mottaghitalab and A. K. Haghi, "A study on electrospinning of polyacrylonitrile nanofibers," Korean Journal of Chemical Engineering, vol. 28, no. 1, pp. 114–118, 2011.

6. D.-G. Yu, X.-Y. Li, X. Wang, J.-H. Yang, S. W. Annie Bligh, and G. R. Williams, "Nanofibers fabricated using triaxial electrospinning as zero order drug delivery systems," ACS Applied Materials & Interfaces, vol. 7, no. 33, pp. 18891–18897, 2015.

7. D.-G. Yu, C. Yang, M. Jin et al., "Medicated Janus fibers fabricated using a Teflon-coated side-by-side spinneret," Colloids and Surfaces B: Biointerfaces, vol. 138, pp. 110–116, 2016.

8. D.-G. Yu, K. White, N. Chatterton, Y. Li, L. Li, and X. Wang, "Structural lipid nanoparticles self-assembled from electrospun core-shell polymeric nanocomposites," RSC Advances, vol. 5, no. 13, pp. 9462–9466, 2015.

9. Heidari, M. M. Mashhadi, and G. Faraji, "A novel approach for preparation of aligned electrospun polyacrylonitrile nanofibers," Chemical Physics Letters, vol. 590, pp. 231–234, 2013.

10. L. W. Ji, A. J. Medford, and X.-W. Zhang, "Effects of process and ambient parameters on diameter and morphology of electrospun polyacrylonitrile nanofibers," Polymer, vol. 50, no. 2, pp. 605–612, 2009.

11. R. Nirmala, K. Jeon, R. Navamathavan, B.-S. Kim, M.-S. Khil, and H. Y. Kim, "Fabrication and characterization of II-VI semiconductor nanoparticles decorated electrospun polyacrylonitrile nanofibers," Journal of Colloid and Interface Science, vol. 397, pp. 65–72, 2013. ·

12. Y. Tong, X.-F. Lu, W.-N. Sun, G. Nie, L. Yang, and C. Wang, "Electrospun polyacrylonitrile nanofibers supported Ag/Pd nanoparticles for hydrogen generation from the hydrolysis of ammonia borane,"Journal of Power Sources, vol. 261, pp. 221–226, 2014. ·

13. K. Yoon, K. Kim, X. Wang, D. Fang, B. S. Hsiao, and B. Chu, "High flux ultrafiltration membranes based on electrospun nanofibrous PAN scaffolds and chitosan coating," Polymer, vol. 47, no. 7, pp. 2434–2441, 2006. ·

14. Y.-Z. Shi, Y.-J. Li, J.-F. Zhang, Z. Yu, and D. Yang, "Electrospun polyacrylonitrile nanofibers loaded with silver nanoparticles by silver mirror reaction," Materials Science and Engineering C, vol. 51, pp. 346–355, 2015.

15. C. Q. Zhang, Q. B. Yang, N. Q. Zhan et al., "Silver nanoparticles grown on the surface of PAN nanofiber: preparation, characterization and catalytic performance," Colloids and Surfaces A: Physicochemical and Engineering Aspects, vol. 362, no. 1–3, pp. 58–64, 2010. ·

16. D.-Y. Lee, K.-H. Lee, B.-Y. Kim, and N.-I. Cho, "Silver nanoparticles dispersed in electrospun polyacrylonitrile nanofibers via chemical reduction," Journal of Sol-Gel Science and Technology, vol. 54, no. 1, pp. 63–68, 2010. ·

17. J. Bai, Q. B. Yang, S. Wang, and Y. Li, "Preparation and characterization of electrospun Ag/polyacrylonitrile composite nanofibers," Korean Journal of Chemical Engineering, vol. 28, no. 8, pp. 1761–1763, 2011. ·

18. M. A. Kudryashov, A. I. Mashin, A. S. Tyurin, A. E. Fedosov, G. Chidichimo, and G. de Filpo, "Morphology of a silver/polyacrylonitrile nanocomposite," Technical Physics, vol. 56, no. 1, pp. 92–96, 2011. ·

19. G. Panthi, S.-J. Park, T. W. Kim et al., "Electrospun composite nanofibers of polyacrylonitrile and Ag2CO3 nanoparticles for visible light photocatalysis and antibacterial applications," Journal of Materials Science, vol. 50, no. 13, pp. 4477–4485, 2015.

20. Q. Shi, N. Vitchuli, J. Nowak et al., "Durable antibacterial Ag/polyacrylonitrile (Ag/PAN) hybrid nanofibers prepared by atmospheric plasma treatment and electrospinning," European Polymer Journal, vol. 47, no. 7, pp. 1402–1409, 2011. ·

21. Mahapatra, N. Garg, B. P. Nayak, B. G. Mishra, and G. Hota, "Studies on the synthesis of electrospun PAN-Ag composite nanofibers for antibacterial application," Journal of Applied Polymer Science, vol. 124, no. 2, pp. 1178–1185, 2012. ·

22. D.-G. Yu, C. J. Branford-White, N. P. Chatterton et al., "Electrospinning of concentrated polymer solutions," Macromolecules, vol. 43, no. 24, pp. 10743–10746, 2010. ·

23. Tyurin, G. de Filpo, D. Cupelli, F. P. Nicoletta, A. Mashin, and G. Chidichimo, "Particle size tuning in silver-polyacrylonitrile nanocomposites," Express Polymer Letters, vol. 4, no. 2, pp. 71–78, 2010. ·

24. Karbownik, O. Rac, M. Fiedot, P. Suchorska-Woźniak, and H. Teterycz, "In situ preparation of silver–polyacrylonitrile nanocomposite fibres," European Polymer Journal, vol. 69, pp. 385–395, 2015.

25. M. A. Kudryashov, A. I. Mashin, A. S. Tyurin, G. Chidichimo, and G. De Filpo, "Metal-polymer composite films based on polyacrylonitrile and silver nanoparticles preparation and properties," Journal of Surface Investigation, vol. 4, no. 3, pp. 437–441, 2010. ·

26. H. Dodiuk-Kenig, K. Lizenboim, S. Roth et al., "Performance enhancement of dental composites using electrospun nanofibers," Journal of Nanomaterials, vol. 2008, Article ID 840254, 6 pages, 2008. ·

27. W.-W. Zhang, D.-S. Zhang, Y.-Y. Chen, and H. Lin, "Hyperbranched polymer functional TiO2nanoparticles: synthesis and its application for the anti-UV finishing of silk fabric," Fibers and Polymers, vol. 16, no. 3, pp. 503–509, 2015.

28. F. Zhang and L. R. Yao, "Preparation of nano-silver solution," China Patent, ZL 201010166358.3, 2011.

29. G. Zhang, Y. Liu, H. Morikawa, and Y. Chen, "Application of ZnO nanoparticles to enhance the antimicrobial activity and ultraviolet protective property of bamboo pulp fabric," Cellulose, vol. 20, no. 4, pp. 1877–1884, 2013. ·

30. E. S. Abdel-Halim and S. S. Al-Deyab, "Utilization of hydroxypropyl cellulose for green and efficient synthesis of silver nanoparticles," Carbohydrate Polymers, vol. 86, no. 4, pp. 1615–1622, 2011. ·

# Chapter 9

# A Review of Electrospun Conductive Polyaniline Based Nanofiber Composites and Blends: Processing Features, Applications, and Future Directions

**Saiful Izwan Abd Razak,[1,2] Izzati Fatimah Wahab,[2] Fatirah Fadil,[3] Farah Nuruljannah Dahli,[4] Ahmad Zahran Md Khudzari,[1,2] and Hassan Adeli[5]**

[1]IJN-UTM Cardiovascular Engineering Centre, Institute of Human Centered Engineering, Universiti Teknologi Malaysia, 81310 Skudai, Johor, Malaysia

[2]Faculty of Bioscience and Medical Engineering, Universiti Teknologi Malaysia, 81310 Skudai, Johor, Malaysia

[3]Department of Chemistry, Faculty of Science, Universiti Teknologi Malaysia, 81310 Skudai, Johor, Malaysia

[4]Department of Bioprocess and Polymer Engineering, Faculty of Chemical and Energy Engineering, Universiti Tun Hussein Onn Malaysia, 86400 Batu Pahat, Johor, Malaysia

[5]Department of Chemical Engineering, University of Mazandaran, Babolsar, Iran

## ABSTRACT

Electrospun polymer nanofibers with high surface area to volume ratio and tunable characteristic are formed through the application of strong electrostatic field. Electrospinning has been identified as a straight forward and viable technique to produce nanofibers from polymer solution as their initial precursor. These nanofiber materials have attracted attention of researchers due to their enhanced and exceptional nanostructural characteristics. Electrospun polyaniline (PANI) based nanofiber is one of the important new materials for the rapidly growing technology development such as nanofiber based sensor devices, conductive tissue engineering scaffold materials, supercapacitors, and flexible solar cells applications. PANI however is relatively hard to process compared to that of other conventional polymers and plastics. The processing of PANI is daunting, mainly due to its rigid backbone which is related to its high level of conjugation. The challenges faced in the electrospinning processing of neat PANI have alternatively led to the development of the electrospun PANI based composites and blends. A review on the research activities of the electrospinning processing of the PANI based nanofibers, the potential prospect in various fields, and their future direction are presented.

# 1. INTRODUCTION

Nanostructures, structures of 100 nanometers (nm) or less in size, have been at the forefront of much modern engineering and science research due to their novel, significantly improved biological, chemical, and physical properties. There are mainly four types of nanostructures: zero- (0D) (e.g., nanoparticles), one- (1D) (e.g., nanowires, nanotubes, and nanofibers), two- (2D) (e.g., films and layers), and three- (3D) dimensional structures (e.g., polycrystals). Among them, one-dimensional (1D) nanostructures have been in the focus of quite extensive studies worldwide, because they represent the smallest dimensional structures with high aspect ratios and degrees of flexibility. Hence, these 1D nanostructures can be exploited as elements in nanodevices with various applications, such as supercapacitors, chemicals, and biosensors, in optoelectronics and tissue engineering [1–7].

Nanoapplication and nanotechnology development continues to stir the scientists' and engineers' interest in exploring the best combination of materials and 1D nanostructures. A few examples of such combinations are carbon nanotubes, inorganic semiconductor nanowire/nanobelts, metallic nanotubes/nanowires, and conjugated polymer nanofibers/nanotubes. The conjugated polymer 1D nanostructures possess certain highly attractive characteristics, such as an easily controllable bandgap, high mechanical flexibility, and greater biocompatibility than that of many inorganic materials. In addition, conjugated polymers are also called conducting polymers; in their neutral states their conductivity typically ranges from $10^{-10}$ to $10^{-5}$ $Scm^{-1}$. However, the conductivity can be enhanced into semiconductor or conductor states via chemical or electrochemical redox reactions. Among the various classes of conducting polymers, polyaniline (PANI) is one of the most widely investigated materials due to its easy synthesis, excellent optical and magnetic properties, and environmental stability. Furthermore, the electrical properties can be controlled by the oxidation and protonation state [8–10].

Up to now, a large number of approaches have already been demonstrated for preparing PANI 1D nanostructures, including chemical routes, such as hard physical template-guided synthesis and soft chemical template synthesis (interfacial polymerization, template-free method, dilute polymerization, and reverse emulsion polymerization), and a variety of lithography techniques [11]. However, the drawbacks of these techniques, such as required postsynthesis process, relatively poor control of the size and morphology uniformity, poorly oriented nanostructure arrays, and high costs, may limit their production on a large scale. Thus, physical approaches such as electrospinning are more advantageous compared to the above-mentioned techniques, especially for the mass production of continuous nanofibers.

## 2. ELECTROSPINNING THEORY AND PRINCIPLE

Electrospinning is a fiber-spinning technology used to produce long, three-dimensional, ultrafine fibers [12]. According to Frenot and Chronakis, nanofibers produced by the electrospinning technique have diameters in the range of 100 nm to 1 µm and lengths of up to kilometers [13]. Electrospun nanofibers present high attractiveness for diverse applications, such as in biomedical engineering and tissue regeneration [14], sensory and electronic devices [15, 16], smart textiles [17], and membrane filtration technology [18]. The viability of this technology is evidenced by the easiness of the spinning procedure and the simplicity in the custom setting of the electrospinning machine [19]. The most amazing characteristics of electrospun nanofibers include high surface area-to-volume ratio of well-interconnected nanofiber webs, flexibility in surface functionalities, tuneable surface properties, high permeability, and good mechanical performance [20].

The principle of the electrospinning technique was first described by Zeleny in his published work on the behavior of fluid droplets at the end of metal capillaries [12, 21]. Later on, in 1934, Formhals established the commercialized system of electrospinning for the fabrication of textile yarns in the U.S. Patent number 195704 [22]. Further development of the electrospinning from a polymer melt rather than a polymer solution using an air-blast to assist the fiber formation was patented by Norton in 1936 [23]. The conceptual principle of electrospinning is based on the employment of the external electric field to induce polymer fluid transformation by the elongation and whipping of the jet. In 1969, Taylor introduced the concept of Taylor cone, which is associated with the deformation of a liquid surface into a characteristic shape induced by electric fields. The theoretical analysis of the disintegration of liquid drops under electric fields, the calculation of the electric force acting on fine jets, of the critical voltage for inducing jet elongation, and of the voltage values at which Taylor cones are formed have been described in detail elsewhere [24]. Recently, the electrospinning process has gained huge popularity and the publication of works related to electrospinning increased over the years. A survey of the publications related to electrospun PANI nanofibers in the past 10 years is given in Figure1(a), whereas the distribution of these published works among the 15 top countries is shown in Figure 1(b). These literature data were obtained based on the Scopus search system.

The data clearly demonstrate the progression of research works associated with PANI nanofibers obtained by means of electrospinning. The increase in the number of patents for diverse applications gives an insight into the recognition of the electrospinning process. A few companies, such as Espin Technologies, Nano Technics, and KATO Tech, are constantly benefited by the utmost features of nanofibers derived by electrospinning, while companies such as Donaldson Company and Freudenberg have already applied the electrospinning process in their air filtration products [25, 26].

In this report, a comprehensive review is made on the research and developments related to PANI electrospun nanofibers, including a discussion of their processing characteristics, physical and mechanical properties, and applications. Other basic issues regarding the processing limitations of the materials and research challenges are also considered in detail.

## 3. TECHNICAL ASPECTS IN ELECTROSPINNING

From the technical point of view, the electrospinning equipment is an inexpensive, robust, and straight forward nanofiber production machine. The standard laboratory electrospinning unit generally consists of a metallic capillary needle, which is connected to a high voltage power supply and a grounded collector. The droplet of the spinning solution is discharged from the capillary needle in a continuous flow and the feed rate of the discharged droplet is adjusted using an infusion syringe pump. The electric field is applied at the tip of the capillary needle attached to the syringe, which is loaded with a spinning polymer solution, to induce an electric charge on the surface of the hanging droplet. The hanging droplet of the spinning solution, which is held by its surface tension, will then develop mutual charge repulsion due to excess pressure. The contraction of the surface charges to the counter electrode is caused by a force directly opposite to the surface tension of the liquid droplet [27].

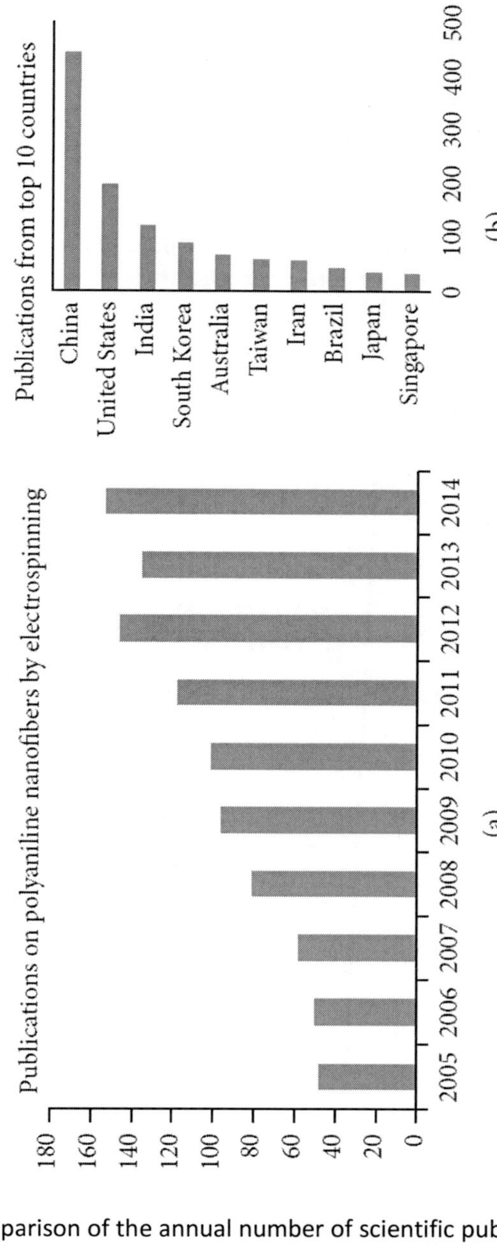


**Figure 1:** (a) Comparison of the annual number of scientific publications related to electrospinning polyaniline nanofibers. (b) Publication distribution around the world based on data analysis carried out using the Scopus search system with the term "polyaniline nanofibers" as of 30 March 2015.

The hemispherical droplet gets distorted into a conical shape, also called Taylor cone, under the simultaneous actions of the repulsive force. Additional increases in the electric field intensity will lead to reaching a critical value, whereby the repulsive electrostatic force overcomes the surface tension of the charged droplet to initiate the polymer jet formation. The initiation of the polymer jet then undergoes an instability and elongation process simultaneously with the evaporation of the solvent.

A grounded collector, which receives the deposition of the electrospun nanofibers from the charged jets, is placed at a certain distance from the metallic capillary needle [20]. A schematic diagram of the electrospinning system with the basic setup of a syringe pump attached to a disposable syringe with a metallic needle, a high voltage power supply unit, and a collector is Figure 2.

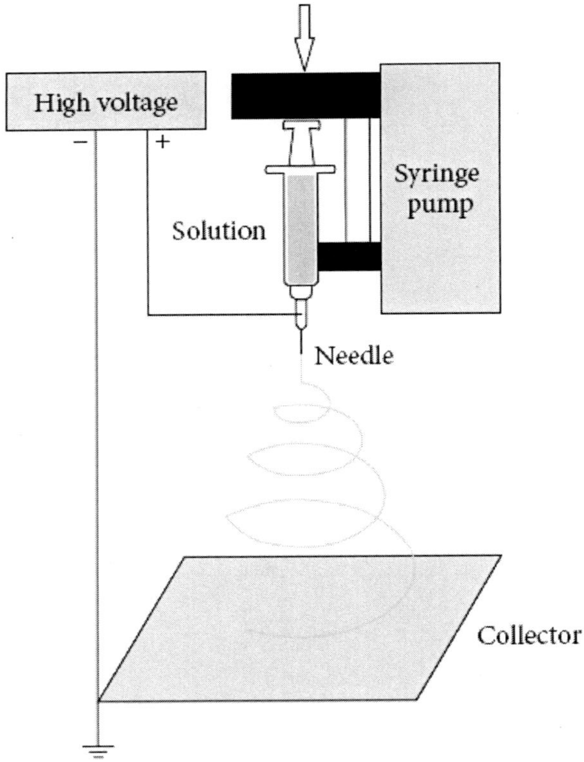


**Figure 2:** Schematic diagram of an electrospinning setup.

Multiple aspects of the electrospinning process have been demonstrated to affect the morphology of the nanofibers [27]. The major factors involved in the electrospinning process, including the polymer solutions, needles, and collectors, are discussed below [20].

## 3.1. Polymer Solutions

When dealing with the spinning solution of various polymers, the spinnability of the polymer solution becomes an utmost issue. The spinnability of a solution depends on several aspects during the electrospinning process [28]. These parameters can be classified into major and minor influences. Major parameters comprise the viscosity, elasticity, and surface tension of the solution, the hydrostatic pressure of the solution in the capillary needle, the electric potential, and the distance from the capillary tip to the collector, while ambient influences, such as solution temperature, humidity, and air velocity, can be classified as minor parameters [29].

The difference in the viscosities and surface tension of the spinning solution will directly influence the morphology of the nanofibers formed either as an ultrathin or as beaded fibrous structure. A spinning solution with viscosity values of 1–20 poises and surface tension values of 35–55 dynes/cm$^2$ is regarded as a suitable solution for inducing fiber formation. Meanwhile, at viscosities of above 20 poises, the electrospinning is hindered because of the flow instability caused by the high cohesiveness of the solution. As the viscosity of the solution is increased, the beads become larger, the average distance between beads is longer, and enlarged fiber diameters are observed. The formation of beaded fibers occurs because the fluid jet breaks up to droplets when the viscosity of the spinning solution is below 1 poise [30].

The need for a volatile solvent as carrier of a particular polymer in solution spinning is one of the fundamental criteria in electrospinning. Solvent selection is of paramount importance in determining the critical minimum solution concentration to allow the transition from electrospraying to electrospinning, thereby significantly affecting solution spinnability and the morphology of the electrospun nanofibers [31]. Bahrami and Gholipour Kanani studied the effects of using solvents of different polarity, such as glacial acetic acid, dimethyl formamide, glacial formic acid, and acetone, on the morphology of polymer nanofibers. It was reported that the use of glacial acetic acid as solvent resulted in the formation of fibers with nonuniform diameter distribution compared to other solvents. This was due to the high dielectric constant value of acetic acid [32]. Generally, the net volume charge density is proportional to the conductivity of the solution. As the net charge density increases, the beads become smaller and change more into a spindle-like shape, while the fibers become thinner. The neutralization of the charge carried by the polymer jet favors the formation of beads, as the tension in the fiber depends on the net charge repulsion and the interaction of the net charge with the electric field [30]. One of the effective approaches to adjust the solution conductivity is by introducing a conductive nanoparticle filler or soluble salt into the solution. The incorporation of salt increased the stretching of the polymer jet and reduced the incident of bead formation by distributing the charges consistently to control the fiber uniformity [33].

## 3.2. Needles

Typical electrospun nanofibers represent solid fibers, with a smooth surface and circular cross section. However, electrospun nanofibers can also be designed and fabricated in hybrid form of either basically random blending with nanoparticles or ordered structures, such as core-sheath and mingled nanofibers, which consist of two distinctive polymers. Various designs of electrospun nanofibers were successfully created by the use of different needles (coelectrospinning). These needles allow the formation of different fiber morphologies that could be exploited for specific uses [34]. Cross-sectional views of electrospun nanofibers in the forms of randomly blended structure, core-shell structure, and mingled structure are illustrated in Figure 3.

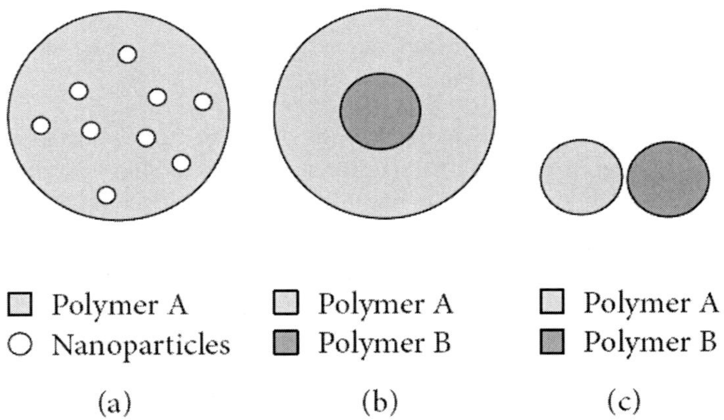


**Figure 3:** Schematic cross-sectional view of different structures of composite nanofibers: (a) randomly blended; (b) core-shell; (c) mingled structure generated from various capillary needle nozzles.

For the randomly blended nanofibers, a standard needle with only one capillary opening is used. However, core-shell and mingle structured nanofibers need to be spun using a special needle, respectively, as shown in Figure 4. For the core-sheath needle, two capillary openings are combined: the outer capillary opening is for the sheath and the inner capillary opening is for the core, while for mingled structured nanofiber formation, the use of a coaxial needle with double or multiple capillary openings in a parallel axis is needed.

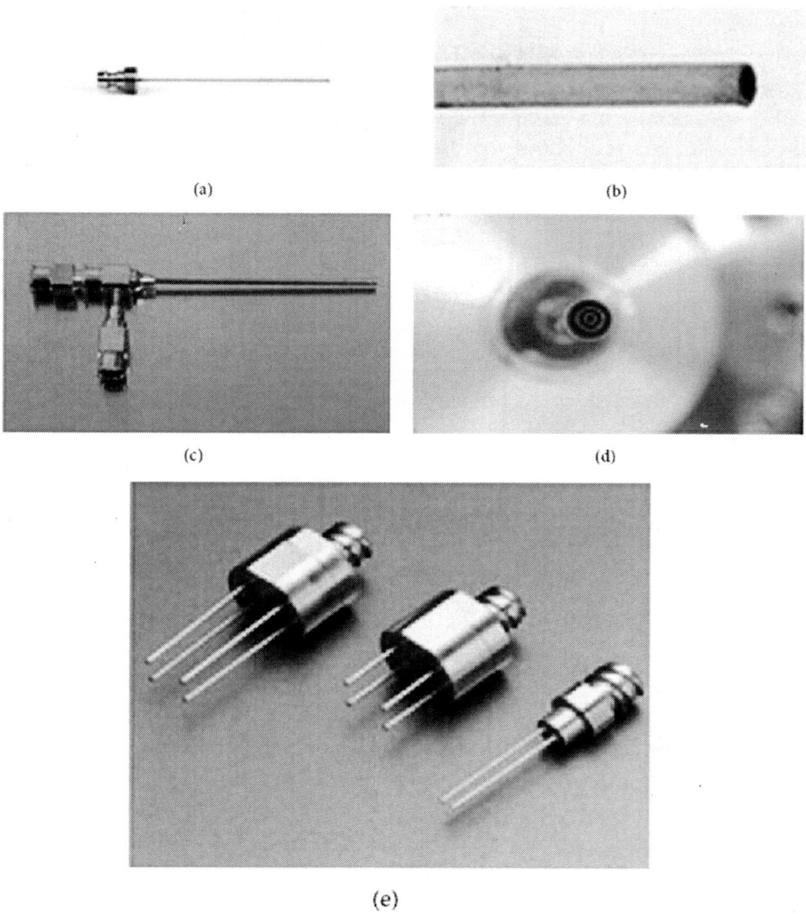


**Figure 4:** Various needles used in electrospinning: (a) standard needle; (b) inlet of standard needle; (c) core-sheath needle; (d) inlet of core-sheath needle; (e) coaxial needle with double or multiple capillary openings.

Nonetheless, the same difficulties in optimizing the solution properties are still present and affecting the multineedle design of the electrospinning system. However, since the throughput is not a primary concern in this case, interesting nanofiber structures can be obtained if the nonuniform deposition of the fiber can be taken under control [35]. Bi-component mingled structured fibers have been obtained by merging side-by-side two spinnerets fed by separate reservoirs [36, 37]. In spite of the fact that the electrospinning of polymer blends can be achieved by spinning a mixture of polymers or polymer solutions through a single needle, the electrospinning setup can be designed as having two separate reservoirs to avoid some limitations related to the physics of polymeric blends. Indeed, with a single spinneret, a homogenous, thermodynamically stable mix of polymers is required and, in the case of solution electrospinning, the interactions between the polymer and the solvent of the opposing pair become crucial. With two separate spinnerets, the polymer solutions will come in contact only at the end of the spinning head, allowing for the formation of blend fibers that are otherwise difficult to produce [35].

## 3.3. Collectors

The collector used during the electrospinning process would also have an influence on the morphology of the electrospun nanofibers. At first, the randomly oriented nanofibers were obtained by collecting the fibers from a stationary collector covered with aluminium foil [38]. An attempt to obtain the deposited nanofibers in aligned orientation has been made as the electrospinning jet was shown to be sensitive to small differences in the electric field profile [39]. A conductive collecting substrate with textured surface, such as wire mesh or grid, has been used to form patterned/textured nanofibers due to its electric field profile. The results show that, in a typical square grid substrate, the nanofibers are preferentially deposited along the wire. As the wires forming the grid are narrow, nanofibers tend to align along the length of the wire. Despite the preferential deposition of fibers on the wire, the presence of numerous neighbouring wires often diverts the electrospinning jet to the adjacent wire. This results in a higher density of the nanofibers that are aligned on the wire, compared to the randomly oriented, scattered nanofibers at the gaps between the wires. Where the distance between the wires gets larger, the difference in the density of the nanofibers on the wires as well as the space between the wires also increases. There are differences in fiber diameter as well whereby the nanofibers deposited on the wire are larger than the nanofibers deposited within the space between the wires. This has been attributed to a greater stretching of the nanofibers when they span across the space [40]. Other forms of patterned substrates have also been used to collect the deposited nanofibers. Grids made out of parallel wires close to one another have been shown to form dense, aligned nanofibers between the wires, while a substrate with arrayed pins gives rise to a membrane made out of radiating nanofibers between the pins [41]. More grid variations, such as having a pin in the center of a round space, form a pattern where fibers radiate from the central pin to the edges.

Furthermore, aligned nanofibers have also been successfully collected by using a rotating disk collector instead of the stationary collector. The adjustment of the take-up velocity of the rotating disk collector proved that the fiber alignment was a direct function of the velocity of the rotating disk. Adding to this, as the take-up velocity increased from 630 m/min to 1890 m/min, the average diameter of aligned nanofibers decreased from 700 to 350 nm [42].

## 3.4. Configurations

There have been established several types of arrangement of the electrospinning setup, such as horizontal, vertical upward, and vertical downward. For each type of arrangement, the positioning of the components is different. Figure 5 illustrates the arrangement of the electrospinning in detail. For a small laboratory scale use, the vertical downward setup of the electrospinning is the most favorable due to the simple optimization and operational monitoring.

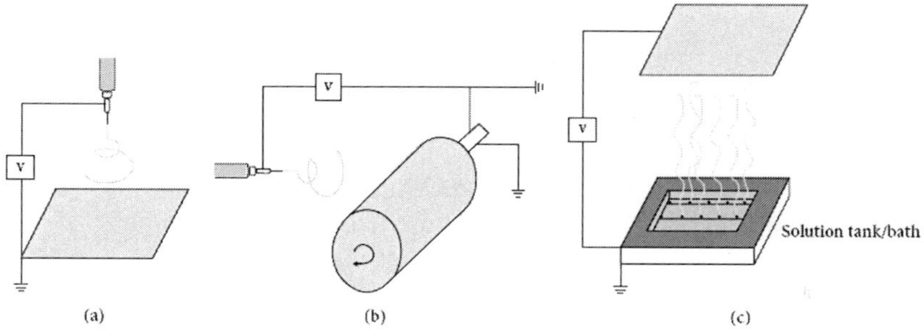


**Figure 5:** Common configuration of electrospinning: (a) vertical; (b) horizontal; (c) needleless.

Most of the research works on the electrospinning nanofibers completed in the universities are based on the conventional single needle. One of the main issues to overcome in this case is the limited throughput obtainable by the use of such a system. For industrial applications, despite the commonly high added value of the products, the low production rate (in the order of tenths of grams per hour) of the basic electrospinning apparatus represents a severe restraint. Therefore, the mass production of the nanofibers by using the conventional single needle is tough and challenging. For this reason, the vertical upward setup has been selected for the industrial production scale. This employs a new design for supplying the solution to a metal roller spinneret. The advantage of this setup is its ease of scaling-up for increased output. By using the needleless electrospinning system, the possibilities to increase the production rate of nanofibers are higher due to the use of multiple jets. The productivity rate has been significantly enhanced within 30 times higher than that of the single needle electrospinning [43].

The setup of the needleless electrospinning system was firstly introduced in 2010 by Tang et al., whereby the nanofibers could be produced in abundance by applying the method of splashing the polymer solution onto the surface of a metal roller spinneret [44]. In this method, the setup consisted of a metal roller spinneret as the positive electrode connected to a high voltage power supply. The polymer solution droplets were splashed onto the surface of the metal roller spinneret through the holes of the solution distributor, which was located above the spinneret. Niu et al. [45] improvised the system by inventing a spiral coil setup and proved that this method had a higher fiber production rate and allowed better control of fiber morphology. Considering the above information, one can conclude that various designs and dimensions of nanofibers can be tailored by using different kinds of electrospinning setup.

# 4. BASICS OF PANI

Conducting polymers are the fourth generation of polymeric materials. This type of nanofibers has become the source for the development of films, membranes, and nanoelectrodes for sensor applications. Their electrical conductivities can be increased by many orders of magnitude from $10^{-10}$–$10^{-5}$ to $10^{2}$–$10^{5}$ $Scm^{-1}$ upon doping [8, 46], which covers the whole insulator-semiconductor-metal range. A variety of conducting polymers, such as polyaniline (PANI), polypyrrole (PPY), poly(p-phenylene-vinylene) (PPV), poly(3,4-ethylene dioxythiophene) (PEDOT), and other polythiophene derivatives, which can be synthesized into 1D nanostructures, such as nanotubes and nanowires, have recently attracted attention in the areas of nanoscience and nanotechnology due to their special conduction mechanism, unique electrical properties, reversible doping/dedoping process, controllable chemical and electrochemical properties, and processability [46–48].

PANI has the longest history among the intrinsically conducting polymers. It is one of the oldest artificial conducting polymers and its high electrical conductivity among organic compounds has attracted continuing attention. PANI has become one of the most attractive conducting polymers due to its high stability, ease of synthesis, feasibility of electrical conductivity control by changing either the protonation state or the oxidation state, and the low cost of the aniline monomer [49–51]. PANI is either totally insulating or electrically conductive, depending on the oxidation state and protonation level. Only in the intermediate oxidation state, the protonated emeraldine form is conductive. The fully reduced leucoemeraldine and fully oxidized pernigraniline are insulating materials [52]. PANI nanofibers have received much attention due to their superior properties compared to the conventional bulk PANI or PANI film [53–55]. PANI nanofibers show enhanced water processability [56] and improved acid-base sensitivity and response time when they are exposed to chemical vapor, due to their large surface area. On the other hand, PANI nanofibers have numerous potential applications, including electric devices, flash welding [55], sensors and actuators [57, 58], rechargeable batteries [59], electromagnetic shielding devices, and anticorrosion coating [60].

The neutral intrinsic redox states of PANI can vary from that of the fully reduced leucoemeraldine to that of the fully oxidized pernigraniline. The 50% intrinsically oxidized polymer has been termed emeraldine and the 75% intrinsically oxidized polymer is called nigraniline [61]. Unlike most other polyaromatics, the fully oxidized state of PANI is not conducting. PANI becomes conducting when the emeraldine base is protonated and charge carriers are generated. The process is generally termed protonic acid doping. The green protonated emeraldine salt has conductivity by many orders of magnitude higher than that of common polymers, but lower than that of typical metals.

PANI is composed of the repeating units of the aniline monomer connected to form a backbone. The existence of a nitrogen atom lying between the phenyl rings allows the formation of different oxidation states, which can affect its physical properties. PANI has an inherently unstable backbone, resulting from the formation of alternate single and double bonds along the monomer units during polymerization. The delocalized bonding electrons, produced across the conjugated backbone, provide an electrical pathway for mobile charge carriers, which are introduced through doping. Consequently, the electronic properties of PANI, as well as many other physicochemical properties, are determined by the structure of the polymer backbone and the nature and concentration of the dopant ions [62–64].

## 4.1. PANI Nanostructures

Much effort has been put in synthesizing nanostructured PANI, owing to its novel physical properties and potential applications. Generally, PANI nanostructures can be synthesized using hard- and soft-template methods. In the hard-template method, which typically employs an aluminium oxide or track-etched membrane as a template for synthesizing the desired material, polymerization occurs within the pores of the membrane [65]. However, a rather tedious postsynthesis process is often required to remove the template used [66]. Additionally, preformed PANI nanostructures may be destroyed or form undesirable aggregates when released from the template. The soft-template method uses micelles, emulsions, liquid crystals, and surfactant gels to synthesize PANI nanostructures [67].

The soft-template method [68, 69] is a simple self-assembly method. By controlling the synthesis conditions, including the temperature and molar ratio of monomer to dopant, PANI nanostructures can be prepared by in situ doping polymerization in the presence of protonic acids as dopants. In the self-assembled formation mechanism using this approach, the micelles formed by dopant and/or monomer dopant act as soft templates in the process of forming tubes/wires [68]. Up to now, a variety of PANI micro/nanostructures, such as micro/nanotubes [70, 71], nanowires/fibers [72–74], hollow microspheres [75, 76], nanotube junctions, and dendrites [77], have been prepared by the soft-template method. However, this method allows for a limited range of chemicals to be used. Moreover, to date, some novel template-less methods have also been developed to shape PANI into nanofibers, such as electrospinning [78]. The nanofibers can be synthesized by deposition directly onto aluminium foil, wire mesh, cotton, silk, wool, and Si wafer.

Many other methods have also been recognized for the synthesis of PANI nanostructures. PANI films with a thickness of 20 to 100 nm have been prepared using vapor-phase polymerization [79]. Vapor-phase polymerization is one of the self-assembling techniques that do not require a host polymer. It utilizes the organic arrangement of macro molecules to fabricate nanolayered aggregates and thin films [80, 81]. There are also reports on the preparation of nanostructured PANI through polymerization in the presence of functionalized protonic acid, which acts as both surfactant (and emulsifier) and protonating agent. This process is called emulsion polymerization and it combines aniline, protonic acid, and oxidant with a mixture of water and a nonpolar or weakly polar liquid [82–84]. In order to avoid agglomeration of thick fibers into irregularly shaped particles, Jing and coworkers have employed ultrasonic radiation to assist the conventional dropwise addition of the oxidant solution to form PANI nanofibers [85]. Ultrasound assistance through the sonochemical method is an important approach, allowing researchers to obtain uniform dispersion of inorganic nanoparticles in the polymer matrix [86, 87]. Another popular method to obtain PANI nanofibers is through interfacial polymerization, which involves step polymerization of two reactive agents or monomers that dissolve in two immiscible phases. This technique allows the synthesis of high crystallinity PANI nanofibers without depending on specific templates and solvents [60, 88–90].

# 5. PANI NANOFIBERS VIA ELECTROSPINNING

## 5.1. Neat PANI Nanofibers

Producing neat PANI nanofiber by electrospinning has been a great challenge for researchers. The poor solubility of PANI in common solvents interferes with electrospinning into uniform fibers, requiring it to be doped by organic acids in order to increase its solubility in organic solvents [91]. Various methods have been attempted to test suitability of electrospinning PANI solutions without blending with other polymers. Being an intrinsically conducting polymer, PANI is available only in the form of low molecular weight, which thus makes the elasticity of its solution insufficient for being electrospun directly into a fiber form [79]. Previous studies reported the formation of structures like beads and droplets in electrospun fibers with low elasticity [92]. The use of an insulating copolymer with high molecular weight is expected to act as a remedy, imparting greater elasticity. However, the presence of a nonconducting polymer strongly modifies the physical and chemical properties of PANI and eventually limits its applications [93]. The fabrication of one-dimensional polymer field effect transistor using PANI nanofibers blended with polyethylene oxide (PEO) in a 1 : 1 ratio has been reported. The addition of PEO resulted in a drastic decrement of conductivity. The results of the study allow the conclusion that reducing the PEO content will enhance the conductivity of the electrospun PANI by one or several orders of magnitude [94].

In order to improve the electrospinning processing of PANI, a few routes have been developed to find the optimum conditions to produce pure PANI nanofibers with smooth fiber surface and to minimize the limitations caused by nonconducting copolymers. The objective was not only to produce neat PANI, but also to obtain fibers that would raise no concern about side reactions with the doping agents when exposed to external agents, such as radiation or gases [93]. In their study, Cárdenas and coworkers proposed using a bath collector in the fabrication of PANI nanofibers by the electrospinning procedure. In this research work, the fibers were not deposited on the grounded collector, but instead they were collected in an acetone bath placed on an electrode [95]. Although there are a number of literature sources [96–98] describing the preparation of PANI based nanofibers by electrospinning, the reports on the fabrication of neat PANI nanofibers by electrospinning are quite rare. In an earlier study, MacDiarmid et al. [99] produced nanofibers of PANI by a method similar to the one used by Cárdenas et al. and achieved fibers with an average diameter of 139 nm by placing a 20 wt% solution of PANI in 98% sulfuric acid in a glass pipette fixed above a copper cathode immersed in pure water. However, in contrast to this work, Cárdenas and coworkers used a solution with a considerably lower concentration of PANI (1 wt%) in acetone [95].

By using the acetone bath as collector, the excess solvent was diffused into acetone and allowed fiber formation. Thus, the researchers have successfully fabricated neat submicron PANI using the electrospinning technique without incorporation of high molecular weight polymer blending for gaining jet stability to form fibers. The fibers produced by this technique were observed and characterized under scanning electron microscope (SEM). These fibers appeared as isolated on the substrates, which is favorable for electrical characterization. The conductivity of these fibers was measured to be in the order of $10^{-3}$ to $10^{2}$ $Scm^{-1}$, with a significant increasing trend in the conductivity values. Nanofibers with larger diameters have higher values of conductivity due to their increased volume to surface ratio. A higher volume to surface ratio causes a relatively slower loss of solvent by evaporation and consequently the fibers will be more partially doped and conductive.

The method of utilizing the copper cathode immersed in pure water was done by placing 20 wt% of PANI solution with 98% sulfuric acid inside the glass pipette. The fibers were collected either in or at the surface of water. In this study, the authors reported having successfully produced fibers with average diameters of 139 nm, the conductivity of single fibers being of about ~0.1 $Scm^{-1}$. It is noticeable that the morphology of the nanofibers deposited in the solution bath was less smooth as compared to that of the nanofibers collected from the conductive surface [99].

Another study by Shie et al. [100] reported the use of the emeraldine base of poly(o-methoxyaniline) powder dissolved in a mixture of tetrahydrofuran/dimethylformamide (THF/DMF) for the preparation of the electrospinning solution. A 5 wt% solution of PANI was loaded into a plastic syringe connected with a metallic needle, with a rate of 0.02 to 0.04 mL/min. PANI nanofibers were collected on an aluminum plate placed 8 to 14 cm below the tip of the needle. The samples were then heated at 100°C for an hour. The biocompatibility of these fibers was examined through various analyses: agar diffusion test, nitric oxide assay, and cell and protein adsorption. The study found that the PANI nanofibers fabricated by electrospinning possessed negligible cytotoxicity, low serum protein absorption, and low myoblast cell attachment. The percentage of myoblast cell attachment to PANI was measured to be 40.7% ± 6.1%, while cell adhesion to electrically stimulated

PANI was 53.6% ± 9.9%, which was slightly higher. The results indicated that the cells on the PANI fibers exhibited a significantly extended lag phase of growth, while after electrical stimulation, the cells demonstrated higher proliferation on PANI fibers. The decreased cell growth on PANI fibers was probably caused by the hydrophobic surface of the fibers. A significant increase of nitric oxide secretion by PANI nanofibers shows that these fibers may elicit a slight inflammatory response after implantation in the human body. The response index of the toxicity assay of PANI to myoblast cells was nearly zero after electrical stimulation. Thus, it is suggested that PANI nanofibers can be considered to be very suitable for biosensor applications.

A study of electrospun PANI doped with 2-acrylomido-2-ethyl-1-propanesulfonic acid (AMPSA) indicated a conductivity of 0.76 $Scm^{-1}$, suggesting that the fibers were in the metallic regime of charge transport since there was no apparent field effect [101]. Pure PANI fibers were prepared by dissolving 900 mg of PANI-AMPSA into 3 mL concentrated sulfuric acid and manually stirring the solution with a glass rod for 30 min. The solution was then held in a glass pipette placed vertically 3 cm over a beaker of water acting as cathode. A value of 15 kV of spinning voltage was successfully applied, leading to the deposition of fine fibers in an erratic manner due to the spark formation at the end of the pipette. Fibers deposited on the surface of water and were left in the water for 12 h to allow the sulfuric acid to dissolve in the water. The fibers were then collected on a degenerately doped $Si/SiO_2$ wafer. The average diameter of these fibers was measured to be approximately 10 μm.

## 5.2. Blend PANI Nanofibers

Since PANI is difficult to be electrospun, most researchers tend to incorporate it with other polymers to make it electrospinnable. Chen and coworkers studied the incorporation of PANI into poly(ε-caprolactone) and gelatin for orthotopic photothermal treatment of tumors in vivo [102]. Apart from having the right polymer candidates to be blended with, the solvent used during the electrospinning solution preparation is also an important parameter to consider. Fryczkowski et al. investigated the effect of two different solvents in preparing electrospinning solution of poly(3-hydroxybutyrate) (PHB) with PANI [103]. Their study showed potential biodegradability and interesting electrical properties for PHB-PANI. THF gave better nanofiber structure and properties than chloroform. PHB-PANI in THF solvent produces smooth, uniform, and fine nanofibers with enhanced hydrophilicity and low electrical resistance.

Karim utilized dual approach to incorporate PANI aligned nanofibers [104]. The first step was to synthesize PANI copolymer with o-aminobenzenesulfonic acid (PANI-co-PABSA) by in-situ polymerization. Then, polyvinyl alcohol (PVA) and chitosan oligosaccharide (COS) were blended into the PANI copolymer to make the electrospinning solution. XRD and FTIR data from this study displayed the existence of hydrogen bonds between PANI-co-PABSA, COS, and PVA molecules that may possibly cause weak interaction in COS.

Sharma et al. made use of electrospun fibers to fabricate scaffolds for three-dimensional cell culture [105]. They reported fabrication of PANI/poly(N-isopropyl acrylamide-co-methacrylic acid) (PANI-CNT/PNIPAm-co-MAA) composite by 1 : 1 ratio of polymers. This study suggests possible use of conducting nanofibers as scaffolds after obtaining better cell survival than the control and PNIPA-co-MAA without PANI. An effort to replace high cost platinum for oxygen reduction reaction catalyst in polymer electrolyte fuel cells with cheaper material has been done by Zamani et al. by having metal-polymer blend [106]. 10 wt% PANI addition into polyacrylonitrile (PAN) was firstly used as nanofiber catalysts and it provides improvements to half-wave potential and ORR onset potential by 70 and 100 mV.

PANI nanofibers with the least amount of PEO, within the range of 1 to 0%w/w of PEO content, with respect to the amount of PANI, revealed some defects such as bead formation [94]. Such drawbacks could be explained by the capillary instability of the spinning jet by surface tension as PEO concentration was decreased [107]. Compared to PEO, polystyrene (PS) was found to be more suitable for blending with PANI, as it led to the formation of thinner PANI nanofibers, with a diameter below 100 nm. Thus, electrospinning the blended polymer solution produced nanofibers with fewer defects, as compared to PEO, and presented better homogeneity of the blend in chloroform. Since the conductivity of PANI is very sensitive to the amount of defects, it is reasonable to suggest that the overall conductivity of the PANI/PS blends will be higher than that of the PANI/PEO blends [91].

## 5.3. Core/Shell PANI Nanofibers

As mentioned earlier, coaxial spinning can also be used in producing PANI nanofibers. This kind of electrospinning uses two spinnerets, which allows the low elastic fluid to elongate along with the electrospinnable fluid [108]. This process will result in nanofiber with continuous core-sheath morphology. Zhang and Rutledge produced fibers of PANI with a dopant of camphor-10-sulfonic acid (HCSA) blended with poly(methyl methacrylate) (PMMA) in core-sheath form through coaxial electrospinning. 100% doped-PANI fibers were then obtained by immersing the fiber blend into isopropyl alcohol. This was done to remove the PMMA shells and thereby release the doped PANI cores. Due to the removal, the diameters of the fibers decreased from to nm [109].

Coaxial electrospinning is also useful in incorporating conducting material onto nonconducting fibers (or vice versa) with distinct layers. Sarvi et al. applied this approach to develop nanofibers with superior piezoelectricity [110]. The nanofibers consist of poly(vinylidene fluoride) (PVDF) as the core and blends of PVA, PANI, and multiwalled carbon nanotube (MWCNT) as the shell. Low concentration of PVDF was added to PANI solution in order to increase the viscosity of relatively dilute PANI solution so that it is electrospinnable. MWCNT was reported to significantly reduce the electrical percolation concentration in the polymer matrix. It was shown that the presence of PANI component increased the PVA nanofibers mat formation by bridging the nanofibers and eliminating air resistance.

An attempt to develop a material extraction for in vivo, semisolid tissue extraction has been done by preparing polystyrene (PS)/PANI core-sheath electrospun nanofibers [111]. Wu et al. have previously worked on PS/crosslinked collagen core-sheath for same purpose but the PS core showed poor extraction ability to very polar and ionic compounds [112]. The collagen sheath failed to extract and therefore would prevent the permeation of tissue matrix into PS core. Later, they successfully modified the fabrication of core-sheath by having PANI sheath. Neat PANI sheath was obtained after selectively removed collagen content in as-spun collagen/PANI sheath. This approach gave better efficiency for acidic phytohormones and high extraction capacity.

## 5.4. Modified Core/Shell PANI Nanofibers

Various modifications can also be done to produce variations of core/shell structure. Spinneret of coaxial electrospinning can be designed to produce novel ribbon-shaped flexible nanocables [113]. The modified spinneret allows the core for the nanoribbon to be adjusted to achieve preferred diameters by changing the inner needle. This novel technique could fabricate tunable core composed of PANI, magnetic $Fe_3O_4$nanoparticles, and PMMA template with photoluminescent $Tb(BA)_3phen$ (BA = benzoic acid, phen = 1,10-phenanthroline). The nanocable array produced electrically conductive-magnetic bifunctional core with insulating photoluminescent sheath. The conductive cores possessed electrical conductivity at the order of $10^{-2}$ $Scm^{-1}$ and may give response in magnetic field. These conductive nanocables were insulated with each other in the array form and able to emit green light under ultraviolet light due to the sheath properties. The ability to tune the emission colour of their belt-shaped coaxial microcables by changing the europium/terbium ($Eu^+$/$Tb^+$) molar ratio in the sheath was shown by Shao et al. [114]. They were able to tune the emission in wide range of red-yellow-green colour by adjusting the mass ratio of $Eu^{+}/Tb^{+}$ complexes, PANI content, or $Fe_3O_4$ nanoparticles content.

An attempt by Liu et al. has led to an interesting way to fabricate PANI nanofiber with hollow structure through electrospinning [115]. Instead of using the conventional coaxial electrospinning, the technique did not require the preparation of two different electrospun solutions. The one-pot electrospinning requires only one mixture of solution favored for the shell and eventually directly produces hollow electrospun nanofibers. Liu et al. fabricated trifunctional (photoluminescent-electrical-magnetic) flexible $Eu(BA)_3phen$/PANI/$Fe_3O_4$/PVP hollow nanofibers. The hollow nanofibers had outer diameters of 305 nm and inner diameters of about 140 nm. The team also used the same procedure to produce $Tb(BA)_3phen$/PANI/$Fe_3O_4$/PVP hollow nanofibers with 238 nm outer diameters and 80 nm inner diameters [116]. There are also few studies that reported hollow structured PANI nanofibers. However, those nanofibers were obtained by the in situ polymerization of aniline on the surface of as-prepared electrospun nanofibers and followed by removal of the electrospun template [117].

Lv and coworkers applied electrospinning using specially designed parallel dual spinnerets to prepare bistrand-aligned nanobundles [118]. These nanobundles consisted of PANI/PVP as one strand and $Eu(BA)_3phen$/PVP as another strand, where the PVP acted as template. They are the first to record this procedure of obtaining bistrand-aligned nanostructures. They suggested better connectivity of PANI in the nanofibers and effective isolation of rare earth complex.

The mechanical properties of electrospun nanofibers are generally considered important for many applications. In spite of this, nanofibers are often found to be mechanically poorer than their corresponding textile fibers made from the same polymers. Their tensile strength and Young's modulus were reported below 300 MPa and 3 GPa, respectively [119–121]. To produce high-performance electrospun nanofibers, the molecular structures of the nanofibers should be oriented with chain extension along the fiber axis [122]. PANI on the other hand usually acts as conducting particulate filler in a suitable matrix, which provides the required mechanical properties. Valentová and Stejskal [123] characterized the mechanical properties of PANI on compressed circular pellets instead of conventional rectangular or dumbbell shaped samples. It was found out that the dynamic mechanical properties of the compressed pellets depended on the compression pressure. The samples became brittle at a compression pressure above 800 MPa. PANI fibers were wet spun from a solution of PANI containing solvent. It was found that the coagulant temperature had significant effects on the strength and modulus of the spun fibers [124].

# 6. APPLICATIONS OF ELECTROSPUN PANI NANOFIBERS

Researchers have been investigating the potential applications of electrospun PANI nanofibers in various fields. This section highlights opportunities of using electrospun PANI nanofibers for emerging technology development, such as in the conductive scaffolding materials, controlled drug delivery systems, nanofiber based sensor devices, supercapacitors, and flexible solar cells.

## 6.1. Tissue Engineering and Drug Delivery

Tissue engineering is a method to fabricate new tissue from cultured cells. Biodegradable scaffolds can act as transplant vehicles for cultured cells and templates to guide tissue regeneration [125, 126]. The scaffolds should be biocompatible, biodegradable, and highly porous; they should have large surface area and highly interconnected pore structure with suitable pore size [127, 128]. PANI is one of the conducting polymers that have the potential to be applied as conductive substrates for tissue engineering [129–131]. It is only quite recently that the tunable electroactivity of PANI has been explored in the area of diverse biological applications, such as for scaffolds in the tissue engineering field. Still, the investigation of PANI for such applications has already brought evidence of the ability of PANI and blended PANI to support cell growth. The ability of varying the oxidative state of PANI has increased the research development in biomedical research fields [132].

Although, like many conductive polymers, PANI is not inherently biodegradable [133], it can be made so by (1) blending it with a biodegradable polymer to form a composite [134]; (2) modifying the backbone to render degradability [135]; and (3) synthesizing short PANI chains that can undergo gradual erosion [136]. Electrospinning a biocompatible or a biodegradable polymer filled with PANI could produce nanofibers that may be useful in tissue engineering applications. Studies on electrospun PANI blends for tissue engineering include reports on PANI/chitosan [137], PANI/poly(glycerol-sebacate) (PGS) [138], PANI/nanohydroxyapatite/gelatin [139], PANI/PLA [140], PANI/PPY [141], and PANI/PCL [142].

Previous research conducted by Li et al. [143] described the use of composite nanofibers of PANI blended with gelatin as a conductive scaffold in the proliferation of myoblast cell tissue culture. Gelatin is a frequently used biomaterial for tissue engineering applications, especially in cardiac tissue engineering [144, 145]. Gelatin, which is less than 3% compared to the amount of PANI in total weight, leads to an alteration of the physicochemical properties of the nanofiber composites. The decrement in the diameter of the composite nanofibers was found with the increment of the PANI to gelatin volume ratio. The cells initially displayed different morphologies, depending on the concentration of PANI. The cells cultured on the smaller diameter PANI-gelatin nanofiber blend were more spread out and exhibited a smooth muscle-like morphology, similarly to the cells growing on a glass cover surface (control sample). However, cell growth on larger diameter nanofiber substrates presented low density and pseudopodia. Other physicochemical measurements, such as tensile strength, elasticity, conductivity, and electron microscopy analysis, have confirmed the uniformity of the ensued PANI nanofibers.

The preliminary study of PANI nanofibers in tissue engineering [143] was then supported with the research works developed by Ghasemi-Mobarakeh et al. It was explained that the surface modification and functionalization of PANI with different biomolecules or dopants allowed them to be modified with biological sensing elements and to exhibit different signaling pathways required for cellular processes. In this way, significant enhancement in cell proliferation and differentiation was observed through the employment of PANI. Thus, conductive PANI provides an excellent opportunity for the fabrication of highly selective, biocompatible, specific, and stable nanocomposite scaffolds for tissue engineering of different organs [146].

The effect of PANI nanofiber orientation on electrical conductivity has also been reported by Ku et al. [147] who prepared electrospun PANI nanofibers in a PCL blend. Rotor speed was adjusted to 100 rpm for collecting random nanofibers and to 900 rpm for aligned nanofibers. The electrospun PANI/PCL nanofibers with both random and aligned orientation were used as scaffolds for the growth of C2C12 mouse myoblast cells. The diameter of the blended PANI-PCL nanofibers was reduced through the increasing concentration of PANI as a result of the increment in electrical conductivity of the spinning solution.

The seeded cells grown on randomly oriented nanofiber matrixes were observed as having flat and multipolar morphologies. In contrast, the seeded cells grown on aligned nanofibers were found to possess bipolar morphologies. PANI loading significantly affected the number of myotubes formed on the PANI-PCL nanofibers. The number of the myotubes formed increased with increasing PANI concentration. The results suggested that both conductivity and alignment of nanofibers can synergistically stimulate the myogenic differentiation.

The development of effective methods to deliver drugs has attracted wide attention, primarily in the fields of tumor therapy and tissue engineering. A controlled drug delivery system can release quantitatively drugs at a specific time and position, especially the stimuli-response drug delivery systems [148–150]. Conducting polymers exhibit a reversible electrochemical response; they will contract upon reduction and expand after oxidation. This feature permits the controlled release of various kinds of drugs [151, 152]. Electrospun conducting polymer nanofibers provide high surface area, which can be easily loaded with drugs. It has been shown that the release of dexamethasone from electrospun PEDOT/PGA nanofibers can be controlled by using electrical stimulation, which is due to the contraction and expansion of the PEDOT [153]. Up to date, there have been a few works reported focusing on electrospun PANI blends for controlled drug release [154].

## 6.2. Sensors

PANI nanofibers are a promising candidate for designing sensors since their conductivity is highly sensitive to chemical vapors due to their high surface area [155]. Synthetically prepared PANI nanofibers [156, 157] and PANI nanocomposites with carbon nanotubes (CNTs) [158], $WO_3$ [159], and $TiO_2$ [160] have been investigated for gas sensing applications towards humidity and vapors and as biosensor devices, thus making use of the conductivity changes upon oxidation or reduction [161]. This specific feature was targeted to amplify signals transduced by the electrochemical reactions [162–164].

Electrospun PANI nanofibers and their polymeric blends have recently captured the researchers' interest due to their ease of processing, excellent detection, and uniform nanofiber formation, compared to other chemical or template routes. Several types of electrospun PANI nanofiber based sensors have been developed lately. Recent reports include the electrospun nanofiber of graphene/PANI/PS for the detection of Pb and Cd ions [165], PANI/$SnO_2$ for $H_2$ gas sensing [166], multiple detections using PANI/PCL [167], PANI/polyamide 66 for colorimetric sensing [168], PANI/poly(ethylene oxide) gas sensor [169], PANI/ZnO for chemiresistor sensor [170], PANI/PLA to sense alcohol vapors of increasing molecular size [171], PANI/polyvinylidene fluoride as strain sensor [172], high sensitivity gas sensors based on PANI/PMMA [173], PANI/fibroin microfibrous mat as reactive sensor [174], and PANI/polyamide 6/$TiO_2$ nanofibers for ammonia sensor [175, 176].

The humidity response of the PANI nanofiber sensors was determined by measuring their impedance at different humidity levels at room temperature. According to Lin et al. [177], the rheology of PANI nanofibers presenting the formation of some beads exhibited improved adhesion to the electrode receptor and better electrical contact due to the high surface area, which in turn enhanced their sensing properties, compared with the nanofibers without beads. The blending of PEO into PANI greatly modified the hydrophilicity of the PANI-PEO nanofibers, and their humidity response. The PANI-PEO nanofiber based sensors revealed much higher sensitivity than their film counterparts with the impedance increasing by three orders of magnitude from 20% to 90% RH. Humidity sensors based on PANI nanofibers with some beads and a small content of PEO revealed high sensitivity, fast response, and small hysteresis [177].

Lin et al. [178], on the other hand, developed humidity sensors of surface acoustic wave (SAW) based on electrospun core-sheath structured polyaniline/poly(vinyl butyral) PANI/PVB blended nanofibers. It was reported that nanostructured PANI exhibited an excellent chemical sensitive response, with higher sensitivity and faster response compared to those of bulk PANI. As it was not possible to prepare nanostructured PANI directly from its solution, it was blended with PVB. The results indicated that PVB was suitable to be coelectrospun with PANI to form a core-sheath structure.

It was demonstrated that high performance SAW humidity sensors could be successfully fabricated based on electrospun core-sheath PANI-PVB composite nanofibers. The composite nanofiber SAW sensor revealed very high sensitivity of 75 kHz/% RH over a wide humidity range, good sensing linearity, and very fast response in both humidification and desiccation processes. Moreover, PANI-PVB nanofiber based sensors could be used to detect humidity as low as 0.5% RH with good sensitivity. The high surface-to-volume ratio and the improved mass load, electroacoustic load, and viscoelastic load of the core-sheath structured composite nanofibers of PANI blended with PVB were responsible for the attractive humidity sensing properties of the SAW sensors [178].

It is known that the detection stability is an important parameter for a sensor. Doped PANI generally shows unsatisfying stability in the open air, leading to a short lifetime of PANI based sensors. However, the PANI/PVB nanofiber sensor showed good stability, owing to the rheology of the electrospun composite nanofibers, which exhibited a core-sheath structure, with PANI inner core, covered by PVB outer sheath, instead of being directly exposed to the open air. The chemically stable and relatively hydrophobic PVB could effectively protect the PANI core. The investigation on the long-term stability of the SAW sensor based on PANI-PVB nanofibers found that the sensor maintained its high sensitivity after storing for more than seven months [178].

Electrospun PANI nanofibers doped with palladium nanoparticles have been demonstrated efficient for hydrogen sensing application in the fabrication of a chemiresistor [179]. The nanofibers were deposited as single nanofibers across two gold electrodes by means of near-field electrospinning without using the conventional lithography process. These palladium nanoparticle based chemiresistors recorded 1.8% resistance change in the environment with 0.3%

hydrogen concentration. These nanofiber sensors are flexible, have good reversibility, and are potentially versatile by doping with different particles.

Further, the study of Zhang and Rutledge reported the fabrication of electrospun fibers of PANI and poly(3,4-ethylenedioxythiophene) (PEDOT), blended with poly(ethylene oxide) (PEO) and poly(methyl methacrylate) (PMMA) over a range of compositions, using template synthesis. Fibers of neat PANI doped with camphor sulfonic acid (CSA) were successfully fabricated for the first time by coaxial electrospinning and subsequent removal of the PMMA shell by dissolution. This allowed for the electrospun PANI/CSA fibers to be tested for electrical performance and its enhancement, as well as for gas sensing application. The conductivities of the PANI-blend fibers were found to increase exponentially with the weight percent of doped PANI in the fibers, to as high as $Scm^{-1}$ for the as-electrospun fibers of 100% PANI/CSA. The conductivity of the neat doped PANI fibers was found to increase to $Scm^{-1}$ with increasing molecular orientation, achieved through solid state drawing [100].

The experimental results thus support those described previously; that is, enhanced molecular alignment within electrospun fibers, during both the electrospinning process and subsequent posttreatment, contributes positively to increasing the electrical conductivity of conductive polymers. Neat PANI fibers with different levels of doping were also fabricated by coaxial electrospinning and subsequent removal of the shell by dissolution and shown to exhibit a large range of fiber electrical conductivities, which increased exponentially with the increasing ratio of dopant to PANI. These fibers were found to be very effective in nanoscale chemiresistive sensors for both ammonia and nitrogen dioxide gases, thanks to this large range of available electrical conductivities. Both sensitivity and response times were shown to be excellent, with response ratios up to 58 for doped PANI sensing of ammonia and up to more than 105 for nitrogen dioxide sensing by undoped PANI fibers. The characteristic times for the gas sensing were shown to be on the order of 1 to 2 min. Using a model that accounted for the effects of intrinsic fiber conductivity (including both composition and molecular orientation), mat porosity, and fiber orientation distribution within the mat, calculated mat conductivities were obtained in quantitative agreement with the mat conductivities measured experimentally [100]. Electrospun PANI/PS nanofiber showed excellent detection of $H_2O_2$ compared to that of its corresponding thin film form. The $H_2O_2$ is useful because it is often a product of enzymatic reactions. It is generated in the reaction between glucose and oxygen in the presence of glucose oxidase (GOX). The large surface area promotes more GOX immobilization than the film sensor [176].

## 6.3. Supercapacitors

The increasing energy need of the world population has urged for new discoveries and manipulation of renewable energy sources. However, it is crucial to find ways to store energy in order to keep getting sufficient supply and to secure energy distribution. For this reason, the supercapacitor has been a phenomenal discovery with its low internal resistance, fast charge and discharge, high power density, high energy density, and long life cycle, compared to conventional capacitors and batteries [180–184]. The key technology that underlies its performance and has turned it into a research hotspot is the electrode material. Electrode materials mainly include carbon material, metal oxide, and conducting polymers. Among conductive polymers, PANI has become the preferred material due to its high capacitance, desirable chemical stability, and multiple intrinsic redox states [185].

Given all these excellent advantages of PANI, it has become a strong competitor as a material for building powerful supercapacitors. Recently, researchers have recognized this potential of electrospun PANI through their studies on electrospun PANI nanofiber web electrodes [186], three-dimensional porous PANI/polyacrylonitrile core-shell [187], sandwiched symmetric supercapacitor consisting of flexible PANI/carbonized polyimide [188], PANI/GO/PVDF [189], hollow-structured PANI [91], and PANI/vanadium pentoxide as asymmetric supercapacitor [190].

Chaudhari and his coworkers [186] initiated a study on applying PANI nanofiber webs as electrode materials for supercapacitors. Since the direct processing of PANI by electrospinning is a difficult task, they proposed the fabrication of high aspect ratio PANI (>50) by electrospinning a polymeric blend of PANI with polyethylene oxide (PEO). The fabrication of electrodes took 75 wt% of PANI, 15 wt% of carbon black, 10 wt% of polyvinylidene fluoride, and a few drops of N-methylpyrrolidinone to form a uniform slurry. The resulting slurry was then used to coat a graphite substrate, which served as current collector.

The performance of electrospun nanofibers produced in aqueous electrolyte (1 M $H_2SO_4$) and organic (1 M $LiClO_4$ in propylene carbonate (PC)) electrolyte is compared with that of PANI powder. The specific capacitance of a PANI nanofiber web in 1 M $H_2SO_4$ is 267 $F \cdot g^{-1}$ at a current density of 0.35 $A \cdot g^{-1}$, which is higher compared to that of PANI powder (208 $F \cdot g^{-1}$). PANI nanofiber webs demonstrate superior stable performance compared to that of PANI powder and deliver a specific capacitance of 230 $F \cdot g^{-1}$ at the 1000th cycle. Over 86% of its original capacitance is retained after 1000 cycles; meanwhile, the capacitance retention of the PANI powder is observed to be 48% after 1000 cycles. This observation indicated that PANI nanofiber webs have better cycle stability than PANI powder. Smaller individual impedance parameters in the case of PANI nanofiber webs were also observed, indicating the availability of electroactive sites needed for charge transfer reaction and counterion diffusion. Given these characteristics demonstrated by PANI nanofibers synthesized through electrospinning, it was found to be a suitable candidate for being used as electrode material in high performance supercapacitors [186].

## 6.4. Solar Cells

Dye-sensitized solar cells (DSCs) have turned out to be the centre of interest in nowaday's low-carbon economy as a solution to the energy requirements [191, 192]. They have been recognized as a potential alternative to silicon based solar cells due to their simple fabrication procedures, high catalytic performance, and cheap counter electrodes [193]. Counter electrodes of DSCs function as collecting electrons, catalyzing the redox couple regeneration. These are usually made of Pt films, but despite their good performance, Pt based solar cells need high temperature of operation and complex facilities, which limits their application [194–196]. For these reasons, recent decades have witnessed the development of PANI usage in this area in order to enhance cell stability, reduce fabrication cost, and simplify the preparation techniques, compared to previously made solar cells [197]. Duan et al. also proved that a solid PANI electrolyte has the ability of catalyzing tri-iodide species, shortening the charge diffusion path length, and recovering dye molecules at the anode/electrolyte interface [197]. The focus of the research on utilizing PANI as a component in DSCs has widened by the effort of certain research teams in studying electrospun PANI nanocomposites. However, this is still an emerging topic of research; therefore the studies dedicated to it seem scarce, regardless of the potential shown. An investigation on electrospun PLA/PANI nanofibers for rigid and flexible DSCs [198] and another on PANI/MWCNT counter electrodes [199] are among the few studies conducted.

A simple electrospinning method was developed [198] to directly deposit conductive 10-camphorsulfonic acid (CSA) doped PANI blended with polylactic acid (PLA) composite films (PANI/CSA-PLA) on flexible indium tin oxide-coated polyethylenenaphthalate (PEN) and rigid fluorine doped tin oxide (FTO) substrates. The result revealed that the catalytic performance of the PANI/CSA-PLA film was higher than that of the PANI-PLA film, which was ascribed to the contribution of the high surface area of PANI/CSA-PLA nanofibers and relatively high electrical conductivity of the PANI/CSA-PLA film. The preparation process of the PANI/CSA-PLA counter electrodes suggested in this experiment was simple and cost-efficient, and thus feasible for large-scale and flexible application of DSCs.

The flexible and rigid DSCs demonstrate high photovoltaic efficiency, close to that of sputtered Pt based DSCs. It was found that the flexible DSC based on the PANI/CSA-PLA counter electrode can exhibit a conversion efficiency of 3.1%, while the PANI/CSA-PLA film on FTO glass as counter electrode exhibits a reasonably high photoelectrical conversion efficiency of 5.30%. Although these results are a little lower than those for the sputtered Pt based DSC (flexible = 4.39%, rigid = 6.51%), they demonstrate that PANI can be used as counter electrode material for DSCs, providing efficiency values comparable to those of the Pt based device [198].

# 7. CONCLUSIONS

(1)Electrospinning is a simple, versatile, and rapid technology advancement, which can generate nonwoven nanofibers with high surface area to volume ratio and tunable porosity and can be designed into various fiber forms. Due to these interesting properties, the electrospinning process seems to be a promising method for various applications.(2)The polymer solution, the needle, and the collector used in electrospinning have significant effects on the fiber morphology; thus by the manipulation of these parameters one can get nanofibers with desired properties for targeted applications.(3)The introduction of polymer blends and fillers into PANI solution is an alternative for the electrospinning of PANI, apart from template coating, which can enhance the processability of PANI.(4)Electrospun PANI nanofiber blends and composites are being tested to develop new smart materials for a variety of applications, such as tissue engineering scaffolds, sensors, supercapacitors, and solar cells. The materials are designed to respond by changing the shape and size of the cells and have demonstrated high sensitivity of detection toward specific environments.(5)In general, electrospun PANI nanofibers show an excellent potential for future applications, particularly in biomedical devices and sensors. However, the studies on biocompatibility for long-term use are still inadequately carried out and are rather limited. It is important that the toxicology profile of the electrospun PANI nanofibers and their health impacts be evaluated before the actual industrial production begins.(6)The core-sheath structure of PANI hybrids with other polymers has improved the usability of PANI. The limitation of PANI stability over long periods of time due to dopant leaching, which may have a negative impact on the ambient environment, should be thus resolved.(7)In addition to issues on extended response/recovery capabilities, the viability of PANI nanofibers is also being held back by poor mechanical properties as PANI tends to be a brittle material with poor ductility. Although films of PANI nanofibers can be deposited onto many substrates, it is not possible to produce PANI nanofiber films, which are mechanically robust enough to be free-standing. A new method towards free-standing PANI nanofibers by improving the mechanical properties is another way to enhance the usability of PANI nanofibers for a wide range of future biomedical, energy, and biotechnology applications.

## CONFLICT OF INTERESTS

The authors declare that there is no conflict of interests regarding the publication of this paper.

## ACKNOWLEDGMENTS

The authors would to express their gratitude to Universiti Teknologi Malaysia for the Potential Academic Staff (PAS) scheme, IJN-UTM Cardiovascular Engineering Centre, Faculty of Bioscience and Medical Engineering, and also Department of Chemistry, Faculty of Science, Universiti Teknologi Malaysia.

## REFERENCES

1. V. M. Agranovich, Y. N. Gartstein, and M. Litinskaya, “Hybrid resonant organic-inorganic nanostructures for optoelectronic applications,” Chemical Reviews, vol. 111, no. 9, pp. 5179–5214, 2011.

2. R. Li, W. Hu, Y. Liu, and D. Zhu, “Micro- and nanocrystals of organic semiconductors,” Accounts of Chemical Research, vol. 43, no. 4, pp. 529–540, 2010.

3. H. Qu, S. Wei, and Z. Guo, “Coaxial electrospun nanostructures and their applications,” Journal of Materials Chemistry A, vol. 1, no. 38, pp. 11513–11528, 2013.

4. L. Xia, Z. Wei, and M. Wan, “Conducting polymer nanostructures and their application in biosensors,”Journal of Colloid and Interface Science, vol. 341, no. 1, pp. 1–11, 2010.

5. Y. Goryacheva, E. S. Speranskaya, V. V. Goftman, D. Tang, and S. De Saeger, “Synthesis and bioanalytical applications of nanostructures multiloaded with quantum dots,” TrAC—Trends in Analytical Chemistry, vol. 66, pp. 53–62, 2015.

6. T. M. Hsieh, A. C. A. Wan, and J. Y. Ying, "Applications of fabricated micro- and nanostructures in biomedicine," MRS Bulletin, vol. 36, no. 12, pp. 990–997, 2011.

7. H. Liu, X. Ding, G. Zhou, P. Li, X. Wei, and Y. Fan, "Electrospinning of nanofibers for tissue engineering applications," Journal of Nanomaterials, vol. 2013, Article ID 495708, 11 pages, 2013. · s

8. G. MacDiarmid, "'Synthetic metals': a novel role for organic polymers (Nobel lecture)," Angewandte Chemie—International Edition, vol. 40, no. 14, pp. 2581–2590, 2001.

9. G. A. Snook, P. Kao, and A. S. Best, "Conducting-polymer-based supercapacitor devices and electrodes,"Journal of Power Sources, vol. 196, no. 1, pp. 1–12, 2011. View at Publisher · ·s

10. S. I. A. Razak, N. F. A. Sharif, and N. H. M. Nayan, "Electrically conductive paper of polyaniline modified pineapple leaf fiber," Fibers and Polymers, vol. 15, no. 6, pp. 1107–1111, 2014. View at Publisher· · s

11. Y.-Z. Long, M.-M. Li, C. Gu et al., "Recent advances in synthesis, physical properties and applications of conducting polymer nanotubes and nanofibers," Progress in Polymer Science, vol. 36, no. 10, pp. 1415–1442, 2011.

12. J. Zeleny, "Instability of electrified liquid surfaces," Physical Review, vol. 10, no. 1, pp. 1–6, 1917.

13. Frenot and I. S. Chronakis, "Polymer nanofibers assembled by electrospinning," Current Opinion in Colloid & Interface Science, vol. 8, no. 1-2, pp. 64–75, 2003. View at Publisher · ·s

14. J. Dias, F. Antunes, and P. Bartolo, "Influence of the rheological behavior in electrospun PCL nanofibers production for tissue engineering applications," Chemical Engineering Transactions, vol. 32, pp. 1015–1020, 2013.

15. X. Y. Wang, C. Drew, S. H. Lee, K. J. Senecal, J. Kumar, and L. A. Sarnuelson, "Electrospunna nofibrous membranes for highly sensitive optical sensors," Nano Letters, vol. 2, pp. 1273–1275, 2002.

16. R. Phani, R. M. T. De Britto, and S. Srinivasan, "Polyaniline nanofibers obtained by electrospin process for hydrogen storage applications," International Journal of Environmental Research and Development, vol. 4, pp. 375–386, 2014.

17. F. K. Ko and H. Yang, “Functional nanofibers: enabling material for the net generations smart textiles,”Journal of Fiber Bioengineering and Informatics, vol. 1, no. 2, pp. 81–92, 2008. View at Publisher ·

18. T. Jaroszczyk, S. Petrik, and K. Donahue, “The role of nanofiber filter media in motor vehicle air filtration,” in Proceedings of the 4th Biennial Conference on Emissions Solutions in Transportation, AFS, Ann Arbor, Mich, USA, October 2009.

19. N. Khan, “Applications of electrospun nanofibers in the biomedical field,” Studies by Undergraduate Researchers at Guelph, vol. 5, pp. 63–73, 2012.

20. Z.-M. Huang, Y.-Z. Zhang, M. Kotaki, and S. Ramakrishna, “A review on polymer nanofibers by electrospinning and their applications in nanocomposites,” Composites Science and Technology, vol. 63, no. 15, pp. 2223–2253, 2003.

21. J. Zeleny, “The electrical discharge from liquid points, and a hydrostatic method of measuring the electric intensity at their surfaces,” Physical Review, vol. 3, no. 2, pp. 69–91, 1914.

22. Formhals, “Process and apparatus for preparing artificial threads,” U.S. Patent no. 1975704, 1934.

23. L. Norton, “Method and apparatus for producing fibrous or filamentary material,” U.S. Patent No. 2408651, 1936.

24. G. S. Taylor, “Electrically driven jets,” Proceedings of Royal Society of London A, vol. 31, pp. 353–475, 1969. View at Publisher ·

25. P. Bhattarai, K. B. Thapa, R. B. Basnet, and S. Sharma, “Electrospinning: how to produce nanofibers using most inexpensive technique? An insight into the real challenges of electrospinning such nanofibers and its application areas,” International Journal of Biomedical and Advance Research, vol. 5, no. 9, pp. 401–405, 2014. View at Publisher ·

26. S. Ramakrishna, K. Fujihara, W.-E. Teo, T. Yong, Z. Ma, and R. Ramaseshan, “Electrospun nanofibers: solving global issues,” Materials Today, vol. 9, no. 3, pp. 40–50, 2006.

27. X. Fang and D. H. Reneker, “DNA fibers by electrospinning,” Journal of Macromolecular Science—Physics, vol. 36, no. 2, pp. 169–173, 1997.

28. F. Fadil, N. A. Buang, M. P. Prabhakaran, and S. Ramakrishna, “The diameter size reduction of composited poly ε-caprolactone-co-lactide

(PLCL) nanofibers using multi-walled carbon nanotubes (MWCNTs) as the nanofiller," Digest Journal of Nanomaterials and Biostructures, vol. 10, no. 1, article A2, pp. 11–17, 2015. · s

29. J. Doshi and D. H. Reneker, "Electrospinning process and applications of electrospun fibers," Journal of Electrostatics, vol. 35, no. 2-3, pp. 151–160, 1995.

30. H. Fong, I. Chun, and D. H. Reneker, "Beaded nanofibers formed during electrospinning," Polymer, vol. 40, no. 16, pp. 4585–4592, 1999.

31. C. J. Luo, M. Nangrejo, and M. Edirisinghe, "A novel method of selecting solvents for polymer electrospinning," Polymer, vol. 51, no. 7, pp. 1654–1662, 2010.

32. S. H. Bahrami and A. Gholipour Kanani, "Effect of changing solvents on poly(ε-Caprolactone) nanofibrous webs morphology," Journal of Nanomaterials, vol. 2011, Article ID 724153, 10 pages, 2011.

33. F. Yalcinkaya, B. Yalcinkaya, and O. Jirsak, "Influence of salts on electrospinning of aqueous and nonaqueous polymer solutions," Journal of Nanomaterials, vol. 2015, Article ID 134251, 12 pages, 2015.View at Publisher ·

34. C. Migliaresi, G. A. Ruffo, F. Z. Volpato, and D. Zeni, Advanced Electrospinning Setups and Special Fibre and Mesh Morphologies Advanced Electrospinning Setups and Special Fibre and Mesh Morphologies, Smithers Rapra, Akron, Ohio, USA, 2012.

35. W. Bonani, D. Maniglio, A. Motta, W. Tan, and C. Migliaresi, "Biomedical gradient in micropatterned nanofibrous scaffold for spatiotemporal release," Journal of Biomedical Materials Research Part B: Applied Biomaterials, vol. 96, no. 2, pp. 276–286, 2011.

36. T. Lin, H. Wang, and X. Wang, "Self-crimping bicomponent nanofibers electrospun from polyacrylonitrile and elastomeric polyurethane," Advanced Materials, vol. 17, no. 22, pp. 2699–2703, 2005.

37. P. Gupta and G. L. Wilkes, "Some investigations on the fiber formation by utilizing a side-by-side bicomponent electrospinning approach," Polymer, vol. 44, no. 20, pp. 6353–6359, 2003. View at Publisher· · s

38. M. K. Leach, Z.-Q. Feng, S. J. Tuck, and J. M. Corey, "Electrospinning fundamentals: optimizing solution and apparatus parameters," Journal of Visualized Experiments, vol. 47, pp. 1–4, 2011. · s

39. Y. Wu, Z. Dong, S. Wilson, and R. L. Clark, "Template-assisted assembly of electrospun fibers," Polymer, vol. 51, no. 14, pp. 3244–3248, 2010.

40. C. Vaquette and J. J. Cooper-White, "Increasing electrospun scaffold pore size with tailored collectors for improved cell penetration," Acta Biomaterialia, vol. 7, no. 6, pp. 2544–2557, 2011. · s

41. D. Zhang and J. Chang, "Electrospinning of three-dimensional nanofibrous tubes with controllable architectures," Nano Letters, vol. 8, no. 10, pp. 3283–3287, 2008.

42. T. Kongkhlang, K. Tashiro, M. Kotaki, and S. Chirachanchai, "Electrospinning as a new technique to control the crystal morphology and molecular orientation of polyoxymethylene nanofibers," Journal of the American Chemical Society, vol. 130, no. 46, pp. 15460–15466, 2008.

43. B. Lu, Y. Wang, Y. Liu et al., "Superhigh-throughput needleless electrospinning using a rotary cone as spinneret," Small, vol. 6, no. 15, pp. 1612–1616, 2010.

44. S. Tang, Y. Zeng, and X. Wang, "Splashing needleless electrospinning of nanofibers," Polymer Engineering and Science, vol. 50, no. 11, pp. 2252–2257, 2010.

45. H. Niu, X. Wang, and T. Lin, "Upward needleless electrospinning of nanofibers," Journal of Engineered Fibers and Fabrics, vol. 7, no. 3, pp. 17–22, 2012. · s

46. J. Heeger, "Semiconducting polymers: the third generation," Chemical Society Reviews, vol. 39, no. 7, pp. 2354–2371, 2010.

47. X. F. Lu, W. J. Zhang, C. Wang, T.-C. Wen, and Y. Wei, "One-dimensional conducting polymer nanocomposites: synthesis, properties and applications," Progress in Polymer Science, vol. 36, no. 5, pp. 671–712, 2011.

48. J. Stejskal, I. Sapurina, and M. Trchová, "Polyaniline nanostructures and the role of aniline oligomers in their formation," Progress in Polymer Science, vol. 35, no. 12, pp. 1420–1481, 2010. · s

49. X. Ding, D. Han, Z. Wang, X. Xu, L. Niu, and Q. Zhang, "Micelle-assisted synthesis of polyaniline/magnetite nanorods by in situ self-assembly process," Journal of Colloid and Interface Science, vol. 320, no. 1, pp. 341–345, 2008.

50. B.-J. Kim, S.-G. Oh, M.-G. Han, and S.-S. Im, "Preparation of polyaniline nanoparticles in micellar solutions as polymerization medium," Langmuir, vol. 16, no. 14, pp. 5841–5845, 2000. · s

51. Y. Guo and Y. Zhou, "Polyaniline nanofibers fabricated by electrochemical polymerization: a mechanistic study," European Polymer Journal, vol. 43, no. 6, pp. 2292–2297, 2007.

52. L. I. Dan, J. Huang, and R. B. Kaner, "Polyaniline nanofibers: a unique polymer nanostructure for versatile applications," Accounts of Chemical Research, vol. 42, no. 1, pp. 135–145, 2009. View at Publisher· · s

53. S. Virji, J. Huang, R. B. Kaner, and B. H. Weiller, "Polyaniline nanofiber gas sensors: examination of response mechanisms," Nano Letters, vol. 4, no. 3, pp. 491–496, 2004.

54. J. Wang, S. Chan, R. R. Carlson et al., "Electrochemically fabricated polyaniline nanoframework electrode junctions that function as resistive sensors," Nano Letters, vol. 4, no. 9, pp. 1693–1697, 2004.

55. J. Huang, "Syntheses and applications of conducting polymer polyaniline nanofibers," Pure and Applied Chemistry, vol. 78, no. 1, pp. 15–27, 2006.

56. D. Li and R. B. Kaner, "Processable stabilizer-free polyaniline nanofiber aqueous colloids," Chemical Communications, no. 26, pp. 3286–3288, 2005.

57. N. Densakulprasert, L. Wannatong, D. Chotpattananont, P. Hiamtup, A. Sirivat, and J. Schwank, "Electrical conductivity of polyaniline/zeolite composites and synergetic interaction with CO," Materials Science and Engineering B: Solid-State Materials for Advanced Technology, vol. 117, no. 3, pp. 276–282, 2005.

58. J. Gao, J.-M. Sansiñena, and H.-L. Wang, "Chemical vapor driven polyaniline sensor/actuators," Synthetic Metals, vol. 135-136, pp. 809–810, 2003.

59. K. Ghanbari, M. F. Mousavi, M. Shamsipur, and H. Karami, "Synthesis of polyaniline/graphite composite as a cathode of Zn-polyaniline rechargeable battery," Journal of Power Sources, vol. 170, no. 2, pp. 513–519, 2007.

60. J. Huang and R. B. Kaner, "A general chemical route to polyaniline nanofibers," Journal of the American Chemical Society, vol. 126, no. 3, pp. 851–855, 2004.

61. T. S. Najim and A. J. Salim, "Polyaniline nanofibers and nanocomposites: preparation, characterization, and application for Cr(VI) and phosphate ions removal from aqueous solution," Arabian Journal of Chemistry, 2014.

62. R. A. Green, S. Baek, L. A. Poole-Warren, and P. J. Martens, "Conducting polymer-hydrogels for medical electrode applications," Science and Technology of Advanced Materials, vol. 11, no. 1, Article ID 014107, 2010.

63. S. I. A. Razak, N. F. A. Sharif, and I. I. Muhamad, "Polyaniline-coated halloysite nanotubes: effect of para-hydroxybenzene sulfonic acid doping," Composite Interfaces, vol. 21, no. 8, pp. 715–722, 2014.

64. S. I. A. Razak, W. A. W. A. Rahman, and M. Y. Yahya, "Electrically conductive nanocomposites of epoxy/polyaniline nanowires doped with formic acid: effect of loading on the conduction and mechanical properties," Nano, vol. 7, no. 5, Article ID 1250039, 2012.

65. C. R. Martin, "Nanomaterials—a membrane-based synthetic approach," Science, vol. 266, no. 5193, pp. 1961–1966, 1994.

66. X. Sun, S. Dong, and E. Wang, "Formation of o-phenylenediamine oligomers and their self-assembly into one-dimensional structures in aqueous medium," Macromolecular Rapid Communications, vol. 26, no. 18, pp. 1504–1508, 2005.

67. G. Chang, Y. Luo, W. Lu et al., "Ag nanoparticles decorated polyaniline nanofibers: synthesis, characterization, and applications toward catalytic reduction of 4-nitrophenol and electrochemical detection of H2O2 and glucose," Catalysis Science and Technology, vol. 2, no. 4, pp. 800–806, 2012.

68. M. X. Wan, "A template-free method towards conducting polymer nanostructures," Advanced Materials, vol. 20, no. 15, pp. 2926–2932, 2008.

69. M. X. Wan, J. Huang, and Y. Q. Shen, "Microtubes of conducting polymers," Synthetic Metals, vol. 101, no. 1, pp. 708–711, 1999.

70. L. J. Zhang, Y. Z. Long, Z. J. Chen, and M. X. Wan, "The effect of hydrogen bonding on self-assembled polyaniline nanostructures," Advanced Functional Materials, vol. 14, no. 7, pp. 693–698, 2004.

71. H. J. Ding, J. Y. Shen, M. X. Wan, and Z. J. Chen, "Formation mechanism of polyaniline nanotubes by a simplified template-free method," Macromolecular Chemistry and Physics, vol. 209, no. 8, pp. 864–871, 2008.

72. H. J. Ding, M. X. Wan, and Y. Wei, "Controlling the diameter of polyaniline nanofibers by adjusting the oxidant redox potential," Advanced Materials, vol. 19, no. 3, pp. 465–469, 2007.

73. H. Ding, Y. Long, J. Shen, and M. Wan, "Fe2(SO4)3 as a binary oxidant and dopant to thin polyaniline nanowires with high conductivity," Journal of Physical Chemistry B, vol. 114, no. 1, pp. 115–119, 2010.

74. Y. Yan, K. Deng, Z. Yu, and Z. X. Wei, "Tuning the supramolecular chirality of polyaniline by methyl substitution," Angewandte Chemie: International Edition in English, vol. 48, no. 11, pp. 2003–2006, 2009.View at Publisher ·

75. L. J. Zhang and M. X. Wan, "Self-assembly of polyaniline—from nanotubes to hollow microspheres,"Advanced Functional Materials, vol. 13, no. 10, pp. 815–820, 2003.

76. L. J. Zhang, M. X. Wan, and Y. Wei, "Hollow polyaniline microspheres with conductive and fluorescent function," Macromolecular Rapid Communications, vol. 27, no. 11, pp. 888–893, 2006. · s

77. Z. Zhang, Z. Wei, L. Zhang, and M. Wan, "Polyaniline nanotubes and their dendrites doped with different naphthalene sulfonic acids," Acta Materialia, vol. 53, no. 5, pp. 1373–1379, 2005.

78. Greiner and J. H. Wendorff, "Functional self-assembled nanofibers by electrospinning," Advances in Polymer Science, vol. 219, no. 1, pp. 107–171, 2008.

79. J.-Y. Kim, J.-H. Lee, and S.-J. Kwon, "The manufacture and properties of polyaniline nano-films prepared through vapor-phase polymerization," Synthetic Metals, vol. 157, no. 8-9, pp. 336–342, 2007.

80. T.-S. Jun, C.-K. Kim, and Y. S. Kim, "Vapor phase polymerization of polyaniline nanotubes using Mn3O4nanofibers as an oxidant," Materials Letters, vol. 133, pp. 17–19, 2014.

81. J. M. D'Arcy, M. F. El-Kady, P. P. Khine et al., "Vapor-phase polymerization of nanofibrillar poly(3,4-ethylenedioxythiophene) for supercapacitors," ACS Nano, vol. 8, no. 2, pp. 1500–1510, 2014.

82. J.-E. Österholm, Y. Cao, F. Klavetter, and P. Smith, "Emulsion polymerization of aniline," Polymer, vol. 35, no. 13, pp. 2902–2906, 1994.

83. P. J. Kinlen, J. Liu, Y. Ding, C. R. Graham, and E. E. Remsen, "Emulsion polymerization process for organically soluble and electrically conducting polyaniline," Macromolecules, vol. 31, no. 6, pp. 1735–1744, 1998.

84. S. Palaniappan and A. John, "Polyaniline materials by emulsion polymerization pathway," Progress in Polymer Science, vol. 33, no. 7, pp. 732–758, 2008.

85. X. Jing, Y. Wang, D. Wu, and J. Qiang, "Sonochemical synthesis of polyaniline nanofibers," Ultrasonics Sonochemistry, vol. 14, no. 1, pp. 75–80, 2007.

86. B. A. Bhanvase, N. S. Darda, N. C. Veerkar, A. S. Shende, S. R. Satpute, and S. H. Sonawane, "Ultrasound assisted synthesis of PANI/ZnMoO4 nanocomposite for simultaneous improvement in anticorrosion, physico-chemical properties and its application in gas sensing," Ultrasonics Sonochemistry, vol. 24, pp. 87–97, 2015.

87. E. J. V. Rani, K. Karthika, N. Saranya, and R. Padmavathy, "The influence of temperature and sonication on the preparation of doped polymeric nano particles," International Journal of ChemTech Research, vol. 6, no. 13, pp. 5168–5173, 2014. · s

88. D. D. Sawall, R. M. Villahermosa, R. A. Lipeles, and A. R. Hopkins, "Interfacial polymerization of polyaniline nanofibers grafted to Au surfaces," Chemistry of Materials, vol. 16, no. 9, pp. 1606–1608, 2004.

89. X. B. Yan, Z. J. Han, Y. Yang, and B. K. Tay, "NO2 gas sensing with polyaniline nanofibers synthesized by a facile aqueous/organic interfacial polymerization," Sensors and Actuators B: Chemical, vol. 123, no. 1, pp. 107–113, 2007.

90. W. Yang, Z. Gao, N. Song, Y. Zhang, Y. Yang, and J. Wang, "Synthesis of hollow polyaniline nano-capsules and their supercapacitor application," Journal of Power Sources, vol. 272, pp. 915–921, 2014.

91. P. Frontera, C. Busacca, P. Antonucci et al., "Polyaniline nanofibers: towards pure electrospun PANI," AIP Conference Proceedings, vol. 1459, pp. 253–255, 2012.

92. J. H. Yu, S. V. Fridrikh, and G. C. Rutledge, "The role of elasticity in the formation of electrospun fibers,"Polymer, vol. 47, no. 13, pp. 4789–4797, 2006.

93. H. J. Yu, S. V. Fridrikh, and G. C. Rutledge, "Production of submicrometer diameter fibers by two-fluid electrospiiming," Advanced Materials, vol. 16, no. 17, pp. 1562–1566, 2004.

94. N. J. Pinto, A. T. Johnson Jr., A. G. MacDiarmid et al., "Electrospun polyaniline/polyethylene oxide nanofiber field-effect transistor," Applied Physics Letters, vol. 83, no. 20, pp. 4244–4246, 2003.

95. J. R. Cárdenas, M. G. O. de França, E. A. de Vasconcelos, W. M. de Azevedo, and E. F. da Silva Jr., "Growth of sub-micron fibres of pure polyaniline using the electrospinning technique," Journal of Physics D: Applied Physics, vol. 40, pp. 1068–1071, 2007.

96. H. Liu, J. Kameoka, D. A. Czaplewski, and H. G. Craighead, "Polymeric nanowire chemical sensor," Nano Letters, vol. 4, no. 4, pp. 671–675, 2004.

97. Y. Zhou, M. Freitag, J. Hone et al., "Fabrication and electrical characterization of polyaniline-based nanofibers with diameter below 30 nm," Applied Physics Letters, vol. 83, no. 18, pp. 3800–3802, 2003.

98. D. Norris, M. M. Shaker, F. K. Ko, and A. G. MacDiarmid, "Electrostatic fabrication of ultrafine conducting fibers: polyaniline/polyethylene oxide blends," Synthetic Metals, vol. 114, no. 2, pp. 109–114, 2000.

99. G. MacDiarmid, W. E. Jones Jr., I. D. Norris et al., "Electrostatically-generated nanofibers of electronic polymers," Synthetic Metals, vol. 119, no. 1–3, pp. 27–30, 2001.

100. M. F. Shie, W. T. Li, C. F. Dai, and J. M. Yeh, "In vitro biocompatibility of electrospinning polyaniline fibers," IFMBE Proceedings, vol. 25, no. 10, pp. 211–214, 2009.

101. M. J. Díaz-de León, "Electrospinning nanofibers of polyaniline and polyaniline/(polystyrene and polyethylene oxide) blends," in Proceeding of the National Conference on Undergraduate Research (NCUR '01), Lexington, Kentucky, March 2001.

102. Y. Chen, C. Li, Z. Hou et al., "Polyaniline electrospinning composite fibers for orthotopic photothermal treatment of tumors in vivo," New Journal of Chemistry, vol. 39, no. 6, pp. 4987–4993, 2015. View at Publisher ·

103. R. Fryczkowski, M. Gorczowska, B. Fryczkowska, and J. Janicki, "The effect of solvent on the properties of nanofibres obtained by electrospinning from a mixture of poly(3-hydroxybutyrate) and polyaniline,"Synthetic Metals, vol. 166, no. 1, pp. 14–21, 2013.

104. M. R. Karim, "Fabrication of electrospun aligned nanofibers from conducting polyaniline copolymer/polyvinyl alcohol/chitosan oligossacaride in aqueous solutions," Synthetic Metals, vol. 178, pp. 34–37, 2013.

105. Y. Sharma, A. Tiwari, S. Hattori et al., "Fabrication of conducting electrospun nanofibers scaffold for three-dimensional cells culture," International Journal of Biological Macromolecules, vol. 51, no. 4, pp. 627–631, 2012.

106. P. Zamani, D. Higgins, F. Hassan et al., "Electrospun iron-polyaniline-polyacrylonitrile derived nanofibers as non-precious oxygen reduction reaction catalysts for PEM fuel cells," Electrochimica Acta, vol. 139, pp. 111–116, 2014.

107. Moayeri and A. Ajji, "Fabrication of polyaniline/poly(ethylene oxide)/non-covalently functionalized graphene nanofibers via electrospinning," Synthetic Metals, vol. 200, pp. 7–15, 2015.

108. Z. Sun, E. Zussman, A. L. Yarin, J. H. Wendorff, and A. Greiner, "Compound core-shell polymer nanofibers by co-electrospinning," Advanced Materials, vol. 15, no. 22, pp. 1929–1932, 2003.

109. Y. Zhang and G. C. Rutledge, "Electrical conductivity of electrospun polyaniline and polyaniline-blend fibers and mats," Macromolecules, vol. 45, no. 10, pp. 4238–4246, 2012.

110. Sarvi, V. Chimello, A. B. Silva, R. E. S. Bretas, and U. Sundararaj, "Coaxial electrospun nanofibers of poly(vinylidene fluoride)/polyaniline filled with multi-walled carbon nanotubes," Polymer Composites, vol. 35, no. 6, pp. 1198–1203, 2014.

111. Q. Wu, D. Wu, and Y. Guan, "Polyaniline sheathed electrospun nanofiber bar for in vivo extraction of trace acidic phytohormones in plant tissue," Journal of Chromatography A, vol. 1342, pp. 16–23, 2014.

112. Q. Wu, D. Wu, and Y. Guan, "In vivo fast equilibrium microextraction by stable and biocompatible nanofiber membrane sandwiched in microfluidic device," Analytical Chemistry, vol. 85, no. 23, pp. 11524–11531, 2013.

113. Q. Ma, J. Wang, X. Dong, W. Yu, and G. Liu, "Flexible ribbon-shaped coaxial electrical conductive nanocable array endowed with magnetism and photoluminescence," RSC Advances, vol. 5, no. 4, pp. 2523–2530, 2015.

114. H. Shao, Q. Ma, X. Dong et al., "Novel flexible belt-shaped coaxial microcables with tunable multicolor luminescence, electrical conductivity and magnetism," Physical Chemistry Chemical Physics, vol. 17, no. 34, pp. 21845–21855, 2015. View at Publisher ·

115. Y. Liu, Q. Ma, X. Dong, W. Yu, J. Wang, and G. Liu, "A novel strategy to directly fabricate flexible hollow nanofibers with tunable luminescence–electricity–magnetism trifunctionality using one-pot electrospinning," Physical Chemistry Chemical Physics, vol. 17, no. 35, pp. 22977–22984, 2015. View at Publisher ·

116. Y. Liu, Q. Ma, M. Yang et al., "Flexible hollow nanofibers: novel one-pot electrospinning construction, structure and tunable luminescence-electricity-magnetism trifunctionality," Chemical Engineering Journal, vol. 284, pp. 831–840, 2016. View at Publisher ·

117. Y. E. Miao, W. Fan, D. Chen, and T. Liu, "High-performance supercapacitors based on hollow polyaniline nanofibers by electrospinning," ACS Applied Materials & Interfaces, vol. 5, no. 10, pp. 4423–4428, 2013.View at Publisher ·

118. N. Lv, X. Dong, Q. Ma, J. Wang, W. Yu, and G. Liu, "Parallel spinnerets electrospinning construct and properties of electrical-luminescent bifunctional bistrand-aligned nanobundles," Journal of Materials Science, vol. 49, no. 5, pp. 2171–2179, 2014.

119. K. H. Lee, H. Y. Kim, Y. J. Ryu, K. W. Kim, and S. W. Choi, "Mechanical behavior of electrospun fiber mats of poly(vinyl chloride)/polyurethane polyblends," Journal of Polymer Science Part B: Polymer Physics, vol. 41, no. 11, pp. 1256–1262, 2003. View at Publisher ·

120. L. S. Nair, S. Bhattacharyya, J. D. Bender et al., "Fabrication and optimization of methylphenoxy substituted polyphosphazene nanofibers for biomedical applications," Biomacromolecules, vol. 5, no. 6, pp. 2212–2220, 2004.

121. K. H. Lee, H. Y. Kim, M. S. Khil, Y. M. Ra, and D. R. Lee, "Characterization of nano-structured poly(ε-caprolactone) nonwoven mats via electrospinning," Polymer, vol. 44, no. 4, pp. 1287–1294, 2003.

122. J. Yao, C. W. M. Bastiaansen, and T. Peijs, "High strength and high modulus electrospun nanofibers,"Fibers, vol. 2, no. 2, pp. 158–186, 2014. View at Publisher ·

123.H. Valentová and J. Stejskal, "Mechanical properties of polyaniline," Synthetic Metals, vol. 160, no. 7-8, pp. 832–834, 2010.

124.S. J. Pomfret, P. N. Adams, N. P. Comfort, and A. P. Monkman, "Electrical and mechanical properties of polyaniline fibres produced by a one-step wet spinning process," Polymer, vol. 41, no. 6, pp. 2265–2269, 2000.

125.W. Zhu, F. Masood, J. O'Brien, and L. G. Zhang, "Highly aligned nanocomposite scaffolds by electrospinning and electrospraying for neural tissue regeneration," Nanomedicine: Nanotechnology, Biology, and Medicine, vol. 11, no. 3, pp. 693–704, 2015. View at Publisher ·

126.K. T. Shalumon and J. P. Chen, "Scaffold-based drug delivery for cartilage tissue regeneration," Current Pharmaceutical Design, vol. 21, no. 15, pp. 1979–1990, 2015. View at Publisher ·

127.H. Chen, Y. Liu, Z. Jiang, W. Chen, Y. Yu, and Q. Hu, "Cell-scaffold interaction within engineered tissue,"Experimental Cell Research, vol. 323, no. 2, pp. 346–351, 2014.

128.B. Guo and P. X. Ma, "Synthetic biodegradable functional polymers for tissue engineering: a brief review,"Science China Chemistry, vol. 57, no. 4, pp. 490–500, 2014. View at Publisher · ·s

129.D. Chen, Y.-E. Miao, and T. Liu, "Electrically conductive polyaniline/polyimide nanofiber membranes prepared via a combination of electrospinning and subsequent in situ polymerization growth," ACS Applied Materials & Interfaces, vol. 5, no. 4, pp. 1206–1212, 2013.

130.M.-C. Chen, Y.-C. Sun, and Y.-H. Chen, "Electrically conductive nanofibers with highly oriented structures and their potential application in skeletal muscle tissue engineering," Acta Biomaterialia, vol. 9, no. 3, pp. 5562–5572, 2013.

131.T. H. Qazi, R. Rai, and A. R. Boccaccini, "Tissue engineering of electrically responsive tissues using polyaniline based polymers: a review," Biomaterials, vol. 35, no. 33, pp. 9068–9086, 2014.

132.M. Mattioli-Belmonte, G. Giavaresi, G. Biagini et al., "Tailoring biomaterial compatibility: in vivo tissue response versus in vitro cell behavior," International Journal of Artificial Organs, vol. 26, no. 12, pp. 1077–1085, 2003. · s

133.T. J. Rivers, T. W. Hudson, and C. E. Schmidt, "Synthesis of a novel, biodegradable electrically conducting polymer for biomedical applications," Advanced Functional Materials, vol. 12, no. 1, pp. 33–37, 2002.

134. N. K. Guimard, N. Gomez, and C. E. Schmidt, "Conducting polymers in biomedical engineering,"Progress in Polymer Science, vol. 32, no. 8-9, pp. 876–921, 2007.

135. Guiseppi-Elie, "Electroconductive hydrogels: synthesis, characterization and biomedical applications,"Biomaterials, vol. 31, no. 10, pp. 2701–2716, 2010.

136. N. Zelikin, D. M. Lynn, J. Farhadi, I. Martin, V. Shastri, and R. Langer, "Erodible conducting polymers for potential biomedical applications," Angewandte Chemie—International Edition, vol. 41, no. 1, pp. 141–144, 2002.

137. K. Gu, M. S. Kim, C. M. Kang, J.-I. Kim, S. J. Park, and C.-H. Kim, "Fabrication of conductive polymer-based nanofiber scaffolds for tissue engineering applications," Journal of Nanoscience and Nanotechnology, vol. 14, no. 10, pp. 7621–7626, 2014.

138. T. H. Qazi, R. Rai, D. Dippold et al., "Development and characterization of novel electrically conductive PANI-PGS composites for cardiac tissue engineering applications," Acta Biomaterialia, vol. 10, no. 6, pp. 2434–2445, 2014.

139. F. F. Azhar, A. Olad, and R. Salehi, "Fabrication and characterization of chitosan–gelatin/nanohydroxyapatite–polyaniline composite with potential application in tissue engineering scaffolds," Designed Monomers and Polymers, vol. 17, no. 7, pp. 654–667, 2014.

140. P. H. S. Picciani, E. S. Medeiros, Z. Pan, W. J. Orts, L. H. C. Mattoso, and B. G. Soares, "Development of conducting polyaniline/poly(lactic acid) nanofibers by electrospinning," Journal of Applied Polymer Science, vol. 112, no. 2, pp. 744–753, 2009.

141. M. P. Prabhakaran, L. Ghasemi-Mobarakeh, and S. Ramakrishna, "Electrospun composite nanofibers for tissue regeneration," Journal of Nanoscience and Nanotechnology, vol. 11, no. 4, pp. 3039–3057, 2011.

142. L. Shadi, M. Karimi, and A. A. Entezami, "Preparation of electroactive nanofibers of star-shaped polycaprolactone/polyaniline blends," Colloid and Polymer Science, vol. 293, no. 2, pp. 481–491, 2015.

143. M. Li, Y. Guo, Y. Wei, A. G. MacDiarmid, and P. I. Lelkes, "Electrospinning polyaniline-contained gelatin nanofibers for tissue engineering applications," Biomaterials, vol. 27, no. 13, pp. 2705–2715, 2006.

144. W.-H. Zimmermann and T. Eschenhagen, "Cardiac tissue engineering for replacement therapy," Heart Failure Reviews, vol. 8, no. 3, pp. 259–269, 2003.

145. T. Shimizu, M. Yamato, A. Kikuchi, and T. Okano, "Cell sheet engineering for myocardial tissue reconstruction," Biomaterials, vol. 24, no. 13, pp. 2309–2316, 2003.

146. L. Ghasemi-Mobarakeh, M. P. Prabhakaran, M. Morshed et al., "Application of conductive polymers, scaffolds and electrical stimulation for nerve tissue engineering," Journal of Tissue Engineering and Regenerative Medicine, vol. 5, no. 4, pp. e17–e35, 2011.

147. S. H. Ku, S. H. Lee, and C. B. Park, "Synergic effects of nanofiber alignment and electroactivity on myoblast differentiation," Biomaterials, vol. 33, no. 26, pp. 6098–6104, 2012.

148. P. Gentile, D. Bellucci, A. Sola, C. Mattu, V. Cannillo, and G. Ciardelli, "Composite scaffolds for controlled drug release: role of the polyurethane nanoparticles on the physical properties and cell behaviour," Journal of the Mechanical Behavior of Biomedical Materials, vol. 44, pp. 53–60, 2015.

149. X. Wang, Y. Yuan, X. Huang, and T. Yue, "Controlled release of protein from core-shell nanofibers prepared by emulsion electrospinning based on green chemical," Journal of Applied Polymer Science, vol. 132, no. 16, Article ID 41811, 2015.

150. H. Hezaveh and I. I. Muhamad, "Controlled drug release via minimization of burst release in pH-response kappa-carrageenan/polyvinyl alcohol hydrogels," Chemical Engineering Research and Design, vol. 91, no. 3, pp. 508–519, 2013.

151. Seyfoddin, A. Chan, W. Chen, I. D. Rupenthal, G. I. N. Waterhouse, and D. Svirskis, "Electro-responsive macroporous polypyrrole scaffolds for triggered dexamethasone delivery," European Journal of Pharmaceutics and Biopharmaceutics, vol. 94, pp. 419–426, 2015. View at Publisher ·

152. L. Weng and J. Xie, "Smart electrospun nanofibers for controlled drug release: recent advances and new perspectives," Current Pharmaceutical Design, vol. 21, no. 15, pp. 1944–1959, 2015.

153.M. R. Abidian, D.-H. Kim, and D. C. Martin, "Conducting-polymer nanotubes for controlled drug release," Advanced Materials, vol. 18, no. 4, pp. 405–409, 2006.

154.C. Martin and M. R. Abidian, "Conducting polymer nanotube actuators for precisely controlled release of medicine and bioactive molecules," U.S. Patent no. 8,936,794, U.S. Patent and Trademark Office, Washington, DC, USA, 2015.

155.J. Huang, S. Virji, B. H. Weiller, and R. B. Kaner, "Nanostructured polyaniline sensors," Chemistry, vol. 10, no. 6, pp. 1314–1319, 2004.

156.Z. Sadek, C. O. Baker, D. Powell, W. Wlodarski, R. B. Kaner, and K. Kalantar-Zadeh, "Polyaniline nanofiber based surface acoustic wave gas sensors—effect of nanofiber diameter on H2 response," IEEE Sensors Journal, vol. 7, no. 2, pp. 213–218, 2007. View at Publisher ·

157.Z. Sadek, W. Wlodarski, K. Kalantar-Zadeh, C. Baker, and R. B. Kaner, "Doped and dedoped polyaniline nanofiber based conductometric hydrogen gas sensors," Sensors and Actuators, A: Physical, vol. 139, no. 1-2, pp. 53–57, 2007.

158.Y. Liao, C. Zhang, Y. Zhang et al., "Carbon nanotube/polyaniline composite nanofibers: facile synthesis and chemosensors," Nano Letters, vol. 11, no. 3, pp. 954–959, 2011.

159.Z. Sadek, W. Wlodarski, K. Shin, R. B. Kaner, and K. Kalantar-zadeh, "A polyaniline/WO3 nanofiber composite-based ZnO/64° YX LiNbO3 SAW hydrogen gas sensor," Synthetic Metals, vol. 158, no. 1-2, pp. 29–32, 2008.

160.H. Tai, Y. Jiang, G. Xie, J. Yu, and X. Chen, "Fabrication and gas sensitivity of polyaniline-titanium dioxide nanocomposite thin film," Sensors and Actuators, B: Chemical, vol. 125, no. 2, pp. 644–650, 2007.

161.Ofer, R. M. Crooks, and M. S. Wrighton, "Potential dependence of the conductivity of highly oxidized polythiophenes, polypyrroles, and polyaniline: finite windows of high conductivity," Journal of the American Chemical Society, vol. 112, no. 22, pp. 7869–7879, 1990.

162.Mulchandani and N. V. Myung, "Conducting polymer nanowires-based label-free biosensors,"Current Opinion in Biotechnology, vol. 22, no. 4, pp. 502–508, 2011.

163. M. Gerard, A. Chaubey, and B. D. Malhotra, "Application of conducting polymers to biosensors,"Biosensors and Bioelectronics, vol. 17, no. 5, pp. 345–359, 2002.

164. Dhand, M. Das, M. Datta, and B. D. Malhotra, "Recent advances in polyaniline based biosensors,"Biosensors & Bioelectronics, vol. 26, no. 6, pp. 2811–2821, 2011.

165. N. Promphet, P. Rattanarat, R. Rangkupan, O. Chailapakul, and N. Rodthongkum, "An electrochemical sensor based on graphene/polyaniline/polystyrene nanoporous fibers modified electrode for simultaneous determination of lead and cadmium," Sensors and Actuators B: Chemical, vol. 207, part A, pp. 526–534, 2015.

166. H. J. Sharma, N. D. Sonwane, and S. B. Kondawar, "Electrospun SnO2/polyaniline composite nanofibers based low temperature hydrogen gas sensor," Fibers and Polymers, vol. 16, no. 7, pp. 1527–1532, 2015.View at Publisher ·

167. K. Low, C. B. Horner, C. Li et al., "Composition-dependent sensing mechanism of electrospun conductive polymer composite nanofibers," Sensors and Actuators B: Chemical, vol. 207, pp. 235–242, 2014.

168. Y. Wen, Y. Li, Y. Si et al., "Ready-to-use strip for l-ascorbic acid visual detection based on polyaniline/polyamide 66 nano-fibers/nets membranes," Talanta, vol. 144, pp. 1146–1154, 2015. View at Publisher ·

169. Li, N. Chartuprayoon, W. Bosze et al., "Electrospun polyaniline/poly(ethylene oxide) composite nanofibers based gas sensor," Electroanalysis, vol. 26, no. 4, pp. 711–722, 2014.

170. S. B. Kondawar, P. T. Patil, and S. P. Agrawal, "Chemical vapour sensing properties of electrospun nanofibers of polyaniline/ZnO nanocomposites," Advanced Materials Letters, vol. 5, no. 7, pp. 389–395, 2014.

171. W. Serrano, A. Meléndez, I. Ramos, and N. J. Pinto, "Sensor response of electrospun poly(lactic acid)/polyaniline nanofibers to aliphatic alcohol vapors of varying sizes," in Proceedings of the IEEE 9th Ibero-American Congress on Sensors (IBERSENSOR '14), pp. 1–4, IEEE, Bogotá, Colombia, October 2014.

172. R. Huang, Y. Z. Long, C. C. Tang, and H. D. Zhang, "Fabrication of nano-branched coaxial polyaniline/polyvinylidene fluoride fibers via electrospinning for strain sensor," Advanced Materials Research, vol. 853, pp. 79–82, 2013.

173.H. D. Zhang, C. C. Tang, Y. Z. Long et al., "High-sensitivity gas sensors based on arranged polyaniline/PMMA composite fibers," Sensors and Actuators A: Physical, vol. 219, pp. 123–127, 2014.View at Publisher ·

174.Y. A. Ismail, J. G. Martínez, and T. F. Otero, "Fibroin/polyaniline microfibrous mat. preparation and electrochemical characterization as reactive sensor," Electrochimica Acta, vol. 123, pp. 501–510, 2014.

175.Z. Pang, J. Fu, L. Luo, F. Huang, and Q. Wei, "Fabrication of PA6/TiO2/PANI composite nanofibers by electrospinning-electrospraying for ammonia sensor," Colloids and Surfaces A: Physicochemical and Engineering Aspects, vol. 461, no. 1, pp. 113–118, 2014.

176.Aussawasathien, J.-H. Dong, and L. Dai, "Electrospun polymer nanofiber sensors," Synthetic Metals, vol. 154, no. 1–3, pp. 37–40, 2005.

177.Q. Lin, Y. Li, and M. Yang, "Polyaniline nanofiber humidity sensor prepared by electrospinning," Sensors and Actuators B: Chemical, vol. 161, no. 1, pp. 967–972, 2012.

178.Q. Lin, Y. Li, and M. Yang, "Highly sensitive and ultrafast response surface acoustic wave humidity sensor based on electrospun polyaniline/poly(vinyl butyral) nanofibers," Analytica Chimica Acta, vol. 748, pp. 73–80, 2012.

179.Chen, X. Guo, Z. Wang, P. Wang, Y. Chen, and L. Lin, "Polyaniline nanofiber gas sensors by direct-write electrospinning," in Proceedings of the 24th IEEE International Conference on Micro Electro Mechanical Systems (MEMS '11), pp. 1369–1372, IEEE, Cancun, Mexico, January 2011. View at Publisher· · s

180.K. Singh and K. Mandal, "Engineering of high performance supercapacitor electrode based on Fe-Ni/Fe2O3-NiO core/shell hybrid nanostructures," Journal of Applied Physics, vol. 117, Article ID 105101, 2015. View at Publisher ·

181.Z. Tang, C.-H. Tang, and H. Gong, "A high energy density asymmetric supercapacitor from nano-architectured Ni(OH)2/carbon nanotube electrodes," Advanced Functional Materials, vol. 22, no. 6, pp. 1272–1278, 2012.

182.Z. B. Lei, L. Lu, and X. S. Zhao, "The electrocapacitive properties of graphene oxide reduced by urea,"Energy and Environmental Science, vol. 5, no. 4, pp. 6391–6399, 2012.

183. S. Chen, J. Zhu, X. Wu, Q. Han, and X. Wang, "Graphene oxide-Mno2 nanocomposites for supercapacitors," ACS Nano, vol. 4, no. 5, pp. 2822–2830, 2010.

184. C. Yuan, L. Yang, L. Hou et al., "Flexible hybrid paper made of monolayer Co3O4 microsphere arrays on rGO/CNTs and their application in electrochemical capacitors," Advanced Functional Materials, vol. 22, no. 12, pp. 2560–2566, 2012.

185. T. Kang, K. G. Neoh, and K. L. Tan, "Polyaniline: a polymer with many interesting intrinsic redox states," Progress in Polymer Science, vol. 23, no. 2, pp. 277–324, 1998.

186. S. Chaudhari, Y. Sharma, P. S. Archana et al., "Electrospun polyaniline nanofibers web electrodes for supercapacitors," Journal of Applied Polymer Science, vol. 129, no. 4, pp. 1660–1668, 2013.

187. Miao, C. Shao, X. Li et al., "Flexible solid-state supercapacitors based on freestanding electrodes of electrospun polyacrylonitrile@polyaniline core-shell nanofibers," Electrochimica Acta, vol. 176, pp. 293–300, 2015. View at Publisher ·

188. Y. Miao, J. Yan, Y. Huang, W. Fan, and T. Liu, "Electrospun polymer nanofiber membrane electrodes and an electrolyte for highly flexible and foldable all-solid-state supercapacitors," RSC Advances, vol. 5, no. 33, pp. 26189–26196, 2015. View at Publisher ·

189. Rose, N. Raghavan, S. Thangavel, B. U. Maheswari, D. P. Nair, and G. Venugopal, "Investigation of cyclic voltammetry of graphene oxide/polyaniline/polyvinylidene fluoride nanofibers prepared via electrospinning," Materials Science in Semiconductor Processing, vol. 31, pp. 281–286, 2015.

190. W. F. Mak, G. Wee, V. Aravindan, N. Gupta, S. G. Mhaisalkar, and S. Madhavi, "High-energy density asymmetric supercapacitor based on electrospun vanadium pentoxide and polyaniline nanofibers in aqueous electrolyte," Journal of the Electrochemical Society, vol. 159, no. 9, pp. A1481–A1488, 2012.

191. J. Burschka, N. Pellet, S.-J. Moon et al., "Sequential deposition as a route to high-performance perovskite-sensitized solar cells," Nature, vol. 499, no. 7458, pp. 316–319, 2013.

192. S. Pan, Z. Yang, P. Chen, J. Deng, H. Li, and H. S. Peng, "Wearable solar cells by stacking textile electrodes," Angewandte Chemie—International Edition, vol. 53, no. 24, pp. 6110–6114, 2014.

193. M. K. Nazeeruddin, F. De Angelis, S. Fantacci et al., "Combined experimental and DFT-TDDFT computational study of photoelectrochemical cell ruthenium sensitizers," Journal of the American Chemical Society, vol. 127, no. 48, pp. 16835–16847, 2005. View at Publisher · ·s

194. N. Papageorgiou, W. F. Maier, and M. Grätzel, "An iodine/triiodide reduction electrocatalyst for aqueous and organic media," Journal of the Electrochemical Society, vol. 144, no. 3, pp. 876–884, 1997. View at Publisher ·

195. Y.-L. Lee, C.-L. Chen, L.-W. Chong, C.-H. Chen, Y.-F. Liu, and C.-F. Chi, "A platinum counter electrode with high electrochemical activity and high transparency for dye-sensitized solar cells," Electrochemistry Communications, vol. 12, no. 11, pp. 1662–1665, 2010.

196. X. Yin, Z. S. Xue, and B. Liu, "Electrophoretic deposition of Pt nanoparticles on plastic substrates as counter electrode for flexible dye-sensitized solar cells," Journal of Power Sources, vol. 196, no. 4, pp. 2422–2426, 2011.

197. Y. Duan, Y. Chen, Q. Tang et al., "A dye-sensitized solar cell having polyaniline species in each component with 3.1%-efficiency," Journal of Power Sources, vol. 284, pp. 178–185, 2015. View at Publisher·

198. S. Peng, P. Zhu, Y. Wu, S. G. Mhaisalkar, and S. Ramakrishna, "Electrospun conductive polyaniline-polylactic acid composite nanofibers as counter electrodes for rigid and flexible dye-sensitized solar cells," RSC Advances, vol. 2, no. 2, pp. 652–657, 2012.

199. S. Ebrahim, M. Soliman, M. Anas, M. Hafez, and T. M. Abdel-Fattah, "Dye-sensitized solar cell based on polyaniline/multiwalled carbon nanotubes counter electrode," International Journal of Photoenergy, vol. 2013, Article ID 906820, 6 pages, 2013.

# Chapter 10

# Tailoring Crystallinity of Electrospun Plla Fibres by Control of Electrospinning Parameters

**Olubayode Ero-Phillips, Mike Jenkins and Artemis Stamboulis** *

Biomaterials Group, School of Metallurgy & Materials, College of Engineering and Physical Sciences, University of Birmingham, Birmingham B15 2TT, UK

## ABSTRACT

Poly(L-lactic acid) (PLLA) fibers were fabricated by electrospinning. The effects of various electrospinning process parameters on the thermal properties, especially the crystallinity of the electrospun fibers were investigated. Thermal analysis of the fibers revealed that they exhibited degree of crystallinity ranging from 23% to 46% while that for the as-received granules was approximately 37%, suggesting that the crystallinity of electrospun PLLA fibres can be controlled by optimizing the electrospinning process. This finding is very important because crystallinity affects polymer properties such as degradation, stiffness, yield stress, modulus and tensile strength, solubility, optical and electrical properties which will in turn affect the behavior of these materials when they are utilized in energy, environment, defense and security applications. The results presented in this paper show that the degree of crystallinity of the electrospun fibers decreased with increasing the polymer solution concentration. Furthermore, an optimum electrospinning voltage at which maximum degree of crystallinity can be obtained was observed. At voltages higher or lower than the optimum electrospinning voltage, the degree of crystallinity will decrease or increase, respectively. The effect of the needle tip to collector distance (NTCD) on the degree of crystallinity follows no predictable and consistent pattern.

## KEYWORDS

poly(L-lactic acid); electrospinning; electrospinning parameters; degree of crystallization

## 1. INTRODUCTION

Interest in electrospinning has grown exponentially over the last two decades due to the versatility of the technique. Electrospinning is capable of producing fibers with high strength [1] and large surface area-to-volume ratios [2]. An advantage of the process over conventional fiber forming methods is that only a small amount of polymer, as low as 50 mg, is required for electrospinning, in comparison to tens of kilogrammes of polymer required in conventional spinning methods [3]. However, in spite of the amount of research that has been carried out on electrospinning and electrospun fibers, studies on the effect of electrospinning on polymer properties have been limited. Most studies have focused on the effect of electrospinning parameters on electrospun fibre diameter [4,5,6,7] and the development of controlled morphology [8,9,10,11,12]. However, very few have focused on the effect of electrospinning on polymer properties, such as molecular orientation of polymer chains [2], polymer crystallization chain conformation [3] and polymer degradation [3,13].

Although electrospun fibers have potential applications in many different areas, their use as tissue engineering scaffold materials is probably the most popular application of electrospun fibers [14]. Furthermore, because most electrospun fibers are proposed for use in biomedical applications, it is imperative that an understanding of their properties, such as crystallinity and degradation kinetics, which will impact their functionality when in service, be properly developed.

Although non-biodegradable materials may also be used as tissue engineering scaffolds [15,16], biodegradable materials with controllable degradation rates are more commonly considered [17]. Specifically, poly(lactic acid), poly(glycolic acid) (PGA) and their copolymers are the most commonly researched polymers for tissue engineering because of their biocompatibility, bioresorbability, controllable degradation rates, and also because they have been approved by the Food and Drug Administration for biomedical applications [18]. Crystallinity has been reported to vary directly proportionally to the rate of degradation [19]. It is therefore crucial to have an understanding of the effect of electrospinning on polymer crystallinity.

Stephens *et al.* reported that the crystalline structure of Nylon-6 changed from α- to γ-crystalline structure due to the high stresses present during the electrospinning process, but no information was given about the effect of electrospinning on the degree of polymer crystallinity [3]. Regarding the effect of electrospinning on polymer crystallinity, it has been generally reported that electrospun fibers possess lower crystallinity than the degree of crystallinity of the as-received polymers and films cast from polymer solutions or melts. Bognitzki *et al.* reported that electrospun PLLA fibers had lower degree of crystallinity than fibers from melt or films from dilute solutions. This was attributed to the rapid deposition rate of the fibers during electrospinning, which allows very little time for crystallization to occur [20]. Zong *et al.* also reported lower degree of crystallinity in electrospun PLLA fibers as compared to as-received PLLA resin and attributed this also to the rapid solidification of the stretched polymer chains during the electrospinning process [21].

Other researchers have also compared the crystallinity of various electrospun fibers to the crystallinity of the pristine materials which they were electrospun from. Ramdhanie *et al.* electrospun blends of PLA and PGA to produce both aligned and random fibers [22]. They reported generally lower values of crystallinity [measured by differential scanning calorimetry (DSC)] for the PLA fibers than for the pristine PLA pellets, while some of the PGA electrospun fibers showed an increase in crystallinity. Deitzel *et al.* reported a wide angle X-ray diffraction (WAXD) study of electrospun polyethylene oxide (PEO) fibers and pristine PEO powder [23,24]. The (120) and (112) reflections of the electrospun fibres were weak and broad in comparison to the sharp peaks obtained for the powder, indicating that the crystalline microstructure of the electrospun fibres was not well developed. In their work on silk-like polymer with fibronectin functionality (SLPF), Buchko *et al.* also observed lower crystallinity for electrospun fibres than in the pristine material using wide-angle X-ray scattering (WAXS) and attributed this also to the rapid solidification of the electrospun fibres [25]. Ayutsede *et al.* obtained crystallinities of 48% and 39% for pristine *Bombyx mori* silk and electrospun fibres, respectively, using WAXD [26]. Similarly, Ribeiro *et al.* recently reported that electrospun PLLA fibres were nearly amorphous, and they attributed this to the rapid solidification of the electrospinning jet [27].

In the contrary, Zhao *et al.* suggested, that the degree of crystallinity of electrospun ethyl-cyanoethyl cellulose fibres was greatly influenced by the crystallization time as the jet travels from the capillary tip to the collector and the molecular orientation was brought about by the electrostatic field [28].

Specifically, the degree of crystallinity of the electrospun fibres increased with increasing voltage up to a maximum after which the degree of crystallinity began to decline.

More recently, Kongkhlang *et al*. reported the electrospinning of random fibres of polyoxymethylene (POM). The authors observed molecular orientation in the electrospun fibres, as well as a chain structure in which fully extended molecules aligned parallel to the needle axis. The presence of molecular orientation in the electrospun fibres was attributed to the substantial stretching forces imposed on the polymer solution by the electric field during the electrospinning process [2].

The high elongation strains and shear stresses that are exerted on polymer chains within the electric field during electrospinning are reported to cause the macromolecular chains to align along the fibre axis, leading to high degree of molecular orientation in the fibres [1,2,3,29,30]. Molecular orientation has been shown to be directly proportional to the degree of crystallinity [1,30]. It is therefore reasonable to suggest, that if electrospinning is able to cause molecular orientation of polymer chains, then it may be possible to influence the degree of crystallinity by carefully controlling the electrospinning parameters.

The crystalline properties of a material are important when considering the material for medical applications because the crystalline properties of a polymer have a direct bearing upon the degradation behavior *in vivo* [19,31,32,33]. A high or low degradation rate may be desirable, depending on the application of the material. It would therefore be beneficial if the degree of crystallinity of electrospun polymer fibres can be tailored by controlling the electrospinning process/parameters.

Although poly(lactic acid) is one of the most researched biodegradable polymers [34,35], there is no comprehensive systematic study in the literature about the effect of varying different electrospinning parameters on the degree of crystallinity of its electrospun fibres. The previous work on the study of crystallinity of electrospun PLLA in the literature [20,21,36] has all been carried out within a fixed set of parameters which limits their results to the parameters used. In a recent paper by Ribeiro *et al*. where a wider set of parameters (*i.e.*, voltage and flow rate) was used, the degree of crystallinity of the electrospun fibres was controlled by an additional annealing process after electrospinning. In that case, a higher degree of crystallinity was achieved, but not directly during electrospinning [27]. This paper aims to investigate the possibility of controlling the degree of crystallinity of electrospun PLLA fibres in a single process by varying the electrospinning parameters.

# 2. EXPERIMENTAL SECTION

## 2.1. Materials

PLLA granules of molecular weight ≈220,000 g/mol were supplied by Goodfellow Chemicals, Cambridge, UK. Dichloromethane (DCM) (puriss. p.a., ACS reagent, ≥99.9%) was supplied by Sigma Aldrich, UK, while *N,N*-Dimethylformamide (DMF) (>99% assay) was supplied by Fisher Chemicals, UK. Four different solutions of 2, 5, 8 and 10 wt% concentration were made by dissolving PLLA granules in a 75:25 solution of Dichloromethane/*N,N*-Dimethylformamide.

## 2.2. Methods

### 2.2.1. Kinematic Viscosity

The kinematic viscosity of the different PLLA-DCM/DMF solutions described above was obtained using an Ostwald U-tube viscometer (Cannon Instrument Company, USA) enclosed in a water bath at 25 °C. The viscosities were calculated using the following formula supplied by the manufacturer

$$\nu = t \times k \tag{1}$$

Where $\nu$ = kinematic viscosity (cSt), $t$ = efflux time (s), and $k$ = viscometer constant (0.01 cSt/s at 25 °C).

### 2.2.2. Electrospinning

The polymer solutions were electrospun using a Model ES1a electrospinning machine manufactured by Electrospinz Ltd., New Zealand. The electrospinning apparatus is shown in Figure 1. A static aluminium foil target was used to collect the non-woven electrospun fibres. A systematic variation of three electrospinning parameters, namely: polymer solution concentration, applied voltage and needle tip to collector distance (NTCD) was conducted. Briefly, during each electrospinning experiment, two of these parameters were held constant while the third one was varied. The electrospinning apparatus has been designed by the manufacturers to function with a gravity feed system which ensures the internal pressure of the droplet at the needle tip is constant

[37]. Therefore, the system does not require a syringe pump to control the flow rate. The electrospinning was carried out at room temperature. The electrospun fibres were stored in a dessicator for one week to allow for complete evaporation of solvent. The 2 and 10 wt% solutions could not be electrospun under all varied electrospinning parameters due to their low and high viscosities, respectively (details discussed in Section 3.1). Additionally, a low yield of fibres from the 5 wt% and 8 wt% solution concentrations was also observed using 10 kV applied voltage, showing that 10 kV was too low for successful electrospinning of the polymer solution.

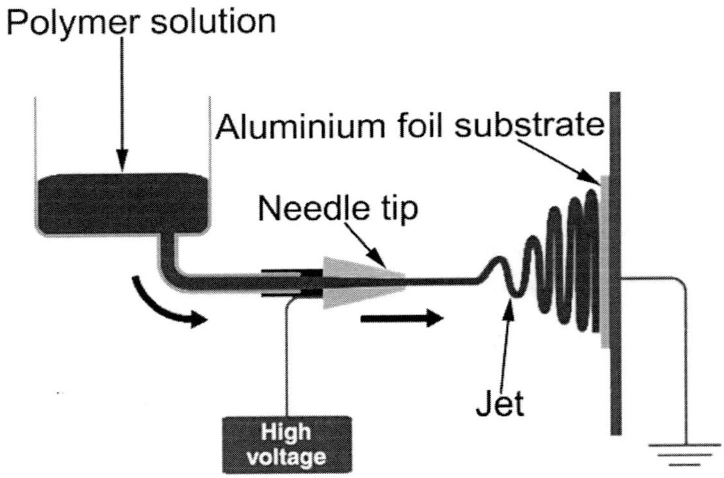


**Figure 1.** Schematic representation of ES1a electrospinning machine.

## 2.2.3. Characterisation of Electrospun Polymer Fibres

The morphology of the electrospun fibres was obtained using a Philips XL-30 FEG SEM (SE mode) at an accelerating voltage of 20 kV. Four different specimens were cut from each sample and then mounted on an aluminium stub using a carbon adhesive film. The specimens were sputter coated with platinum using an Emscope SC500 (Emscope laboratories, Kent, UK) Sputter Coater at a current of 25 mA for 3 min.

A Perkin Elmer Differential Scanning Calorimeter (DSC 7) was used for the thermal analysis of the elctrospun polymer fibres. Approximately 6 mg of samples were crimped into standard DSC aluminium pans. The fibres were taken through a temperature scan of 25 °C to 200 °C at a heating rate of 10 °C/min, held at 200 °C for 2 min, followed by cooling to 25 °C at a cooling rate of 10 °C/min. With the aid of the Pyris™ software (Perkin Elmer) installed on the DSC equipment, the glass transition temperature ($T_g$) was calculated from the endothermic transition relating to the glass transition on DSC scans using the half-$C_p$ (heat capacity) extrapolated method.

The values for the degree of crystallinity were calculated using the formula [38,39,40] below:

$$X_c\,(\%) = \frac{\Delta H_f^m - \Delta H_c^m}{\Delta H_f^{100\%}} \times 100\% \tag{2}$$

where $X_c$ (%) = Degree of crystallinity (%), $\Delta H_f^m$ = Heat of fusion (J·g$^{-1}$), $\Delta H_c^m$ = Heat of cold crystallization (J·g$^{-1}$) and, $\Delta H_f^{100\%}$ = Heat of fusion for 100% crystalline material (93.6 J·g$^{-1}$) [39].

# 3. RESULTS AND DISCUSSION

## 3.1. Effect of Polymer Solution Parameters and Processing Conditions on the Electrospinning Process

Several attempts were made to electrospin fibres from the 2 wt% PLLA solution (kinematic viscosity of 2.22 cSt), however a mixture of electrospraying and electrospinning was observed (Figure 2A). The lower polymer solution concentration resulted in a lower polymer chain entanglement density [41], which caused the chains to break up at the tip of the capillary due to the high surface tension [24]. Fibres, as well as some beads were obtained when the polymer solution concentration was increased to 5 wt% (Figure 2B), however, when the polymer solution concentration was increased to 8 wt%, bead-free fibres were obtained (Figure 2C). The 10 wt% polymer solution was too viscous with a kinematic viscosity of 142.57 cSt (Figure 3) and it was difficult to electrospin, even at very high electrospinning voltages.

The polymer solution concentration has a direct bearing upon the number of polymer chain entanglements within the solution and is one of the major factors which determine if fibres can be formed during electrospinning. As has been previously reported in the literature, there is an optimum range of polymer solution concentration within which uniform fibres can be electrospun when all other parameters are held constant [42], and in our study, this range of polymer solution concentration was found to between >2 wt% and <10 wt%.

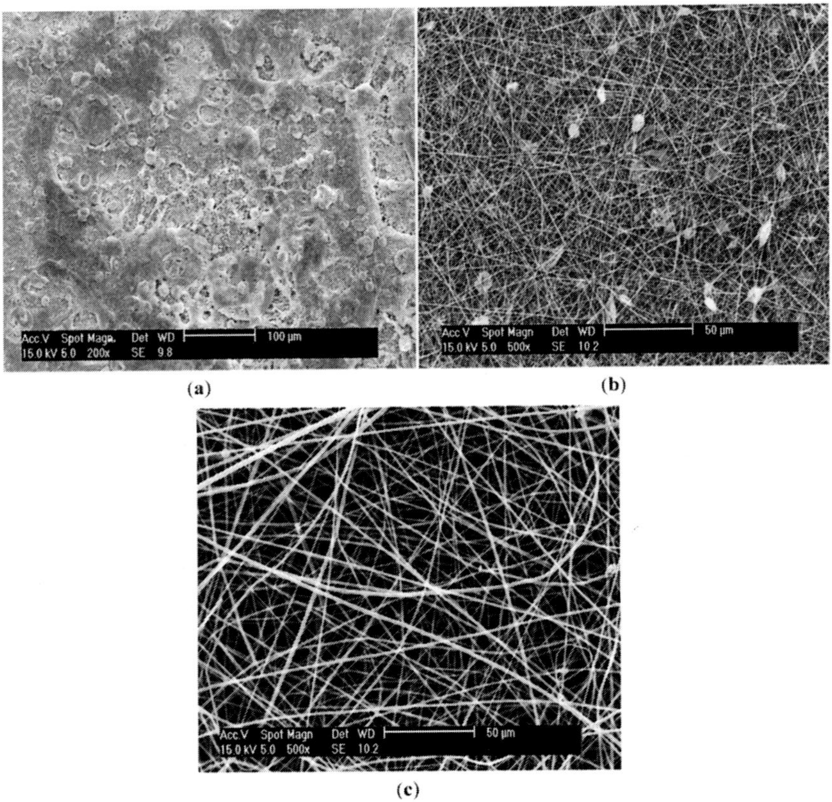


**Figure 2.** SEM micrographs of electrospun polymer fibres obtained when electrospinning from (**a**) 2 wt%; (**b**) 5 wt%; and (**c**) 8 wt% polymer solutions at 20 kV and NCTD of 12.5 cm, respectively.

When electrospinning was carried out at 10 kV, the amount of fibres produced was very small. The voltage applied during electrospinning induced charges into the polymer solution/melt and it was when the electrostatic forces created by these charges overcame the surface tension of the solution that electrospinning occurred. Hence, when the applied voltage was too low, the electrostatic forces were too low to initiate electrospinning. Similar to polymer solution concentration, the applied voltage has been shown to have a window within which electrospinning can be successful [4]. As mentioned above, electrospinning at 10 kV was not possible and therefore this is considered to be the minimum possible value.

Figure 3 shows the variation of viscosity with polymer solution concentration. The viscosity of the polymer solution increased with increasing polymer solution concentration as a result of increased number of polymer chain entanglements [41]. The increase in viscosity is dependent on polymer solution concentration [43].

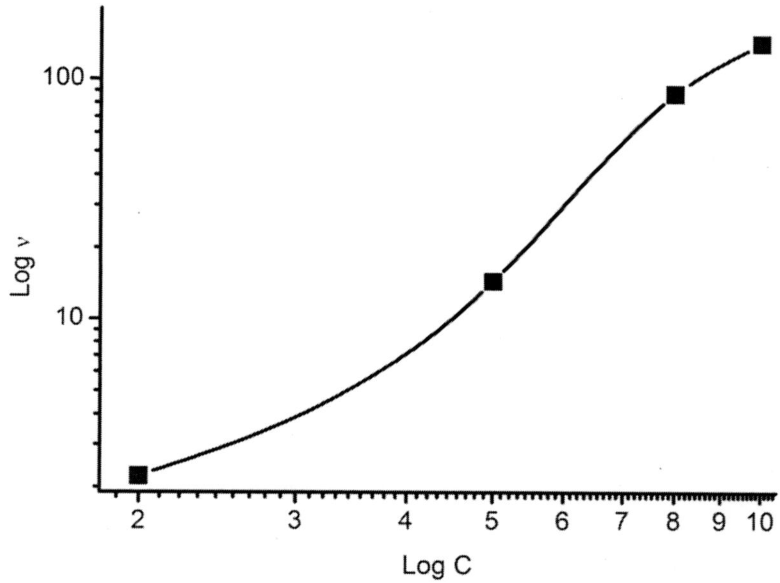


**Figure 3.** Variation of kinematic viscosity with polymer solution concentration (where *v* is kinematic viscosity and C is polymer solution concentration).

## 3.2. Effect of Electrospinning on the Glass Transition Temperature ($T_g$)

The glass transition temperature, $T_g$, of the electrospun fibres under any of the experimental parameters and for all the polymer solutions was lower than that of the as-received granules (67 °C) and ranged from 52 °C to 63 °C (Table 1). Varying the voltage or the NTCD had no effect on the $T_g$ of the electrospun fibres. The $T_g$ generally increased when the polymer concentration was increased from 5 wt % to 8 wt%; this was due to increased polymer chain interaction, but the $T_g$ remained lower compared to the as-received polymer. A decrease in $T_g$ values has previously been also reported for electrospun fibres [21,44,45]. Zong *et al.* attributed the decrease in $T_g$ in electrospun PLLA fibres to the plasticisation of the nanofibres by air and the large surface to volume ratio [21]. Similarly, Carrizales *et al.* reported a decrease of 20 °C for $T_g$ of electrospun poly(methyl methacrylate) (PMMA) nanofibres during the 1st heating cycle of a DSC study compared to the $T_g$ of the as-received PMMA powder. However, during the 2nd heating cycle of the electrospun nanofibres, it was reported that the $T_g$ of the electrospun nanofibres was equal to that of the as-received PMMA powder. Carrizales *et al.* attributed this reduction in the $T_g$ values of electrospun PMMA nanofibres to the plasticisation of the fibres by absorbed water, although their work was not clear about the source of the absorbed water [46]. We did not expect this to be the case in our studies because the electrospun fibres were properly dried prior to DSC characterisation. In addition, the glass transition temperatures measured during the 2nd DSC cycle (Table 1) were not much different to the 1st DSC cycle. Unfortunately, it is not clear from this work and the literature why there is a reduction in $T_g$ after epectrospinning. Zong *et al.*, Kim *et al.* and Cui *et al.* suggested that the reduction in $T_g$ is possibly due to three factors: 1. High surface area [21], 2. Increase in segmental mobility as a result of decreased chain entanglements [13], and 3. Higher molecular orientation and stress within polymer chains [44]. It is possible that it is not only one reason that causes the reduction in $T_g$. From our results it is clear that the $T_g$ increased with increasing the polymer concentration. This could possibly be related to segmental mobility however, this is still only a speculation.

## 3.3. Effect of Polymer Solution Concentration on Degree of Crystallinity

Figure 4 presents DSC scans of polymer fibres electrospun from 5 wt% and 8 wt% PLLA solutions at 20 kV and 10 cm NTCD. Just above the $T_g$, a hysteresis endothermic peak was observed (Figure 4) characteristic of volume relaxation due to the rapid structure formation of the fibres during electrospinning as has also been reported by Bognitzki *et al.* [20]. An exothermic peak observed around 75 °C was associated with "cold crystallization". Cold crystallization has been reported the literature [21,36]. Cold crystallization is a transition that occurs at low temperature due to enhanced primary nucleation when heating the polymer above its glass transition temperature when polymer chains have enough energy to become mobile and achieve an ordered state via an exothermic reaction [22,40]. The occurrence and extent of this reaction is dependent on the processing history of the sample. Mezghani and Spruiell reported that generally, a crystalline or semi-crystalline polymer will have a lower tendency to undergo this transition than a less crystalline or amorphous polymer [38] and this will be reflected by a small $\Delta H_c$ for the crystalline/semi-crystalline polymer and a large $\Delta H_c$ for an amorphous polymer. In this study, the cold crystallization enthalpy ($\Delta H_c$), increased from 2.5 $J{\cdot}g^{-1}$ to 18.3 $J{\cdot}g^{-1}$ when the polymer solution concentration was increased from 5 wt% to 8 wt% and voltage and NTCD were held constant at 15 kV and 10 cm, respectively (Figure 4).

At low polymer solution concentrations, the polymer chains are more mobile due to relatively small number of chain entanglements resulting in higher molecular orientation during electrospinning compared to fibres resulting from higher solution concentrations. Also, the solidification process is expected to be slower for a low polymer solution concentration resulting in higher electric-field assisted molecular orientation during electrospinning [36]. Therefore, it can be said that low polymer concentration solutions can yield fibres with higher molecular orientation and most likely lower degree of cold crystallization. Using Equation 1 therefore the degree of crystallinity after electrospinning is expected to be higher. Conversely, higher polymer concentration solutions can yield fibres with reduced molecular orientation due to higher viscosity and most likely higher degree of cold crystallization and therefore generally lower degree of crystallinity. Based on the above explanation the reduction in the degree of crystallinity with increasing polymer solution concentration can be justified. However, in the exceptional case of fibers electrospun from the 8 wt% polymer solution concentration at a voltage of 25 kV, the degree of crystallinity increased compared to the fibers electrospun from the 5 wt% polymer concentration solution (Table 1). The reason for this will be discussed in Section 3.4.

## 3.4. Effect of Voltage during Electrospinning on the Degree of Crystallinity

As shown in Figure 5, the degree of crystallinity of the fibers electrospun from the 5 wt% PLLA solution generally decreased with increasing voltage from 15 kV to 25 kV albeit in some cases, the reduction was only minimal. For instance, when the voltage was increased from 15 kV to 25 kV at a constant NTCD of 10 cm, the cold crystallization enthalpy, $\Delta H_c$, increased from 2.5 $J{\cdot}g^{-1}$ to 6.7 $J{\cdot}g^{-1}$ (Figure 6). Consequently, the degree of crystallinity, $X_c$, calculated using Equation 1, decreased with increasing the voltage from 15 to 25 kV (Figure 5). On the contrary, the degree of crystallinity for the fibers electrospun from the 8 wt% solution concentration increased as the voltage was increased from 15 kV to 25 kV (Figure 7).

From Figure 5 and Figure 7, it is obvious that there was a voltage at which a maximum degree of crystallinity was obtained. We have called this voltage the optimum electrospinning voltage. For the fibers electrospun from the 5 wt%

solution concentration, the maximum degree of crystallinity was obtained when electrospinning at 15 kV, while for the fibers electrospun from the 8 wt% solution concentration, the maximum degree of crystallinity was obtained when electrospinning at 25 kV.

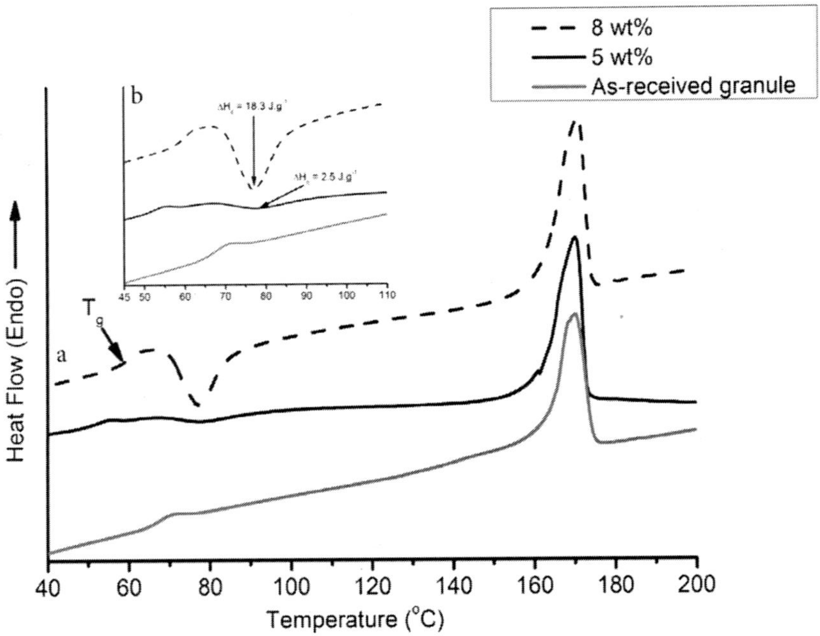


**Figure 4.** (**a**) DSC scans of fibers electrospun from 5 wt% and 8 wt% solutions at 15 kV and 10 cm needle tip to collector distance (NTCD); (**b**) Cold crystallization region of scans shown in A.

During the electrospinning process, as the jet travels from the needle tip to the collector, both molecular orientation induced by the electric field, and crystallization of the polymer occur [28]. Also, as the voltage is increased during electrospinning, the number of charges in the polymer solution increases and the effect of this is that there will be more stretching of the jet [4]. Furthermore, increasing the voltage will lead to an increase in the alignment of the polymer chains (*i.e.*, molecular orientation) during electrospinning and consequently an increase in the degree of crystallinity of the electrospun fibers as reported by Zhao *et al.* [28]. However, this increase in the degree of crystallinity with an increase in electrospinning voltage is not expected to continue indefinitely, on the contrary, above a certain voltage (the optimum electrospinning voltage), the degree of crystallinity will reduce. This is because the degree of crystallinity is also influenced by the flight time of the jet, which at high voltages is short enough to leave insufficient time for the polymer to crystallize [47]. Increasing the voltage beyond the optimum electrospinning voltage will lead to a reduction in the flight time due to the higher acceleration of the jet, resulting in decreased crystallinity [28,47]. Generally, the degree of crystallinity will increase with increasing electrospinning voltage until the optimum electrospinning voltage is reached; above the optimum electrospinning voltage, the degree of crystallinity is reduced due to the effect of the acceleration imposed on the spun fibers by the electric field.

For the fibers electrospun from the 5 wt% solution concentration, as the maximum degree of crystallinity ($X_c$) was obtained at 15 kV, after which $X_c$ began to decrease, it is likely that if a voltage lower than 15 kV was used to electrospin the polymer solution, the degree of crystallinity would increase with voltage up to a maximum at *circa* 15 kV. Above this voltage, reduction of the flight time of the jet occurs with a consequent decrease in the degree of crystallinity. It was attempted to electrospin the polymer solutions at lower voltages than 15 kV but this was not possible as has been explained in the Experimental Section. Consequently, it is believed that the range of voltages used in this study may have been above the optimum electrospinning voltage. Within this range, it is likely that increases in voltage will lead to increased acceleration of the jet to the extent that it reduces its flight time; the result of this is that there was insufficient time for crystallization and hence lower degree of crystallinity was observed beyond 15 kV.

For the fibers electrospun from the 8 wt% solution concentration, the degree of crystallinity reached a maximum at 25 kV when the voltage was increased from 15 kV to 25 kV (Figure 7). This suggests that the optimum voltage at which maximum degree of crystallinity can be obtained is at or close to 25 kV. It is therefore believed that the range of voltages used in this study fall below the optimum electrospinning voltage. It is very likely that further increases in voltage beyond 25 kV will lead to a reduction in the degree of crystallinity.

It was interesting to note, however, that the optimum voltage at which maximum degree of crystallinity can be obtained for fibers electrospun from the 5 wt% solution concentration was lower than that for fibers electrospun from 8 wt% solution concentration, *i.e.*, 15 kV and 25 kV for the 5 wt% and 8 wt% solutions, respectively. As has been explained in the previous section, the polymer chains are more mobile in polymer solutions of low concentration; it is therefore believed, that this will make them more susceptible to influences of voltage at low voltage values. As the polymer solution concentration is increased however, there are more polymer chain entanglements, hence higher solution viscosity, resulting in the polymer chains exhibiting a higher resistance to stretching [48] during electrospinning. As a result, the probability of the electric field inducing molecular orientation in the electrospun fibers at low voltages will be small, except at higher voltages (in this study, this was observed to be approximately ≥ 25 kV).

**Table 1.** Variation of $T_g$ and $X_c$ with electrospinning parameters. (n/a–not applicable; $X_c$ was calculated during 1st heating cycle).

| PLLA electrospinning solution | Electrospinning parameters | | $T_g$ (°C) | | $X_c$ (%) | PLLA electrospinning solution | Electrospinning parameters | | $T_g$ (°C) | | $X_c$ (%) |
|---|---|---|---|---|---|---|---|---|---|---|---|
| | NTCD (cm) | Voltage (kV) | 1st heating cycle | 2nd heating cycle | | | NTCD (cm) | Voltage (kV) | 1st heating cycle | 2nd heating cycle | |
| Granule | n/a | n/a | 67 | n/a | 37.4 | n/a | n/a | n/a | - | - | - |
| | | 15 | 52 | 56 | 45.3 | | | 15 | 61 | 61 | 37.3 |
| | 7.5 | 20 | 54 | 54 | 45.1 | | 7.5 | 20 | 64 | 61 | 43.8 |
| | | 25 | 52 | 54 | 44.8 | | | 25 | 63 | 61 | 46.3 |
| | | 15 | 53 | 54 | 43.9 | | | 15 | 60 | 61 | 30.5 |
| | 10 | 20 | 52 | 55 | 42.2 | | 10 | 20 | 60 | 61 | 36.7 |
| 5% | | 25 | 53 | 54 | 37.9 | 8% | | 25 | 62 | 60 | 43.9 |
| | | 15 | 55 | 54 | 45.5 | | | 15 | 60 | 61 | 33.4 |
| | 12.5 | 20 | 53 | 55 | 43.7 | | 12.5 | 20 | 58 | 61 | 36.9 |
| | | 25 | 54 | 55 | 42.4 | | | 25 | 64 | 61 | 46.6 |
| | | 15 | 52 | 55 | 46.6 | | | 15 | 60 | 61 | 31.2 |
| | 15 | 20 | 55 | 54 | 28.4 | | 15 | 20 | 60 | 61 | 34.7 |
| | | 25 | 56 | 54 | 26.2 | | | 25 | 63 | 61 | 45.8 |

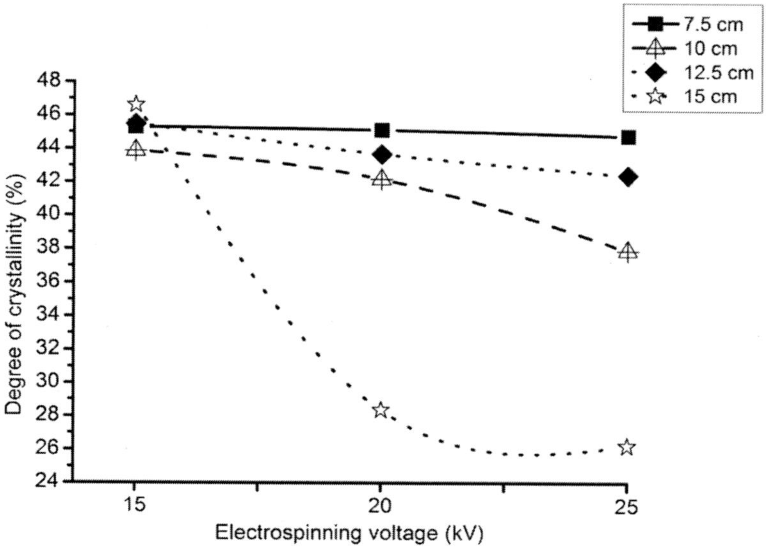


**Figure 5.** Effect of voltage on the degree of crystallinity of fibers electrospun from 5 wt% solution concentration at different NTCDs.

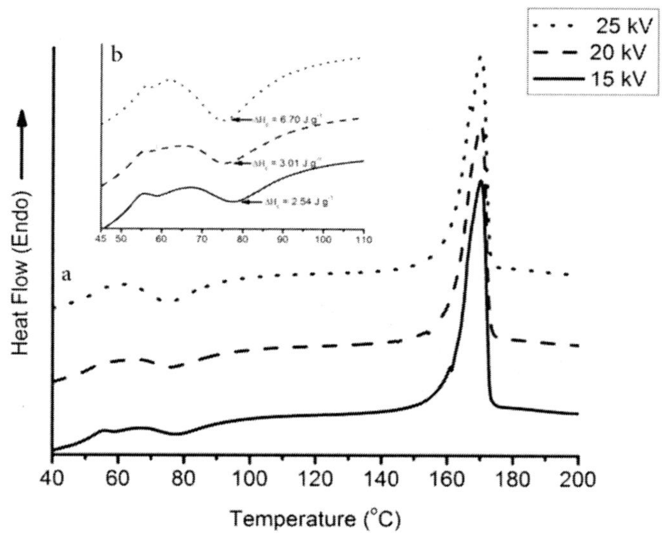


**Figure 6.** (**a**) DSC scans of fibers electrospun from 5 wt% solution at a constant NTCD of 10 cm and varying voltages; (**b**) Cold crystallization region of scans shown in A.

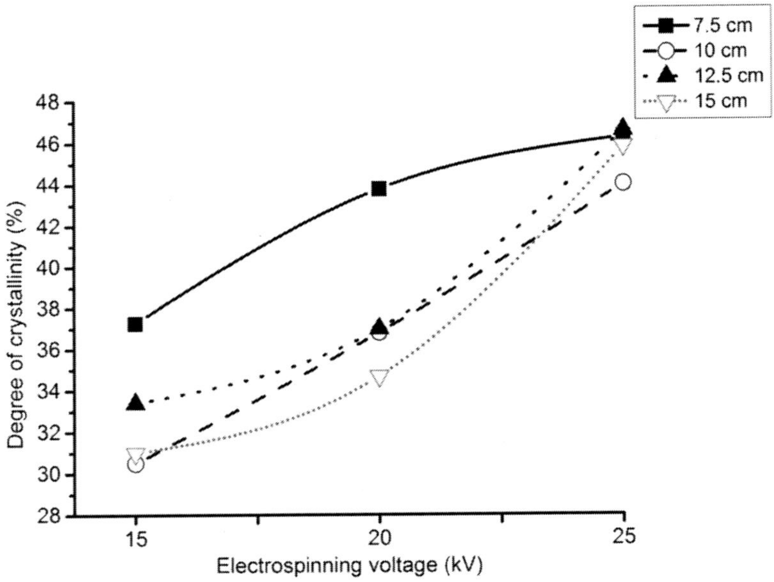


**Figure 7.** Effect of voltage on the degree of crystallinity of fibers electrospun from 8 wt% solution concentration at different NTCDs.

A comparison of the degree of crystallinity of fibers electrospun from the 8 wt% solution at 25 kV with the degree of crystallinity of fibers electrospun from the 5 wt% solution at the same voltage shows that the degree of crystallinity was slightly higher for the first case. As explained above, at voltages above the optimum voltage at which maximum degree of crystallinity can be obtained the degree of crystallinity of the fibers will decrease. As the optimum voltage for fibers electrospun from the 5 wt% solution was at or close to 15 kV, any further increase in voltage will lead to a reduction in the flight time of the jet and consequently a decrease in the degree of crystallinity. However, for the fibers electrospun from the 8 wt% solution, because the optimum voltage was at *circa* 25 kV, the maximum degree of crystallinity was observed at this voltage.

## 3.5. Effect of NTCD on the Degree of Crystallinity

Ideally, increasing the NTCD should result in an increase in the flight time of the jet [4]. Since molecular orientation can be induced by the electric field as the jet travels from the needle tip to the collector, if the flight time of the jet is increased by increasing the NTCD [28], theoretically, it would be reasonable to suggest that the molecular orientation of the electrospun fibers could increase with a consequent increase in the degree of crystallinity. The recent work of Tong and Wong however, show that increasing the NTCD may not necessarily increase the flight time of the jet due to the complicated three-dimensional spiralling trajectory of the jet [4]. In this study, for fibers electrospun from the 5 wt% solution, no constant trend was observed in the variation of the degree of crystallinity ($X_c$) with changing NTCD, suggesting that the NTCD influences the degree of crystallinity of the electrospun fibers but not in a predictable and consistent way (Figure 5). Although the reason for this is not particularly clear, we suggest that it could be due to the inconsistency of trajectory of the polymer jet as Tong and Wang suggested [4]. However, it is interesting to note, that for the fibers electrospun from the 5 wt% solutions, at 15 kV which was the optimum electrospinning voltage for this solution, the degree of crystallinity remained fairly constant in the whole range of NTCD used in this study suggesting that the degree of crystallinity is not affected by variations in NTCD when electrospinning at or close to the optimum voltage. Similar observations as above were also made for fibers electrospun from the 8 wt% solution concentration (Figure 7). This has not been suggested before in the literature and therefore it is not possible to compare with any other similar example, however this is an indication that the acceleration voltage has a main role in the jet trajectory perhaps more important than the NTCD.

# 4. CONCLUSIONS

PLLA polymer fibers were electrospun from two different polymer solution concentrations, *i.e.*, 5 wt% and 8 wt%, at varying voltages and NTCDs using an ES1a electrospinning machine manufactured by Electrospinz Ltd, New Zealand. The thermal characterisation as conducted by DSC revealed that the electrospun fibers exhibited crystallinities ranging from 23% to 46% while that for the as-received granules was approximately 37%. This suggests that by optimizing the solution and processing parameters during the electrospinning process, it may be possible to tailor the degree of crystallinity of PLLA. This can be a very important factor not only when considering an electrospun material for biomedical applications, but in other areas as well, because crystallinity affects polymer properties such as stiffness, yield stress, modulus and tensile strength, solubility, optical and electrical properties [49,50] which will in turn affect the behaviour of these materials when they are utilized in energy, environment, defence and security applications.

It was also observed that factors such as polymer solution concentration, electrospinning voltage and NTCD influenced the degree of crystallinity of the electrospun fibers. More specifically, the degree of crystallinity decreased with increasing the polymer solution concentration. In addition, an optimum electrospinning voltage at which maximum degree of crystallinity can be obtained was observed. At voltages higher or lower than the optimum electrospinning voltage, the degree of crystallinity will decrease or increase, respectively. It is believed that each polymer solution concentration would exhibit its individual optimum electrospinning voltage. The effect of the NTCD on the degree of crystallinity followed no predictable and consistent pattern, possibly due to inconsistency of trajectory of the polymer jet. However, it was interesting to observe that the degree of crystallinity was always maximum irrespective of the NTCD when electrospinning at the optimum electrospinning voltage.

In summary, in order to tailor the crystallinity of electrospun fibers, the polymer solution concentration and voltage should be carefully controlled.

## REFERENCES

1. Baji, A.; Mai, Y.W.; Wong, S.C.; Abtahi, M.; Chen, P. Electrospinning of polymer nanofibers: Effects on oriented morphology, structures and tensile properties. *Compos. Sci. Technol.* **2010**, *70*, 703–718.

2. Kongkhlang, T.; Tashiro, K.; Kotaki, M.; Chirachanchai, S. Electrospinning as a new technique to control the crystal morphology and molecular orientation of polyoxymethylene nanofibers. *J. Am. Chem. Soc.* **2008**, *130*, 15460–15466.

3. Stephens, J.S.; Chase, D.B.; Rabolt, J.F. Effect of the electrospinning process on polymer crystallization chain conformation in nylon-6 and nylon-12. *Macromolecules* **2004**, *37*, 877–881.

4. Tong, H.W.; Wang, M. An investigation into the influence of electrospinning parameters on the diameter and alignment of poly(hydroxybutyrate-co-hydroxyvalerate) fibers. *J. Appl. Polym. Sci.* **2011**, *120*, 1694–1706.

5. Boland, E.D.; Wnek, G.E.; Simpson, D.G.; Pawlowski, K.J.; Bowlin, G.L. Tailoring tissue engineering scaffolds using electrostatic processing techniques: A study of poly(glycolic acid) electrospinning. *J. Macromol. Sci. Pure Appl. Chem.* **2001**,*38*, 1231–1243. ]

6. Sukigara, S.; Gandhi, M.; Ayutsede, J.; Micklus, M.; Ko, F. Regeneration of bombyx mori silk by electrospinning-part 1: Processing parameters and geometric properties. *Polymer* **2003**, *44*, 5721–5727.

7. Tan, S.H.; Inai, R.; Kotaki, M.; Ramakrishna, S. Systematic parameter study for ultra-fine fiber fabrication via electrospinning process. *Polymer* **2005**, *46*, 6128–6134.

8. He, W.; Ma, Z.W.; Teo, W.E.; Dong, Y.X.; Robless, P.A.; Lim, T.C.; Ramakrishna, S. Tubular nanofiber scaffolds for tissue engineered small-diameter vascular grafts. *J. Biomed. Mater. Res. A* **2009**, *90A*, 205–216.

9. Katta, P.; Alessandro, M.; Ramsier, R.D.; Chase, G.G. Continuous electrospinning of aligned polymer nanofibers onto a wire drum collector. *Nano Lett.* **2004**, *4*, 2215–2218.

10. Kidoaki, S.; Kwon, I.K.; Matsuda, T. Mesoscopic spatial designs of nano- and microfiber meshes for tissue-engineering matrix and scaffold based on newly devised multilayering and mixing electrospinning techniques. *Biomaterials* **2005**, *26*, 37–46.

11. Saraf, A.; Lozier, G.; Haesslein, A.; Kasper, F.K.; Raphael, R.M.; Baggett, L.S.; Mikos, A.G. Fabrication of nonwoven coaxial fiber meshes by electrospinning. *Tissue Eng. C Methods* **2009**, *15*, 333–344.

12. Sun, Z.C.; Zussman, E.; Yarin, A.L.; Wendorff, J.H.; Greiner, A. Compound core-shell polymer nanofibers by co-electrospinning. *Adv. Mater.* **2003**, *15*, 1929–1932.

13. Kim, J.S.; Lee, D.S. Thermal properties of electrospun polyesters. *Polym. J.* **2000**, *32*, 616–618.

14. Zhang, Y.Z.; Su, B.; Venugopal, J.; Ramakrishna, S.; Lim, C.T. Biomimetic and bioactive nanofibrous scaffolds from electrospun composite nanofibers. *Int. J. Nanomed.* **2007**, *2*, 623–638.

15. Jagur-Grodzinski, J. Polymers for tissue engineering, medical devices, and regenerative medicine. Concise general review of recent studies. *Polym. Adv. Technol.* **2006**, *17*, 395–418.

16. Vacanti, J.; Vacanti, C.A. The History and Scope of Tissue Engineering. In *Principles of Tissue Engineering*, 3rd; Lanza, R., Langer, R., Vacanti, J., Eds.; Elsevier Academic Press: Burlington, VT, USA, 2007; pp. 1–6.

17. Ma, P.X. Scaffolds for tissue fabrication. *Mater. Today* **2004**, *7*, 30–40.

18. Jones, J.R. Scaffolds for Tissue Engineering. In *Biomaterials, Artificial Organs and Tissue Engineering*; Hench, L.L., Jones, J.R., Eds.; Woodhead Publishing Limited: Cambridge, UK, 2005; pp. 201–213.

19. Middleton, J.C.; Tipton, A.J. Synthetic biodegradable polymers as orthopedic devices. *Biomaterials* **2000**, *21*, 2335–2346.

20. Bognitzki, M.; Czado, W.; Frese, T.; Schaper, A.; Hellwig, M.; Steinhart, M.; Greiner, A.; Wendorff, J.H. Nanostructured fibers via electrospinning. *Adv. Mater.* **2001**, *13*, 70–72.

21. Zong, X.H.; Kim, K.; Fang, D.F.; Ran, S.F.; Hsiao, B.S.; Chu, B. Structure and process relationship of electrospun bioabsorbable nanofiber membranes. *Polymer* **2002**, *43*, 4403–4412.

22. Ramdhanie, L.I.; Aubuchon, S.R.; Boland, E.D.; Knapp, D.C.; Barnes, C.P.; Simpson, D.G.; Wnek, G.E.; Bowlin, G.L. Thermal and mechanical characterization of electrospun blends of poly(lactic acid) and poly(glycolic acid). *Polym. J.* **2006**, *38*, 1137–1145.

23. Deitzel, J.M.; Kleinmeyer, J.D.; Hirvonen, J.K.; Beck Tan, N.C. Controlled deposition of electrospun poly(ethylene oxide) fibers. *Polymer* **2001**, *42*, 8163–8170.

24. Deitzel, J.M.; Kleinmeyer, J.; Harris, D.; Beck Tan, N.C. The effect of processing variables on the morphology of electrospun nanofibers and textiles. *Polymer* **2001**, *42*, 261–272.

25. Buchko, C.J.; Chen, L.C.; Shen, Y.; Martin, D.C. Processing and microstructural characterization of porous biocompatible protein polymer thin films. *Polymer* **1999**, *40*, 7397–7407.

26. Ayutsede, J.; Gandhi, M.; Sukigara, S.; Micklus, M.; Chen, H.-E.; Ko, F. Regeneration of bombyx mori silk by electrospinning. Part 3: Characterization of electrospun nonwoven mat. *Polymer* **2005**, *46*, 1625–1634.

27. Ribeiro, C.; Sencadas, V.; Miguel Costa, C.; Gomez Ribelles, J.L.; Lanceros-Mendez, S. Tailoring the morphology and crystallinity of poly(L-lactide acid) electrospun membranes. *Sci. Technol. Adv. Mater.* **2011**, *12*, 015001:1–015001:9.

28. Zhao, S.; Wu, X.; Wang, L.; Huang, Y. Electrospinning of ethyl-cyanoethyl cellulose/tetrahydrofuran solutions. *J. Appl. Polym. Sci.* **2004**, *91*, 242–246.

29. Arinstein, A.; Burman, M.; Gendelman, O.; Zussman, E. Effect of supramolecular structure on polymer nanofibre elasticity.*Nat. Nanotechnol.* **2007**, *2*, 59–62.

30. Wong, S.C.; Baji, A.; Leng, S.W. Effect of fiber diameter on tensile properties of electrospun poly(epsilon-caprolactone).*Polymer* **2008**, *49*, 4713–4722.

31. Kister, G.; Cassanas, G.; Bergounhon, M.; Hoarau, D.; Vert, M. Structural characterization and hydrolytic degradation of solid copolymers of D,L-lactide-co-epsilon-caprolactone by raman spectroscopy. *Polymer* **2000**, *41*, 925–932.

32. Dong, Y.X.; Liao, S.; Ngiam, M.; Chan, C.K.; Ramakrishna, S. Degradation behaviors of electrospun resorbable polyester nanofibers. *Tissue Eng. B Rev.* **2009**, *15*, 333–351.

33. Hollinger, J.O.; Battistone, G.C. Biodegradable bone repair materials-synthetic-polymers and ceramics. *Clin. Orthop. Relat. Res.* **1986**, 290–305.

34. Gomes, M.E.; Reis, R.L. Biodegradable polymers and composites in biomedical applications: From catgut to tissue engineering—part 1—available systems and their properties. *Int. Mater. Rev.* **2004**, *49*, 261–273.

35. Rasal, R.M.; Janorkar, A.V.; Hirt, D.E. Poly(lactic acid) modifications. *Prog. Polym. Sci.* **2010**, *35*, 338–356.

36. Inai, R.; Kotaki, M.; Ramakrishna, S. Structure and properties of electrospun plla single nanofibres. *Nanotechnology* **2005**, *16*, 208–213.

37. Electrospinz. Frequently Asked Questions. 2008. Available online: http://www.electrospinz.co.nz/faq.php (accessed on 12 January 2012).

38. Mezghani, K.; Spruiell, J.E. High speed melt spinning of poly(L-lactic acid) filaments. *J. Polym. Sci. B Polym. Phys.* **1998**, *36*, 1005–1012.

39. Fambri, L.; Pegoretti, A.; Fenner, R.; Incardona, S.D.; Migliaresi, C. Biodegradable fibres of poly(l-lactic acid) produced by melt spinning. *Polymer* **1997**, *38*, 79–85.

40. Wunderlich, B. *Thermal Analysis of Polymeric Materials*; Springer: Berlin/Heidelberg, Germany, 2005.

41. Shenoy, S.L.; Bates, W.D.; Frisch, H.L.; Wnek, G.E. Role of chain entanglements on fiber formation during electrospinning of polymer

solutions: Good solvent, non-specific polymer-polymer interaction limit. *Polymer* **2005**, *46*, 3372–3384.

42. Sill, T.J.; von Recum, H.A. Electro spinning: Applications in drug delivery and tissue engineering. *Biomaterials* **2008**, *29*, 1989–2006.

43. Young, R.J. *Introduction to Polymers*; Chapman and Hall: New York, NY, USA, 1981.

44. Cui, W.G.; Li, X.H.; Zhu, X.L.; Yu, G.; Zhou, S.B.; Weng, J. Investigation of drug release and matrix degradation of electrospun poly(D,L-lactide) fibers with paracetanol inoculation. *Biomacromolecules* **2006**, *7*, 1623–1629.

45. Zong, X.H.; Ran, S.F.; Kim, K.S.; Fang, D.F.; Hsiao, B.S.; Chu, B. Structure and morphology changes during *in vitro*degradation of electrospun poly(glycolide-co-lactide) nanofiber membrane. *Biomacromolecules* **2003**, *4*, 416–423.

46. Carrizales, C.; Pelfrey, S.; Rincon, R.; Eubanks, T.M.; Kuang, A.X.; McClure, M.J.; Bowlin, G.L.; Macossay, J. Thermal and mechanical properties of electrospun PMMA, PVC, Nylon 6, and Nylon 6,6. *Polym. Adv. Technol.* **2008**, *19*, 124–130.

47. Ramakrishna, S.; Fujihara, K.; Teo, W.; Lim, T.; Ma, Z. *An Introduction to Electrospinning and Nanofibres*; World Scientific Publishing Co. Ltd.: Singapore, 2005.

48. Jarusuwannapoom, T.; Hongrojjanawiwat, W.; Jitjaicham, S.; Wannatong, L.; Nithitanakul, M.; Pattamaprom, C.; Koombhongse, P.; Rangkupan, R.; Supaphol, P. Effect of solvents on electro-spinnability of polystyrene solutions and morphological appearance of resulting electrospun polystyrene fibers. *Eur. Polym. J.* **2005**, *41*, 409–421.

49. Billmeyer, F.W. *Textbook of Polymer Science*, 3rd ed; John Wiley & Sons, Inc.: Singapore, 1984.

50. Hou, X.X.; Yang, X.P.; Zhang, L.Q.; Waclawik, E.; Wu, S.Z. Stretching-induced crystallinity and orientation to improve the mechanical properties of electrospun pan nanocomposites. *Mater. Des.* **2010**, *31*, 1726–1730.

# Chapter 11

# Synthesis of Alumina Nanofibers and Composites

**Mohamad Ridzuan Noordin[1] and Kong Yong Liew[1, 2]**

[1] Faculty of Industrial Science and Technology, Universiti Malaysia Pahang, Kuantan,, Malaysia

[2] College of Chemistry and Materials Science, South Central University for Nationalities, Wuhan, China

## 1. INTRODUCTION

Alumina, $Al_2O_3$, is a ceramic metal oxide of great importance as building material, refractory material, electrical and heat insulator, attributed to its high strength, corrosion resistance, chemical stability, low thermal conductivity, and good electrical insulation. One phase of alumina, the γ-phase is widely used as catalyst, catalyst support and adsorbent because its high porosity and surface area. It is conventionally prepared by heating the hydrated trihydroxides, gibbsite and bayerite, to temperature of > 1000°C to obtain the α-alumina, corundum, a material with great hardness and low surface area. The trihydroxides, gibbsite, bayerite, both are $Al(OH)_3$. $nH_2O$ and the monohydroxyl oxide, boehmite, AlO(OH) $\alpha H_2O$, have the monoclinic crystal structure with similar lattice parameters ($a$=0.866 nm,$b$=0.506 nm, $c$=0.983 nm, β=94 34′ in boehmite, $a$=0.868 nm, $b$=0.507 nm, $c$=0.972 nm, β=94 34′ in gibbsite, and $a$=0.867 nm, $b$=0.506 nm, $c$=0.942 nm, β=90 26′ in bayerite [1, 2]. In the process of heat treatment, the trihydroxide undergoes a series of transformations. It loses the water of hydration, then dehydroxalate at < 300°C to form the monohydroxyl oxide boehmite, AlO(OH), which on further heating to increasingly higher temperatures, changes to the transition aluminas, including the η-, γ-, δ-, and θ-phases, which then transform finally to the α-form. These transition aluminas are crystalline solids with high porosity and surface areas as well as acidic and basic properties, which make them suitable as adsorbents, catalyst, catalyst support and fabricated into filtration membranes as well as used as fillers or components in polymer/ inorganic composite materials with enhanced mechanical properties. The conventional alumina as obtained are usually powder of particulates. These could be fabricated into alumina fibers by various method, through melt growth techniques including the internal crystallization method and extrusion, electrospraying [3] and electrospinning [4-8] have been developed for producing $Al_2O_3$ fibers.

The high surface areas of transition aluminas are due to the presence of pores. These pores are irregular in sizes and size distributions, and are mainly micropores of diameter less than 2 nm. Since the synthesis of mesoporous materials was first reported in the late 1980s, a number of methods for the synthesis of mesoporous alumina have been reported [8,9]. More recently, interest in nano-materials and nanotechnology has spurred a fury of investigations on the preparation and the applications of nano-materials including that of inorganic oxides such as aluminum oxide. Nanoparticles can be of 1, 2 or 3 dimensions. One dimensional nano-structured alumina may be described as fiber, wire, belt, rod, laths and tubes [9-17]. Nanofibers are one dimensional material with high aspect ratio of length to diameter of 1000:1 and diameter of 1 to 100 nm, which may be solid fibers or hollow nanotubes. The final properties of these nanofibers depend on the size and homogeneity of the fibrous structure, which shall enable tailoring $Al_2O_3$ nanofibers to advanced technological usage with the enhanced sintering ability, strength, catalytic activity, and absorption ability. However, for the present report, lower aspect ratio materials are also included.

## 2. SYNTHESIS OF ALUMINA NANOFIBERS

Various routes including hydrothermal or solvothermal, sol-gel techniques, electrospinning and extrusion, template techniques [19,20], chemical vapor deposition in flame[30], and mercury mediated technique [29] have been developed for producing $Al_2O_3$ nanofibers. Many of the nanfibers have been produced using hydrothermal treatment, with or without surfactant, [18-25] and sol-gel [17,27] techniques of various alumina precursors, subsequent calcinations dehydrate the solid from the sol-gel solution produced by the hydrothermal process. The solid, which consists of naonfibrous boehmite, may be calcined to obtain the α-alumina nanofibers. The gel has high viscosity and may be electrospun to obtain the nanofibers. The solid either from the hydrothermal or sol-gel processes or the electrospun fibers is dehydrated by heat treatment leading to transition aluminas and finally to the nanofibrous α-alumina. Other methods such as vapor–liquid–solid deposition to burn aluminum together with silica nanoparticles at high temperature between 1300 and 1700 C, and by using other 1-D materials as hard template have also been reported [13-15, 19, 30].

Most of the earlier reports are on the generation of $Al_2O_3$ fibers of diameters in the submicrons to several tens of microns range. Only more recently did the synthesis of nanofibrial alumina appeared. However, information regarding the growth and particulate properties of $Al_2O_3$ nanofiber still remains very limited. For the hydrothermal and sol-gel processes, the aluminum precursors could be Al metal, the alkoxide or aluminum salts. The size and morphology of aluminum hydrate nanoparticles synthesized by these processes can be controlled by manipulating the pH value of the reaction mixture, the hydrothermal temperature and aging time, the solvent and the surfactant used. [2,3,20-27]

## 2.1. Hydrothermal and Sol Gel Processes

The first step in the synthesis of nanofibrous alumina both these processes is the hydrolysis of the starting aluminum compound or elemental Al powder to the hydrated hydroxide of aluminum, boehmite, which are then aged subsequently. In the hydrothermal process, the aging could be done in an enclosed reactor under its own autogenic pressure, the colloidal solution obtained is either filtered and the solid further processed. In the sol gel process, the sol is transformed to a gel after the aging. The two processes are closely related and the reactions that take place are similar, these processes are discussed together. The boehmite formed in the first step has the formula, $AlOOH.nH_2O$. It may be amourphous particles initially which then transform into network structures linked by hydrogen bonds and forms nanosheets or sheets of crumpled nanostructures. The nanosheets could rolled up and be transformed into hollow tubular structures or nanofibers[17, 31], which may or may not be crystalline. The initial spherical particles of boehmite are sol solution with low viscosity. As it aged, the network structures and fibers are formed and the sol changed into gel with high viscosity. The solid in the gel is initially spherical nanoparticles [17], which on aging gradually transform into the crystalline boehmite. The solid on heating to higher temperature dehydrates and dehydroxylates forming the different phases of transition aluminas in stages at progressively higher temperature than 400°C, which finally transforms to the final α-alumina at temperature of 1000°C or higher. The reactions may be represented as follows [13-16, 18, 19, 21-27, 31]:

$$Al(OH)_3 .\ nH_2O \rightarrow \text{amorphous } Al(OH)_4^- + H_3O^+ + (n-1)\ H_2O \quad (1)$$

$$Al(OH)_4^- \leftrightarrow \gamma\text{-}AlOOH.nH_2O + (1-n)H_2O \quad (2)$$

$$\gamma\text{-}AlOOH \cdot nH_2O \rightarrow \gamma\text{-}Al_2O_3 + 2(1-n)H_2O \quad (3)$$

$$\gamma\text{-}Al_2O_3 \rightarrow \rightarrow\rightarrow \alpha\text{-}\ Al_2O_3 \quad (4)$$

The γ crystalline boehmite is formed under suitable synthesis condition and the morphology is retained in the various transition aluminas as well as in the final α-alumina as demonstrated by the TEM and Xrd characterizations, although the crystal structures are different. The boehmite formed nanosheets in the sol and gelling process and the curled up fibers or tubes were visible in the TEM images of the boehmite.

Alumina nanofibers can be synthesized with metallic aluminum particles [10,22, 43, 59], Al alkoxides [ 36; 46, 17] or its salts [27, 45, 47, 48, 49, 51] as the starting materials. Other starting materials that have been used include Al pentanedionate, [50], aluminum hydroxide, [52], alumina composite, [53] aluminum film, [42] and sodium aluminate. [54] However, whatever the starting materials used in the synthesis, the nanofibrous boehmite intermediate will have to be formed [22] for the formation of the nanofibers as mentioned above. The starting Al precursor is first hydrolyzed with water to form the hydroxide [17, 22, 32, 37, 38], aged to the viscous gel fibers of hydrated AlOOH, which can be calcined directly or electrospun or drawn from the viscous gel and converted to ceramic fibers after firing [16, 27, 31-37]. The metallic aluminum powder used for the process is prepared by exploding the metal wire by a high density current pulse normally produced by discharging a capacitor bank. The produced particles are spherical of about 80-120 nm [22, 43].

For the case in which metallic Al nanoparticles is used the first step of the reactions are depicted as follows:[17, 22]

$$2\ Al + 4\ H_2O \rightarrow 2\ AlO(OH) + 3\ H_2 \quad \text{without stirring} \quad (5)$$

$$2\ Al + 6\ H_2O \rightarrow 2\ Al(OH)_3 + 3\ H_2 \quad \text{with stirring} \quad (6)$$

reaction (5) leads to boehmite, while reaction (6) leads to formation of the trihydroxide, bayerite, which on dehydration via a series of transformation forms the boehmite:

$$Al(OH)_3 \rightarrow AlO(OH) + 2\,H_2O \qquad (7)$$

The boehmite will further dehydrate to convert to the transition aluminas which will further dehydrate to $\gamma$-$Al_2O_3$ up to 700°C without significant change in morphology. With the aluminum salts, hydrolysis occurs forming aluminum hydroxide, which then proceeds as in eq.7. Na aluminate reacts with acid to form the aluminum hydroxide and proceeds to be dehydrated similarly.

With alkoxides, $Al(OR)_3$, as the starting materials, the initial hydrolysis with water yield alcohol and the hydroxylated aluminum centers as follows: [17]

$$Al(OR)_3 + H_2O \rightarrow Al(OR)_2(OH) + ROH \qquad (8)$$

$$Al(OR)_2(OH) + H_2O \rightarrow AlO(OH) + 2\,ROH \qquad (9)$$

$$Al\text{-}OH + Al\text{-}OX\text{-}Al \rightarrow Al\text{-}O\text{-}Al + XOH\ (X{=}H \text{ or } R) \qquad (10)$$

$$2\,AlO(OH) \rightarrow Al_2O_3 + H_2O \qquad (11)$$

In the initial stage of the sol gel reaction, small three dimensional oligomers are formed with the –Al-OH on the outer surface. Linkange of –Al-OH groups occurs via polycondensation and results in oxypolymers. When sufficient interconnected Al-O-Al bonds are formed, they respond cooperatively to form the colloidal particles or a sol. These sol particles aggregate and increase in size with time forming three dimensional network and randomly distributed spherical nanoparticles as shown by TEM in the earlier stage of the process. With further aging, the nanosheets are shown to curl up forming the nanofibers. The morphology was seen to be preserved even after calcinations at 1200°C [17,21, 45,55].

The size and morphology of the end product of the hydrolysis and subsequent dehydration of the intermediate boehmite from the hydrolysis depend on the conditions in which these processes are performed. The primary factors involved the nature of the starting material, whether a surfactant or templating agent is used, the amount of water used, the aging time in the hydrolysis, the pH value of the reaction mixture, and the hydrothermal temperature [17-19, 21-23, 45-47, 55-59]. Even a simple procedure of stirring the hydrolyzing reaction mixture of Al particles in water, affect the outcome of the resulting hydroxylate product. Without stirring, boehmite was produced, while with stirring, the trihydroxide, bayerite, was resulted.[22]. In the sol-gel synthesis using iospropoxide as the starting material [17], the reaction had to be carried out at the range of $R_w$=60-65, outside of this range no fibrous material was formed. Under acidic conditions, hydrolysis was enhanced and that the rate of polycondensation and aggregation to larger particles was increased too high for nanofibers formation. Additionally, the presence of organic solvent influences the reaction rate significantly. Of the solvent tested only isopropanol in the hydrolysis mixture led to nanofibers formation. Although the morphology is retained during the calcinations of the nanofibrous alumina, the physical property such as the pore size increased with calcinations temperature from the as prepared samples to 1000°C, however, the surface area decreased at temperature > 550°C. Electron microscopic observations of microstructure, together with studies of the evolution of X-ray and electron diffraction patterns, have been used to provide mechanistic information on the dehydration of boehmite to $\gamma$-$Al_2O_3$. Calcinations of boehmite at 400 C produces slow development of a fine, lamellar, porous microstructure, oriented parallel to $(001)_\gamma$, corresponding to the loss of oxygen atoms of the boehmite lattice. The mechanism proposed for the dehydration is controlled by diffusion in a direction determined by the hydrogen bond chains in the boehmite structure and involves counter migration of the Al cations and protons with formation of voids in a coherent cubic close-packed matrix. Final reorganization to give the spinel structure of $\gamma$-$Al_2O_3$ is suggested to involve gradual filling of the tetrahedral cation sites [3].

To control the hydrolysis and the polycondensation rates, much works have been reported for the synthesis of the nanofibers with the addition of various surfactants and other additives as templating agents to direct the growth of the fibers. Alumina nanofibers have been successful grown with the use of polyethylene oxide polyethylene glycol, hexamethyltetraamine, and organic solvents [14, 15, 17,25, 41, 45, 48]. The surfactant directs the fibrils growth by forming rodlike micelles. Alumina nanofibers were successfully prepared by electrospinning technique using aluminum acetate and aluminum nitrate as alumina precursor or other different precursors with a combination of PVA and PEO as polymer precursor [48]. Recently, a one pot synthesis of high area alumina nanofibers were synthesized from aluminum isopropoxide and acetic acid in supercritical $CO_2$ (SCF) as the solvent has been described in which the water in the process was limited by the reaction between the alkoxide and the acetic acid [60]. Another report on the use of supercritical CO2 as a drying agent has also been described for the synthesis of unidirectional nanostructures such as nanotubes, nanofibers, and nanorods. The morphology of the final product depends on the nature of the employed surfactants [25].

Another method to ensure the formation of the nanofibrous alumina is by using the hard template route. Aluminum surfaces created by anodic oxidation electrically with nanosize pores could be used as the nanoreactors for the synthesis of the oxide fibers [14, 15, 19].

## 2.2. Other Methods

### 2.2.1. Mercury Mediated Method

Alumina nanofibers were successfully synthesized in mercury media at room temperature. The procedure consisted of immersing high purity aluminum strips $HgCl_2$ solution for a few minutes, removed and exposed to air. After some time, the white alumina powder was formed on the surface of the aluminum strips. The as-grown aluminas were heated and calcined in air at 850°C for 2 h to. The product was initially amorphous with diameters of 5-15 nm and length of several µm but converted to γ-alumina after calcinations. During the process, the following reactions took place:

$$2\,Al + 3\,HgCl_2 \rightarrow 3\,Hg + 2AlCl_3 \quad (12)$$

$$4\,Al + 3\,O_2 + 2n\,H_2O = 2\,Al_2O_3 \cdot nH_2O \quad (13)$$

The Al reacted with mercuric chloride forming Al amalgam by metathesis and got oxidized by air when removed from the solution. The dissolution of the Al into the Hg was a continuous process. The final product is the alumina nanofibers [28].

### 2.2.2. Flame Aerosol Method

Flame aerosol technology is used for large-scale manufacture of ceramic commodities such as pigmentary titania, fumed silica and alumina. The dynamics of formation and growth of these particles are reasonably well understood. The formation of alumina nanofibers in carbon-containing coflow diffusion flames with pure oxygen support, using trimethylaluminum vapor as the precursor has been reported recently. The alumina nanofibers formed were non-crystalline, typically 2–10 nm in diameter and 20–210 nm in length. These fibers were mixed with spherical nanoparticles along with carbon nanotubes formed at the same time. No alumina nanofiber formation was found in flames supported by air or in pure $H_2$ flames. The alumina nanofibers were formed from gas-phase aluminum-containing species in the flame. Gas-phase carbon-containing species such as CO or hydrocarbon were probably crucial to the formation of the alumina nanofibers. The nanofibers were formed in the region past the maximum temperature zone of the flame. Sufficiently high temperature may be due to the higher concentration of aluminum in the gas phase, while the presence carbon nanotubes or nanofibers may serve as templates or mediating agent for the alumina nanofibers formation [29, 64].

### 2.2.3. Vapor-Liquid-Solid Process

Crystalline alumina nanowires were synthesized at elevated temperatures in a catalyst assisted process using iron as catalyst. Nanotrees formed by alumina nanowires were also found. The typical nanowires are crystalline of size around 50 nm in diameter and around 2 μm in length. The tree trunks of the nanotrees are around 100 nm in diameter and around 10 μm in length. The results are explained in terms of growth mechanism based on a vapor–liquid–solid (VLS) process. The process involved mixing commercially available aluminum powder, iron powder and silicon carbide powder in an appropriate ratio and then sprinkled on tungsten (W) boat. The boat was then placed at the center of the vacuum bell jar and evacuated down to about $5.0\times10^{-2}$ Torr. The boat was gradually heated up by passing current through it, and kept at 1700 C for 1 h under flowing argon. The treated powders contained in the W boat were taken out and cooled down to room temperature in flowing argon [64]. Single-crystal $\alpha$-$Al_2O_3$ fibers by vapor–liquid–solid deposition from aluminum and silica powder and $Al_2O_3$ nanowires were synthesized by heating a mixture of Al, $SiO_2$ and $Fe_2O_3$ catalyst. $Al_2O_3$ nanowires and nanotrees have also being grown on silicon carbide particles' surface in thermal evaporation process and using iron as catalyst [20].

## 2.3. Electrospinning

Electrospinning is a old technology for the production of polymer fibers. This process has experienced renewed interest for the synthesis of nanofibers of polymers, ceramics, and their composites. The process uses electrical force to produce fibers with nanometer-scale diameters from its solution. Nanofibers have a large specific surface area and a small pore are being used or finding uses in filtration, protective clothing, biomedical applications including wound dressings and drug delivery systems, structural elements in artificial organs, and in reinforced composites. Recently, this potentially commercially viable process has been much investigated for the production of ceramic including alumina nanofibers [4-7, 51, 53].

The process setup consists of a capillary tube, connected to a reservoir of the colloidal solution or melts, with a small orifice through which the fluid could be ejected and a collector. A high voltage is used to create an electrically charged jet of solution or melt out of the capillary solution fluid. This induces a charge on the surface of the droplet formed at the tip of the capillary and held by its surface tension. As the intensity of the electric field is increased, the hemispherical surface of the droplet at the tip of the capillary tube elongates to form a conical shape known as the Taylor cone. On further increasing the electric field, a critical value is attained with which the repulsive electrostatic force overcomes the surface tension and the charged jet of the fluid is ejected. The ejected colloidal solution jet undergoes an instabilization and elongation process, which causes the jet to become very long and thin. Meanwhile, the solvent evaporates, leaving behind a charged fiber. In the case of the melt the discharged jet solidifies when it travels in the air.

Many reports have appeared in the literature in the past few years on the fabrication of alumina nanofibers as the viscosity of the boehmite gel is well suited for the electrospinning technique [5,34,48,50,51,53]. The gel solution for the spinning consisted of nanofibrous colloid prepared by the sol-gel or hydrothermal processes, however, the precursor mixture may also be used. A typical solution for electrospinning is prepared by mixing suitable concentration of the solution of an aluminum salt with a polymer solution such as polyvinyl alcohol or polyethylene oxide. The distance between the capillary tip and the collector electrode, the flow rate, the voltage were adjusted suitably. Polymer-alumina composite nanofibers were obtained which can then be processed by further drying or calcinations [48].

# 3. ALUMINA NANOFIBER COMPOSITES

Alumina nanoparticles and nanofiber show interesting properties such as ability to be formed into structures that enhance the functions of osteoblast for bone replacement [65, 66], to form ultrathin alumina hollow fiber microfilm membrane [67] for separation processes, to form novel nanofilter for removal and retention of viral aerosols [68] and to serve as high performance turbidity filter [69]. The preparation of the final form of alumina fiber may involve the use of binders such as acid phosphate and silica colloid binder or polymers [70]. It the case where polymer was used it would be burnt off during high temperature treatment of the prepared.

Many composite materials comprising of alumina nanoparticles or nanofibers as minor or major component in the presence of polymer or inorganic substrate also show similar interesting properties. S. Sundarajan [71] for example explored the fabrication of nanocomposite membrane comprising of polymer and alumina nanoparticles for protection against chemical warfare stimulants and found alumina based materials to be prospective candidates. While traditionally alumina film has been used as protective film on metal substrate [72, 73], alumina presence thermoplastic and thermosetting polymeric materials are also gaining wide application as surface coatings. Landry [74] for example studied the preparation of alumina and zirconia acrylate nanocomposites for coating wood flooring. It was found that for both the alumina and zirconia nano-composites, the conversion of acrylate resin is faster and more important when silane is used as the coupling agent.

The addition of the nanoparticles and nanofibers is meant to enhance the mechanical and thermal properties compared to in absence of such constituents. It is generally agreed that the large surface to volume ratio of the nanoscale constituents plays a key role to the improvement.

## 3.1. Alumina in Polymer Substrate

The effect of alumina in polyaniline, diglycidyl ether of bisphenol A type epoxy resin, carbon fiber epoxy resin composite and PA1010 has been studied recently. Generally it has been found that the present of low loading of alumina nanoparticles and nanofibers tend to enhance the thermal and mechanical properties of the polymer matrix. In many instances the strength of the composites are below the strength of neat resin due to non-uniform particle size distribution and particle aggregation. Ash et.al (as noted in [75]) studied the mechanical behavior of alumina particulate/poly(methyl methacrylate) composites. They concluded that when a weak particle matrix interphase exists, the mode of yielding for glassy, amorphous polymers changes to cavitational to shear, which leads to a brittle-to-ductile transition.

Two challenges have been observed to be overcome to facilitate the enhancement of the properties of the polymer substrate. First is the need to disperse the nanoparticles and nanofibers uniformly throughout the polymer substrate and secondly to facilitate the interaction between the nanofibers and the molecules of the polymer substrate. At low loading the nanoparticles could be distributed uniformly across the substrate. However at high loading, there is the tendency for the fibers to clusters together and hence limit the enhancement of the mechanical properties and in fact lowered it. It has also been observed that while the tensile strength increases with reduction in particles size for micron-scale particles, the tensile strength decreased with reduced particle size for nano-scale particles [76].The changes were attributed not to the strength of bonding between the particulate with the matrix but more to the poor dispersion of particles.

Various preparative methods have been adopted to facilitate good dispersion. This include mechanical milling [77] and Mechanical milling followed by hot extrusion [78]. In both studies the alumina nanofibers was found in the as sintered product. M.I Flores-Zamora concluded that the presence of alumina based nanoparticles and nanofibers seemed to be responsible for the reinforcement effect.

Attempts to meet the second challenge involve functionalizing alumina particles such as in the on fiber and epoxy resin composites [79]. The functionalizing of the alumina surfaces is meant to enhance the miscibility of the alumina particles in the polymer substrate and also the facilitate bridging between alumina surfaces with the substrate. It was however observed that where the functionalised alumina, L-alumoxane is miscible with the resin, high loading results in a marked decrease in performance due to an increase in brittleness. This was proposed to be due to weak inter-phase bonding between resin and the alumina fiber.

## 3.2.Alumina In Inorganic and Carbon Based Substrate

The composites comprising of carbon and alumina is also of significant interest. Study on this category of composite material includes fabrication of macroscopic carbon nanofiber (CNF)/alumina composite by extrusion method for catalytic screening [80]. It was found that the synthesized composite possessed a mesoporous structure with a relatively high surface area (340 $m^2/g$) and a narrow particle size distribution, displaying a good thermal stability. A comparison of the surface acidity between the composite and commercial alumina demonstrated that the total number of acid sistes on composite was significantly increased along the distinct decrease in the number of strong acid sites, which may enhance the activity and anti coking property as a promising industrial catalyst support in petroleum industry.

Hirato [81] fabricate carbon nanofiber–dispersed alumina composites by pulsed electric-current pressure sintering and their mechanical and electrical properties. High bending and fracture strength were observed on the composites compared to that of monolithic alumina. The electrical resistivity of the composite material was also observed to reduce by $10^{17}$ order of magnitude. Xia [82] studied the fracture toughness of highly ordered carbon nanotube/alumina nanocomposites. The result of the study demonstrate that nanotube bridging/sliding and nanotube bridging necessary to induce nanoscale toughening, and suggest the feasibility of engineering residual stresses, nanotube structure and composite geometry to obtain high toughness nanocomposites.

## 3.3. Future Development of Alumina Nanocomposite

There are wide applications of alumina nanofiber composites. These potential would continue to drive research interest in this field. In the case of applying alumina composites for application that is dependent on its strength, future work is like to further elucidate the factors that would first facilitate good dispersion of the fiber in the host matrix and second that would enhance the interphase interaction between alumina nanofiber and the matrix. The effect of additive that could facilitate dispersion or/and bonding between alumina fiber and the matrix may be of great interest. The application for catalysis may be of special interest especially the matrix between alumina and carbon nanofibers and similar substances. It may be argued that catalysts are in nearly all cases composite materials, however the interaction between purposely prepared nano size precursors may yield effect not seen when prepared through the traditional mode.

## REFERENCES

1. H. Knoezinger, R. Ratnasamy, 1978 Catal. Rev., 17(1):31

2. V. Saraswati, G. V. V. N. Rao, G. V. Rama Rao, 1987 Structural evolution in alumina gel, J. Mater. Sci., 22(7):2529;

3. S. J. Wilson, 1979 The dehydration of Boehmite, γ-AlOOH, to γ--Al2O3, J. Solid State Chem. 30(2), :247

4. Jawaorek, A. T. Sobczyk, 2008 Electrospraying route to nanotechnology: an overview, J. Electrostatic, 60:197

5. M. Azad, 2006 Fabrication of transparent alumina nanofibers by electrospinning, Mater. Sci. Engineer. A., 435-436:468;

6. T. Subbiah, G. S. Bhat, R. W. Tock, S. Parameswaran, S. S. Ramkumar, 2005 Electrospping of nanofibers, J. Appl. Polym. Sci.,, 96(2):557;

7. D. Li, Y. Xia, 2003 Electrospinning of nanofibers: reinventing the wheel?, Advanced Mater. 16(14):1151

8. K. Niesz, P. Yang, 1986 G. A.Somorjal, 2005Chem. Comm.

9. S. A. Bagshaw, T. J. Pinnavaia, 1996 Mesoporous Alumina Molecular Sieves, Angew. Chem. (Intern.Ed.English), 35 10 1102 1105 .

10. X. S. Peng, L. D. Zhang, G. W. Meng, X. F. Wang, Y. W. Wang, C. Z. Wang, G. S. Wu, 2002 Photoluminescence and Infrared Properties of α-Al2O3 Nanowires and Nanobelts, J. Phys. Chem. B, 106 (43)

11. J. C. Hulteen, 1997 A general template-based metod for the preparation of nanometerials, J. Mater. Chem., 7(7):1075;

12. L. Pu, X. Bao, J. Zou, D. Feng, 2001 Individual alumina nanotubes, Angew Chem., 40(8):1490;

13. H. Hou, Y. Xie, Q. Yang, Q. Guo, C. Tan, 2005 Perapration and characterization of γ-AlOOH nanotubes and nanorods, Nanotechnol.,, 16(6):714;

14. Y. Xia, Y. Yang, P. Sun, Y. Wu, B. Mayers,, Y. Gates, . F. Yin, . H. Kim, H. Yan, 2003 One-dimensional nanostructures: Synthesis,characterization, and applications, Advanced Mater., 15(5),:353

15. J. G. Lu, P. Chang, Z. Fan, 2006 Quasi-one-dimensional metal oxide materials-Synthesis, properties and applications Mater. Sci. Engineer. R: Reports, 52(1-3),: 49-91

16. W. Cai, J. Yu, S. Mann, 2009 Template-free hydrothermal fabrication of hierarchically organized γ-AlOOH hollow microspheres, Microp. Mesop. Mater. 122(1-3),:42

17. G. L. Teoh, K. Y. Liew, W. A. K. Mahmood, 2007 Synthesis and characterization of sol-gel alumina nanofibers, J. Sol-Gel Sci. Technol., 44(3): 177

18. M.G. Ma, J. F. Zhu, 2009 A facile solvothermal route to synthesis of γ-alumina with bundle-like and flower-like morphologies, Mater. Lett. 63: 881;

19. Y. Zuo, Y. Zhao, X. Li, N. Li, X. Bai, S. Qiu, W. Yu, 2006 Synthesis of alumina nanowires and nanorods by anodic oxidation method, Mater. Lett., 60( 24),:2937;

20. Y. Z. Jin, Y. Q. Zhu, K. Brigatti, H. W. Kroto, D. R. M. Walton, 2003 Catalysed growth of novel aluminium oxide nanorods Appl. Phys. A: Mater. Sci. Proc., 77(1),: 113.;

21. H. C. Lee, H. J. Kim, S. H. Chung, K. H. Lee, H. C. Lee, J. S. Lee, 2003 Synthesis of Unidirectional Alumina Nanostructures without Added Organic Solvents J. Am. Chem. Soc.,, 125 (10):2882

22. J. H. Park, M. K. Lee, C. K. Rhee, W. W. Kim, 2004 Control of hydrolytic reaction of aluminum particles for aluminum oxide nanofibers. Mater Sci.Eng. A, 375-377, 1263

23. W. Cai, H. Li, Y. Zhang, 2007 Influences of processing techniques of the H2O2 -precipitated pseudoboehmite on the structural and textural properties of γ-Al2O3. Colloids Surf. A: Physicochem. Engin Asp, 295( 1-3), 185

24. V. Valcárcel, A. Pérez, M. Cyrklaff, F. Guitián, 2000 Novel Ribbon-Shaped α-Al2O3 Fibers, Advanced Mater. 10(16):1370

25. H. Y. Zhu, X. P. Gao, D. Y. Song, Y. Q. Bai, S. P. Ringer, Z. Gao, Y. X. Xi, W. Martens, J. D. Riches, R. L. Frost, 2004 Growth of Boehmite Nanofibers by Assembling Nanoparticles with Surfactant Micelles. J. Phys. Chem. B, 108 (14): 4245

26. W. Cai, H. Li, Y. Zhang, 2006 Azeotropic distillation-assisted preparation of macro-mesostructured γ-Al2O3 nanofibres of crumpled sheet-like morphology, Mater Chem. Phys., 96(1),:136;

27. J. Wang, Y. Wang, M. Qiao, S. Xie, K. Fan, 2007 A novel sol-gel synthetic route to alumina nanofibers via aluminum nitrate and hexamethylenetetramine, Mater. Lett., 61(28),:5074;

28. Q. Yang, Y. Deng, W. Hu, 2009 Synthesis of alumina nanofibers by a mercury-mediated method. Ceram. Intern 35(1),: 531

29. Guo, H. Yim, Z. P. Luo, 2009 Formation of alumina nanofibers in carbon-containing coflow laminar diffusion flames, J. Aerosol Sci., 40(4), 379

30. Jagminas, J. Kuzmarskytė, L. Malferrari, M. Cuffiani, 2007 A new route of alumina template modification into dense-packed fibrilous material, Mater. Lett., 61(14-15),:2896

31. R. Y. Bando, T. Sasaki, 2004 Directly rolling nanosheets into nanotubes, J. Phys. Chem.B, 108(7),:211

32. J. Wang, Y. Wang, M. Qiao, S. Xie, K. Fan, 2007 A novel sol-gel synthetic route to alumina nanofibers via aluminum nitrate and hexamethylenetetramine, Mater. Lett., 61(28), 5074

33. X. Liu, Z. Wu, T. Peng, P. Cai, H.W. Lv, 2009 Lian Fabrication of alumina nanofibers by precipitation reaction combined with heterogeneous azeotropic distillation process, Mater. Res. Bul., 44(1),:160-167

34. V. Maneeratana, W. M. Sigmund, 2008 Continuous hollow alumina gel fibers by direct electrospinning of an alkoxide-based precursor, Chem. Engineer. J., 137(1), 137

35. J. Chandradass, M. Balasubramanian, 2006 Extrusion of alumina fibre using sol-gel precursor J. Mater. Sci.,41(18),:6026

36. E. Elaloui, A. C. Pierre, G. M. Pajonk, 1997 Influence of the Sol-Gel Processing Method on the Structure and the Porous Texture of Nondoped Aluminas, J. Catal., 166(2),:340

37. J. Chandradass, . D. S. D.S. Bae, M. Balasubramanian, 2008 Synthesis and characterization of sol-gel alumina fiber by seeding α-alumina through extended ball milling, Mater. Manuf. Proc. 23(8),:786

38. R. Venkatesh, P. K. Chakrabarty, B. Siladitya, M. Chatterjee, M. D. Ganguli, 1999 Preparation of alumina fibre mats by a sol-gel spinning technique, Ceramics Intern., 25(6), 539-543

39. G. Zhihong, N. Guilin, L. Yuan, C. Yu, 2006 Heat-induced transformation between nanospheres and nanofibers of boehmite, Proc. 1st IEEE-NEMS 2006, Article 4135091 886 889

40. R. Y. Bando, T. Sasaki, 2004 Directly rolling nanosheets into nanotubes, J. Phys. Chem.B, 108(7),:2115

41. Z. Zhu, H. Liu, H. Sun, D. Yang, 2009 PEG-directed hydrothermal synthesis of multilayered alumina microfibers with mesoporous structures, Microp Mesop. Mater, 123( 1-3), 39-44

42. Y. Kobayashi, T. Ishizaka, Y. Kurokawa, (2005 Preparationof.aluminafilms.bythe.solgel.method J. Mater Sci., 40263

43. F. Tepper, L. Kaledin, 2005 Nanosize electropositive fibrous adsorbent, US patent:6838005;

44. K. M. S. Khalil, 1998 Synthesis of Short Fibrous Boehmite Suitable for Thermally Stabilized Transition Aluminas Formation, J Catal.,178(1):198

45. J. Wang, Y. Wang, M. Qiao, S. Xie, K. Fan, 2007 A novel sol-gel route to alumina nanofibers via aluminum nitrate and hexamethylenetetramine, Mater. Lett. 61(28):5074.

46. T. Yogo, H. Iwahara, 1992 Synthesis of Synthesis of α-alumina from modified aluminum alkoxide precursor, J. Mater Sci.,, 27:1499;

47. J. F. Hochepied, P. Nortier, 2002 Influence of precipitation conditions :pH and temperature on the morphology and porosity of boehmite particles, Powder Technol, 128(2-3), 268

48. P. K. Panda, S. Ramakrisna, 2007 Eletrospinning of alumina nanofibers using different precursors, J. Mater Sci,,42: 2189

49. K. Song, K. Woo, Y. Kang, 1999 Preparation of alumina fibers from aluminum salts by the sol gel methods, Korean J. Chem. Eng. 16(1),:75

50. M. Azad, 2006 Fabrication of transparent alumina nanofibers by electrospinning, Meter. Sci. Eng. A,, (435-473):468

51. R. W. Tuttle, A. Chowdury, E. T. Bender, R. D. Ramsier, J. L. Rapp, M. P. Espe, 2008 Electrospun ceramic fibers: Composition, structure and the fate of precursors, Appl. Surf. Sci.,, 254:4928;

52. S. C. Shen, W. K. Ng, Q. Chen, X. T. Zeng, R. B. H. Tan, 2007 Novel synthesis of lace-like nanoribbons of boehmite and γ-alumina by dry gel conversion method, Mater lett,,61(21):4280 steam crystallization

53. K. M. Sawicka, P. Gouma, 2006 Electrospun composite nanofibers for functional applications, J. Nanopart. Res., 8:769

54. H. Y. Zhu, X. P. Gao, D. Y. Song, S. P. Ringer, Y. X. Xi, R. L. Frost, 2005 Manupilating the size and morphology of aluminum hydrous oxide nanoparticles by soft chemistry approaches, Microp. Mesop. Mater. 85(3):226

55. H. Y. Zhu, J. D. Riches, J. C. Barry, 2002 γ-alumina nanofibers prepared from aluminum hydrate with poly(ethylene oxide) surfactant, Chem. Mater., 14, 2086

56. X. Liu, Z. Wu, T. Peng, P. Cai, H.W. Lv, 2009 Lian Fabrication of alumina nanofibers by precipitation reaction combined with heterogeneous azeotropic distillation process, Mater. Res. Bul., 44(1),:160

57. H. Y. Zhu, X. P. Gao, D. Y. Song, P. Simon, Y. X. Ringer, Ray. L. Xi, Frost, 2005 Manipulating the size and morphology of aluminum hydrous oxide nanoparticles by soft-chemistry approaches, Microp. Mesop. Mater. 85(3),:226

58. W. Cai, H. Li, Y. Zhang, 2007 Influences of three post-precipitation techniques, Colloids Surf. A: Physicochem. Engin Asp, 295( 1-3),, 185,

59. M. Thiruchitrambalam, V. R. Palkar, V. Gopinathan, 2004 Hydrolysis of aluminium metal and sol-gel processing of nano alumina, Mater Lett., 58(24),:3063

60. M. B. I. Chowdhury, R. Sui, R. A. Lucky, P. A. Charpentier, 2000 A One-Pot Procedure to Synthesize High Surface Area Alumina Nanofibers using Supercritical Carbon Dioxide, Langmuir (in press)

61. H. C. Lee, H. J. Kim, C. H. Rhee, K. H. Lee, J. S. Lee, S. H. , 2005 Chung, Synthesis of nanostructured γ-alumina with a cationic surfactant and controlled amounts of water, Microp. Mesop. Mater., 79(1-3),: 61-68

62. J. Zhang, F. Shi, J. Lin, S. Y. Wei, D. Chen, J. M. Gao, H. Z., X. X. Ding, C. Tang, 2008 Nanoparticles assembly of boehmite nanofibers without a surfactant, Mater. Res. Bul., 43(7),: 1709

63. S. E. Pratsinis, 1998 Flame aerosol synthesis of ceramic powders, Prog. Energy Comb. Sci., 24(3),:197

64. J. Zhou, S. Z. Deng, J. Chen, J. C. She, N. S. Xu, 2002 Synthesis of crystalline alumina nanowires and nanotrees, Chem. Phys. Lett.,365(5-6).:505

65. R. L. Price, K. M. Heberstroh, T. J. Webster, 2003 Enhanced functions of osteoblasts on nanostructured surfaces of carbon and alumina, Med. Biol. Eng. Comp., (2003) :41, 372.

66. T. J. Webster, E. L. Hellenmeyer, R. L. Price, 2005 Increased osteoblast function on theta + delta nanofiber alumina, Biomaterials 26,: 953

67. H. Choi, I. C. Kim, B. R. Min, K. H. Lee, 2006 Preparation and characterization of ultrathin alumina hollow fiber microfiltration membrane, Desalination, 193:256

68. H. Li, C. Wu, F. Tepper, J. Lee, C. N. Lee, 2009 Removal and retention of viral aerosols by a novel alumina fiber filter, Aerosol Science 40,: 65

69. F. Tepper, L. Kaledin, C. Hartmann, 2005High Performance. TurbidityFilter.February 55 57

70. J. A. Fernando, D. D. L. Chung, 2005 Thermochemical properties of alumina fiber membrane, Ceram. Intern., 31,: 453

71. S. Sundarajam, S. Ramakrishna, 2007 Fabrication of nanocomposite membranes from nanofibers and nanoparticles for protection against chemical warfare stimulants, J. mater. Sci.,,42: 8400

72. Jing, X. Xu, J. Hou, 2007 Preparation of compact Al2O3 film on metal on metal for oxidation resistance by polyvinylpyrrolidone, J Sol-Gel Sci Technol 43: 321

73. Jing, X. Xhao, Y. Zhang, 2007 Elimination of a zero-growth in thickness of Al2O3 protective film deposited by cycles of dip coating method, J Sol-Gel Sci Technol 42: 151

74. V. Landry, B. Riedl, P. Blanchet, 2008 Alumina and zirconia acrylate nanocomposites coatings for wood flooring: Photocalorimetric characterization, Progr.Org. Coatings, 61,:76

75. T. Thostenson, C. Li, T. Chou, 2005 Nanocomposites in context, Composites Sci. Techn., 65: 491

76. L. Sun, R. F. Gibson, F. Gordaninejad, J. Suhr, 2009 Energy absorption capability of nanocomposites: A Review, Composites Sci. techn. 69:2392

77. Santos-Beltran, V. Gallegos-Orozco, I. Estrada-Guel, L. B. Gomez, F. E. Magana, M. M. Yoshida, R. M. Sanchez, 2007 J. Alloys Comp. 434-435, :514

78. M. I. Flores-Zamora, I. Estrada-Guel, J. Gonzalez-Hernandez, M. Miki-Yoshida, R. Martinez-Sanchez, 2007 Aluminium-graphite composites produced by mechanical milling and hot extrusion, J. Alloys Comp. 434-435,:518

79. N. Shahhid, R. G. Villate, A. R. Barron, 2005 Chemically functionalized alumina nanoparticles effect on carbon/epoxy composites, Composites Sci. Techn. 65:2250

80. J. H. Zhou, C. Chen, R. Guo, X. C. Fang, X. G. Zhou, 2009 Synthesis and characterization of carbon nanofiber/alumina composites by extrusion casting, Carbon 47 2077

81. K. Hirota, Y. Takaura, M. Kato, Y. Miyamoto, 2007 J. Mater. Sci, 42: 47920.

82. Z. Xia, W. A. Curtin, B. W. Sheldon, 2004 , Fracture toughness of highly ordered carbon nanotube/alumina nanocomposites, Transaction of ASME, 126,:239

# Index


**B**

Biomaterials, 18, 23, 45, 47, 50, 97, 122, 123, 125, 173, 232, 241, 242, 243, 249, 267, 268, 270, 287

**C**

cell, 20, 21, 49, 50, 77, 78, 79, 81, 92, 93, 94, 95, 96, 98, 99, 100, 193, 216, 218, 221, 222, 227, 233, 241, 243, 248
Collectors, 26, 210
core-shell, 26, 32, 33, 34, 35, 37, 44, 53, 54, 55, 182, 198, 208, 209, 226, 239, 243, 247, 267

**D**

degree of crystallization, 250

**E**

EDX, 79, 82, 86, 87, 96
electrospinning, 2
electrospinning parameters, 39, 46, 250, 252, 253, 262, 266
Energy Storage Materials, 13

**F**

fiber, 2, 8, 11, 25, 26, 27, 28, 29, 30, 31, 32, 33, 34, 35, 37, 38, 39, 269, 272, 279, 280, 281, 282, 285, 287
Flame Aerosol Method, 277
FTIR, 79, 82, 98, 99, 164, 165, 218

**H**

helical, 25, 26, 32, 36, 37, 38, 39, 56, 57, 90, 98, 99, 104, 124
hierarchical, 25, 26, 27, 32, 33, 40, 44, 55, 58, 61
hydrogen sensor, 130, 139, 140, 141

**K**

Keratin-based Nanofibres, 101

**N**

Nanofibers, 1, 4, 6, 8, 13, 16, 18, 26, 55, 63, 75, 130, 146, 182,

198, 215, 216, 217, 218, 219, 221, 271, 272, 278, 284, 287
Nanopores, 40
nanoporous, 26, 32, 44, 52, 59, 98, 245
Nanoprotrusions, 41
Needles, 208

## P

PANI, 153, 202, 203, 204, 212, 213, 214, 215, 216, 217, 218, 219, 220, 221, 222, 223, 224, 225, 226, 227, 228, 237, 242, 246
patterned, 26, 30, 31, 40, 44, 50, 51, 59, 210
physics, 78, 210
poly(L-lactic acid), 46, 250, 269
polypeptide, 78, 80, 82, 91, 95, 97, 120, 124
Porosity, 29

## S

scaffold, 19, 20, 21, 45, 47, 49, 50, 74, 78, 92, 173, 202, 221, 232, 233, 239, 241, 250, 267
sensing mechanism, 129, 130, 138, 139, 245
Sensors, 142, 223, 237, 244, 245, 246
Supercapacitors, 226

## W

Wool Keratin, 103